THE GREAT
FOSSIL ENIGMA

LIFE OF THE PAST

James O. Farlow, editor

THE GREAT
FOSSIL ENIGMA

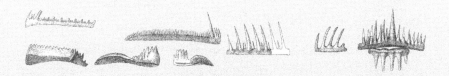

The Search for the Conodont Animal

SIMON J. KNELL

INDIANA UNIVERSITY PRESS   Bloomington & Indianapolis

Illus., p. vi: Pinhead fossils. This scanning electron microscope image of conodont fossils on a dressmaking pin became an iconic image in those years after the animal had been discovered. Photo: Mark Purnell, University of Leicester.

This book is a publication of

INDIANA UNIVERSITY PRESS
601 North Morton Street
Bloomington, Indiana 47404–3797 USA

iupress.indiana.edu

Telephone orders 800-842-6796
Fax orders 812-855-7931

© 2013 by Simon J. Knell

*Manufactured in the
United States of America*

*Library of Congress
Cataloging-in-Publication Data*

Knell, Simon J.
    The great fossil enigma : the search for the conodont animal / Simon J. Knell.
    pages cm – (Life of the past)
    Includes bibliographical references and index.
    ISBN 978-0-253-00604-2 (cloth : alk. paper)
    ISBN 978-0-253-00606-6 (ebook)
    1. Conodonts. 2. Science – Social aspects. I. Title.

QE899.2.C65K59 2012
562'.2 – dc23

                                        2012016799

1 2 3 4 5   18 17 16 15 14 13

FOR THE

HARBINGERS OF SPRING

AND

ALL WHO WONDERED

Contents

The fossil is found – the search begins – Pander's chicks – a friend of Murchison – Pander sees teeth and imagines fishes – others see trilobites, sea cucumbers – Owen sees anything but fish – Harley sees crustaceans – Moore in paleontological ecstasy – the fossil arrives in Cincinnati – Newberry sees fish – Grinnell misses Custer's Last Stand – Ulrich denies the fish and sees a worm – Hinde proves the fish – Smith's surprise – Huxley's encouragement – Zittel and Rohon murder the fish and see a worm.

Oil enters the American soul – fossils take on economic importance – microfossils take center stage – scientific worth of conodonts recognized – the black shales problem – godlike Ulrich vs. Kindle – Bryant's fishes – Hibbard's washing machine – Ulrich and Bassler's proof of method – Stauffer's chance discovery – Gunnell's vision of the future – Chalmer Cooper the disciple – Branson and Mehl's big campaign – Huddle's accidental beginnings – an index discovered – *Icriodus* the ideal – Cooper's cellophane fossils – contamination – Ellison's logic.

The fight for biological paleontology begins – Croneis's influence – the geography of thinking – Macfarlane's conodont oil theory – giant conodonts – Kirk finds bone – Gunnell's imaginings – Eichenberg's novel idea – Schmidt's assemblage – Stadtmüller's influence – Scott finds oil, earthquake and assemblage, and worm – Jones finds assemblages and Denham thinks of sex – Ulrich and Bassler in denial – Loomis and Pilsbry see snails – Demanet finds the fish – Cullison's remarkable jaw.

Illustrations

Preface & Acknowledgments

THIS BOOK HAS TWO BEGINNINGS. THE STORY OF THE "GREAT fossil enigma" begins with the Prelude. However for those expecting an analytical history of science, I suggest the afterword as a useful introduction because it explains how and why I have written this book and what lies beneath the narrative of the chapters. When I began this book in 2003, I envisaged it as the second in a trilogy of monographs revealing the fossil as a cultural object. In the first book, *The Culture of English Geology, 1815–1851*, I tried to show how fossils acted as cultural objects, shaping lives, playing a role in local politics, and leading to the founding of museums and the emergence of geology as a *cultural* field. As a museologist and cultural historian rather than a historian of science, I wanted to explain fossils and their place in society not in terms of a history of ideas but more holistically.

I took these ideas further in explorations of the politics of "English geology," the possibilities of a cultural revolution in science, and cultural changes in geology in late-twentieth-century Britain. In this book, I turn my attention to the workings of a single community of researchers over a period of 150 years or so and consider how these men and women imagined and conceptualized the fossils that bound them together. Once again I have not attempted to write a straightforward history of ideas or to explore the contributions these fossils made to science in any detailed sense. My interest remains predominantly in culture, in the relationships between people, objects, and practices (the things people do, often formalized and institutionalized in some way). While I have long thought of this as a blend of cultural history and museology – as a branch of museum studies – I am aware that it is also a form of ethnology as it has long been understood in northern Europe (as a curiosity about one's own culture). No doubt it is this approach to the history of a science that caused a number of readers of the manuscript to consider it rather "strange." If that is the case, I am delighted, for a study of this kind requires the author and reader to remove themselves sufficiently far from their subject to see the actions of individuals as strange rather than natural.

As I researched and wrote this book, I became increasingly aware that, as one of science's great enigmas, its subject already existed, though not in any coherent form, as a possession of all those who wondered about and worked on these fossils. If I began my research believing that I might be another user of these fossils, this time in my own analytical history, I soon understood that I had to prioritize the needs of those for whom the fossils already held meaning. Having never set out simply to tell a story, I came to understand that accessible narrative was critically important. Indeed, I felt it was an obligation. Consequently, I decided to put a good deal of my analysis into the afterword (which at various times I have considered an explanatory introduction) without making that part of the book oppressively long. I have written the narrative of the chapters broadly in chronological fashion, though some themes prevented me from adhering to this unwaveringly. I hope they retain something of the strangeness of my original outlook. I will leave the reader to decide whether he or she wants to stray into the afterword.

Finally, I should note that nowhere in this book do I assume the authority of a scientist. I am a historian. I listened to my actors and believed, as they did, in what they told me. Consequently, the animal appears in glimpses, as if through a mist. First it seems to be one thing then another, almost coming into view but then disappearing. For a long time the animal did not so much appear in the headlights of science as in its wing mirrors, implicitly present but shrouded in darkness.

Of course, books tend not to be read as they are written, and in the closing weeks before going to press, I became a little concerned that some readers might think that I am making a scientific argument, that I have a scientific opinion on the matters described in this book. I do not. I take no view on what is right or wrong, only in what is believed, and I have no preferred position regarding the biological attributes of the animal myself. I have been fascinated by all the various manifestations of conodont animal, but scientists have to convince each other of the truth of their science, not me. It was inevitable, however, given the focus of this book, that I allocated considerable space to Melton and Scott's, Conway Morris's, and Clarkson and company's animals. Given the sensational significance of the latter, I am sure supporters and critics alike will recognize that British views on the animal have shaped the course of debate over the last thirty years and warrant detailed examination. In all three cases, those who possessed the conodont animal fossils became hugely empowered but also much criticized. I will not say here whether everything has been resolved. That is for the reader to consider at the end of chapter 14.

Among British workers, my main contact has been Dick Aldridge. He gave me, at the outset of my research, the manuscript of a popular account of his scientific work on the animal that he had prepared for publication. Much of this was subsequently dissected by him and turned into papers that have reflected on the scientific investigation of the animals. The presence of Dick's account, with its focus on recent scientific developments and his personal program of research, encouraged me to write instead a long history of conodont research and develop a rather different approach to the discovery of the animal. Dick also gave me unhindered access to the library of conodont papers held by the Geology Department at the University of Leicester and the uncurated and uncensored correspondence between himself, Derek Briggs, Euan Clarkson, and others around the time of the animal's discovery. The manner in which these three men published the first description of the animal, their early ideas, and the criticisms made by friends are there for the reader to see. Dick was among those who read the whole manuscript, though he did so in first draft and not as it appears now. Interestingly, nearly all his comments concerned my use of English; he made no attempt to alter what I had written. My first words of thanks, then, must go to Dick. All readers of the manuscript have been equally generous. Maurits Lindström, who never saw the finished book, expressed it perfectly: "It is your book, write it as you like!"

Maurits was one of the "harbingers of spring" to whom I dedicate this book and who are introduced in full in chapter 6. I studied their work and lives in tremendous detail before meeting many of them. For me, these meetings were the high spots in an epic journey. It is perhaps an artifact of my writing this book at this particular moment – when the full panorama of their scientific lives is available to me – that places them at the center of this study, but one cannot deny their transformative influence. Sadly, three of them – Lindström, Müller, and Walliser – have died since I interviewed them.

I met Maurits Lindström in Stockholm in March 2007. I had flown to Copenhagen and caught the train up to Lund in Sweden as the guest of veteran conodont worker, Lennart Jeppsson, who is perhaps a decade or more younger than members of this 1950s generation. I knew I wanted to write about Lennart's extraordinary Gotland experiences, and we had long conversations about all aspects of Swedish science. I also met members of Lennart's technical team and saw his remarkable laboratory setup – a conodont factory if ever there was one. I then took that long train journey up to Stockholm to meet Maurits Lindström. It had taken a bit of effort to track Maurits down as I had been told it would be difficult to find him. This added a little mystery to the man, which I built up in my mind rather more than was necessary. Finding him was actually not so hard, and Maurits e-mailed me, asking me to meet him on "the southwest corner of Drottninggatan and Tegnergatan. . . . I would [sic] be passing there (parking is precarious) from 9 A M and am driving a grey metallic Renault Megane."

As I waited on the street corner for this man I had never met and with whom I had barely corresponded, my breath appearing in white clouds before me, I felt like Harry Palmer in one of those spy movies set in Berlin in the 1960s. This sense was not dispelled when at lunch time Maurits drove me to a canteen in the middle of an industrial area of the city. The canteen reminded me of those communist-era socialist realist posters featuring the heroic worker; I felt undressed without a gray boiler suit. Maurits himself was tall and thin, sophisticated and clever, but in that Swedish way that always makes others feel comfortable. We chatted all day about his discoveries and thinking, his worldview and his extraordinary life. He told me much more than I could include in this book. It was only on meeting him that I understood his wonderfully playful scientific outlook.

I met Klaus Müller in an up-market retirement complex in Bonn. He was the oldest of this generation, and with his large glasses and those milky eyes of old age, he reminded me a bit of my own father in his final days. Klaus was a proud German (or Berliner, he might say) who had lived an extraordinary life despite chronic health problems. His wife had been central to his being and had played an important role in helping him make the right decisions. In his company, I felt young again, for he had experienced a life during the war that I could barely imagine despite my consumption of so many war histories. From Bonn I caught the train down the picturesque Rhine Valley, a journey I had traveled many times in the 1970s, and then up to Göttingen to meet Otto Walliser. Klaus had talked of the difficulties of his relationship with Otto – the two men fell out almost on first meeting – so I was now even keener to meet him. I found Otto at the University of Göttingen, in a large room crammed with books and fossils, with giant cactuses lining the tall windows. Like Lindström, Otto was still mentally young and full of energy. He was naturally gregarious and enthusiastic, and his smile rarely left his lips. Positive and motivating, he seemed to exude happiness, and it was easy to understand why others might follow his lead. You knew you would have some fun along the way.

In May 2007, I flew out to the United States and spent long days in the science library at the University of Chicago, where I found so much written about these fossils that had rarely escaped the country. I recall that the American Association of Museums was in town at that time, so I spent my evenings partying in another world. From there I flew to Washington to work in the Smithsonian Institution's archives, and there I discovered the political controversy surrounding the black shales that led to the birth of the conodont as a properly scientific fossil. From there I flew to Columbus, Ohio, to meet Walt Sweet at his house. Walt seemed to measure moments in his career in terms of the students he had taught at the time – he was immensely proud of them all. I had corresponded a good deal by e-mail with him, so I knew he was a generous yet pretty straight-talking guy who valued precision. He took me to the Geology Department at Ohio State University, which had been the center of his life for so long. The department's building, which was now quite old and a little cramped, had been constructed with geological principles in mind, and in it every space seemed to tell a scientific story. In the vastness of the Columbus campus, it seemed rather homely.

In December that year, I spent several days in the archives at the University of Illinois in Urbana, looking through the materials Harold Scott had lodged there. It was with great delight that I discovered a big folder filled with correspondence on his and Melton's animal. In those papers I also believed I could see another Scott, not the Scott others remembered. My plan was to fly from there to Chicago and then down to Missouri to look for Branson and Mehl material. But the tiny jet did not want to leave the ground at Urbana, and by the time it did, O'Hare Airport was already under the jurisdiction of snow plows. All flights were canceled, but for how long? No one knew. My bag flew south, and the next day I flew east to Washington, D.C. The TV news that night showed the Midwest ravaged by extraordinary ice storms. Branson and Mehl (and my bag) remained out of reach. On the outskirts of Washington, I visited the U.S. Geological Survey, gathering the last pieces of information.

Unfortunately, I did not interview all the harbingers. Willi Ziegler had died a few years earlier, but I believe I met him briefly at the Senckenberg Museum in Frankfurt, when digging for fossils at Messel in the late 1980s, long before I had any thoughts of writing this book and when I was still a museum curator. I finally corresponded with Frank Rhodes when this book was pretty much finished, and he kindly agreed to read the manuscript. I would have enjoyed interviewing them both.

There are many other people I must thank. I interviewed Ronald Austin, coincidentally Dick Aldridge's former doctoral dissertation supervisor, in Swansea, Wales, quite early in the project and caught Carl Rexroad briefly, another legend from the old days, and still very active, in Leicester, England, in 2006. I exchanged a number of e-mails with Gil Klapper in 2005 and 2007, and I received materials, ideas, information, and so on from Eric Robinson, Mark Purnell, Lyall Anderson, Richard Davis, Danita Brandt, Wendy Cawthorne, Martin Langer, Irena Malakhova, Hannes Theron, Andrew Polaszek, Neil Clark, Peter von Bitter, Paul Smith, and Sandra Dudley. Dick Aldridge and my good friend Mike Taylor read the first complete, but rather different, draft in 2009. Euan Clarkson and Derek Briggs commented on chapter 13 before a final rewrite permitted me to introduce some additional archival material. Derek Briggs and Stig Bergström reviewed the whole book in slightly different versions on behalf of Indiana University Press, and both gave me very useful corrections. Walt Sweet kindly read chapters 8 and 9, Chris Barnes chapter 9, Lennart Jeppsson chapter 10, Mark Purnell chapters 12 to 14, and Paul Smith chapters 13 and 14. Jeff Over, John Repetski, Eberhard Schindler, Stig Bergström, Helje Pärnaste, Dick Aldridge, Mark Purnell, and Debbie Maizels gave me additional help with images.

I would have liked to have met and interviewed many other people discussed in this book, but there simply was not time. Delayed by my increasing management responsibilities, this book was overtaken by a number of other projects that made finishing it something of a struggle. As I type the final words, I am already deeply immersed in a similar kind of study, but this time focused on Europe's national art galleries. It is something of a mental stretch to keep both projects in my mind!

I am grateful to librarians and archivists at the University of Chicago, University of Illinois, Natural History Museum in London, U.S. Geological Survey, Smithsonian Institution, University of Leicester, and Geological Society of London. I would like to thank the Smithsonian Institution for permission to use and quote from archival material in its possession and the University of Illinois at Urbana-Champaign for permission to use and quote from Archives Record Series no. 15/11/25 in the Harold W. Scott Papers. I am very grateful to the following for permission to use illustrations: the American Association of Petroleum Geologists, Cambridge University Press, the Deutsche Gesellschaft für Geowissenschaften, Elsevier, the Geological Association of Canada, the Geological Society of America, the Geological Society of London Publishing House, Geoscience Australia, Kyoto University, the Palaeontological Association, the Royal Society, the Society for Sedimentary Geology (SEPM), The Treatise on Invertebrate Paleontology, Gebruder Borntraeger, and Wiley. I am also very grateful to the British Academy for a travel grant (RA17036) that permitted me to access archives and libraries and allowed me to interview key actors in various parts of the world. I would like to thank the University of Leicester for permitting me two semesters of study leave over the last seven years, in which I managed to make considerable headway with this book, and my colleagues, who have been so encouraging during the long process of writing about these strange "teeny tiny teeth." Finally I must thank Marg, Callum, and Ciaran for their patience – this has been a long journey. All remaining errors are, of course, my own, and because changes were made to the narrative right up to going to the presses, these may not have been present in the versions reviewed by readers.

THE GREAT
FOSSIL ENIGMA

The Impossible Animal

AS STUDENTS, WE DREW AND LABELED A JARGON-RICH palaeontological world, only too ready to be captivated by the objects before us. Our tutor, however, seemed to have other ideas. He evidently had no passion for his subject. To him, ammonites and trilobites were just things to carry names. As each fossil was introduced in sequence, then drawn, annotated, named, and removed, our enthusiasm waned. How could paleontology be so dull? Why would any tutor wish it to be so? Our disapproval turned to disdain. Then, one day, we were greeted by rows of binocular microscopes. Through them, we looked at "microfossils" and, among these, some peculiar tooth-like objects. Immediately, and to our great surprise, everything now changed. Our roles were reversed. We initially thought these new objects dull (they were not the prettiest examples of their kind), but our tutor had woken up! He asked us what they were. We made a few feeble guesses, which he easily rebuffed. He did, however, take our suggestions seriously. That too was new. Then he began to list other possibilities, and one by one he explained that they too were incorrect. Before long, every blackboard in the room – and there were many – was covered with names and sketches of what seemed like the whole animal kingdom, and a few plants besides, and yet still we seemed no nearer the truth. We waited patiently for the answer, but that answer never came. Sporting a smile we had never previously seen, and with obvious relish, this dour Yorkshireman (or so we had thought) admitted he didn't know what they were either. There was a moment of silence. Then we became brave: "What about . . . ?" "If . . . ?" "Couldn't they . . . ?" But our speculation was futile. In every case someone had been there before us.

We looked again at these tiny teeth. They were so evocative. How could no one have any sense of what they were? How could we not even know whether the animal that possessed them also possessed a backbone? How could a natural object exist in this advanced age and yet remain beyond the most general categorization? Undoubtedly enhanced by a perfect prelude – that dull journey through paleontological gems – our tutor's performance had been quite brilliant. For years after, we would recall this impossible thing and dream a little about that magnificent moment when all would be revealed. Many years later, over a cup of coffee with curator Peter Crowther, who was also one of the editors of the journal *Palaeontology,* I recalled the wonder of this little fossil. His face lit up. It was clear that he, too, had experienced a similar moment. It was as though we had shared a religious rite of passage. Then he said, "And have you heard? They have recently found the animal!"

The natives, in order to get rid of their troublesome guests, continually described Dorado as easy to be reached, and situate at no considerable distance. It was like a phantom that seemed to flee before the Spaniards, and to call on them unceasingly. It is in the nature of man, wandering on the earth, to figure to himself happiness beyond the region which he knows. El Dorado, similar to Atlas and the islands of the Hesperides, disappeared by degrees from the domain of geography, and entered that of mythological fictions.

ALEXANDER VON HUMBOLDT,
Personal Narrative of Travels to the Equinoctial Regions of America During the Years 1799 to 1804 (1853)

ONE

The Road to El Dorado

THEY WERE JEWEL-LIKE THINGS: LUSTROUS, COLORFUL, AND perfect. Their evocative shape suggested they had fallen from the mouths of living fish, but Christian Pander knew this was just a wonderful illusion, for he had not found them in any river, lake, or sea, but in some of the oldest rocks then known.[1] Oblivious to the chemistry of their surroundings, they had survived as objects of beauty when all around them had turned to stone or not survived at all. So small that several would fit on the head of a pin, these tooth-like things were also older than any known trace of vertebrate life. From the very moment of their discovery, then, they were quite extraordinary objects. Evocative, ambiguous, contradictory, and secretive, they had the capacity to mesmerize, to compel mind and body to go in search of the animal that had once possessed them. For more than a century and a half this animal was pursued, its assailants acquiring little more than glimpses as the animal repeatedly concealed itself in illusions. Before long it became science's El Dorado.

We, too, will go in search of the animal, but our journey will not take us into dense jungles, a distant past, or much into the arcane world of rocks and fossils. Instead we shall journey through the minds of those who looked and believed, for only in the scientific imagination was this animal clothed in flesh and made to breathe. The animal was real enough – be assured of that – but no human ever saw it alive.

So to begin this journey, we must cast aside our fishes, fossils, and teeth – indeed, we must put out of our minds all preconceptions of what these things are or how they might be understood. The geologists and paleontologists discussed here needed to believe that these objects existed in, and came from, a distant past. That is a necessity of their discipline. We, however, are not interested in the real world but in what these scientists experienced and thought. Consequently, we must distance ourselves from their outlook and consider that such things as fossils just appear, born into the world of known things. One moment they had never entered a human thought; the next, they had. Indeed, unlike fossil ammonites, oysters, and sea lilies, Pander's fossils had not existed in folklore or prehistory. They made their first appearance in a world that was already scientifically mature and ready to make sense of them. Since then, they have only existed in the enclosed world of science. And despite their enigmatic status, they have never spawned the kind of romance and fantasy that has been so important to the making of *Tyrannosaurus rex*. So with all preconceptions put to one side, we are ready to return to that moment of discovery when the animal first entered the human imagination.

It just so happens that Pander was peculiarly equipped to discover these tiny fossils, for his eyes had been trained to notice the minute anatomical details of unhatched chicks. Born in 1794, he came from that wealthy, German-speaking merchant class that had for centuries dominated his native city of Riga, the capital of modern-day Latvia but then in Livonia, a province of Russia. The city's official language and many of its intellectual ties remained German. It was natural, then, for Pander to seek an education in Germany, and so, in 1814, he took his studies to Berlin and then to Göttingen. On this southward migration, his intention had been to train for a career in medicine, but that ambition was soon displaced by a fascination with nature itself. That he could make this subject his life became a reality when, in March 1816, he caught up with his good friend Karl Ernst von Baer in Jena. Since their last meeting Baer had fallen under the spell of the distinguished anatomist Professor Ignaz Döllinger at Würzburg and become intoxicated with embryology and the opportunities it presented for understanding how organisms are made. Baer now recruited Pander to the cause, convincing him to take up Döllinger's proposal for a new study of the first five days of the chick embryo's life. Baer would have accepted this challenge himself, but for his impecunity and the enormous costs of experimentation and illustration. So instead Pander found himself in Baer's shoes and on a journey into the very origins of life itself. "With bewilderment we saw ourselves transported to the strange soil of a new world," Pander later remarked. Two thousand eggs later, he emerged from his studies, crowned with a "laurel of eggshells,"[2] doctorate in hand and placed in the pantheon of pioneering embryologists. In a single stroke he had risen from student to distinguished man of science.

Baer continued this search for the origins of life in the world of the unborn and soon eclipsed Pander as an embryologist. Pander, by contrast, began to turn his attentions to the long dead, joining his illustrator and naturalist friend Eduard d'Alton on a tour of the great natural history museums of Europe. It was in these bizarre menageries of fossils, animal corpses, and dismembered bodies that these two men saw an opportunity for a gigantic work they called *Comparative Osteology*. Laying the groundwork for this fourteen-volume series during their travels in 1818 and 1819, these books revealed Pander to be an early evolutionist envisaging the development of life as an ongoing transformation of species in response to environmental factors. In this, of course, Pander was not alone; he knew well the early evolutionary literature then being published across continental Europe, particularly in France.[3] Pander's views were shared by Baer and, when he eventually made his contribution to the evolutionary literature, some forty years later, acknowledged by Charles Darwin.

On his return to Russia, Pander was elected to the Academy of Sciences at St. Petersburg, beginning his field study of fossils in the 1820s with an examination of the older rocks that outcrop along the river valleys around that city and form the picturesque coastal cliffs of modern-day Estonia. Part of a major geological structure known today as the Baltic Klint, these rocks run westward some twelve hundred kilometers to the Swedish island of Öland. In all this distance they are undisturbed by Earth movements and show little lithification despite their extraordinary age. In the 1820s, the new "geologists" were still in the early stages of working out the order in which these rocks had been laid down, work that would enable them to figure out the passage of geological time.

The rocks that interested Pander were simply known as the "transition formation." They were unexplored. Or so Pander thought. And soon he understood why: Laboring long and hard, he could only turn up mere fragments of fossils. At this low ebb in his research, he chanced upon a local community of fossil collectors who had been far more successful than he had. It was a turning point. Now he could exploit the curiosity and impecuniousness of children and local villagers to build a collection overflowing with fine specimens.

Many of these fossils found their way into the 940 hand-colored illustrations in his book *Contributions to the Geology of the Russian Empire: The Environs of St. Petersburg,* published in 1830. But Pander was still not happy. Suffering repeated bouts of malaria and having to foot the bill for the plates himself – the academy being unwilling to do so – he resigned from that august body in 1827. Leaving St. Petersburg in 1833, he returned to his father's estate of Zarnikau near Riga, there to be – perhaps unwillingly – a gentleman farmer with only a leisure interest in paleontology.[4]

Nevertheless, Pander's book – which was published before British geologists Roderick Murchison and Adam Sedgwick had packed their "knapsacks" to begin their own investigations of rocks of equivalent age in Britain – would in time give him some recognition. In the early nineteenth century, no country offered greater geological opportunities than Great Britain. Its extraordinary rocks – diverse in age and type, rich in fossils, and exposed in mountains, coasts, and the countless quarries and excavations produced by its Industrial Revolution – gave the country a huge advantage in the new science. Britain fostered individualism and social ambition and at that time possessed a rapidly expanding middle class only too ready to elevate themselves in the new science of geology. That science had by the 1820s worked out its methods and was beginning to locate its "great men," as these British geologists increasingly wished to see themselves. The science was becoming white hot and deeply entangled in controversy and dispute. By 1830, the "transition formation" marked the geological frontier, and all who sought fame looked in its direction. Only a few of them, however, knew anything of Pander's book.[5]

Pander did not live in such a competitive world, though he may have experienced it on his trips to Britain, France, and Spain. But it was not simply that he lived beyond the reach of this world that prevented his 1830 study achieving for him the fame reserved for Murchison and Sedgwick; the geology itself was also to blame. The rocks Pander studied were arranged simply one upon another and were unchanged over huge distances. He needed no complex terminological inventions to describe them and simply named what he saw: a basal blue clay overlain by a sandstone rich in a brachiopod he called *Ungulites* (better known today as *Obolus*), then a black shale (named after the fossil *Dictyonema*, now *Rhabdinopora*, which it contains) and a green sandstone, its greenness caused by the presence of the mineral glauconite. It was in this green sandstone, many years later, that Pander would discover his strange tiny teeth. This succession of rocks was topped off by an out-jutting limestone crammed with straight-shelled nautiluses.

The rocks that confronted Murchison and Sedgwick could not have been more different: folded, faulted, and metamorphosed strata in mountainous Wales and sod-covered Cornwall and Devon. Using fossils as time indicators, and with considerable effort, these men managed to connect rocks in different regions and of different ages as if assembling a great jigsaw. In order to do so, both men, together and alone, conceptualized great swathes of rocks, and thus vast blocks of geological time, in new abstract "systems" they named Cambrian, Silurian, Devonian, and Permian. They, like Pander, exploited the knowledge and specimens of local collectors. However, working on their own projects in different parts of the country, in overlapping sequences of rocks, it was inevitable that Murchison's Silurian and Sedgwick's Cambrian would come into conflict. This dispute is, from our perspective, still in the future and not of great concern to our story, but it says something of the personal investment involved in this new science. Murchison, who had once been a military man and whose Silurian was the first invention of its kind, wished to see his system as an international standard, and to this end he marched into Russia painting the geological map of continental Europe in the colors of his own precious system. When he did so, he was delighted to discover that Pander had done some of the groundwork for him and had, indeed, already compared the Russian strata with rocks in Sweden and Norway. He was even happier when Pander offered to support his scheme.

It was on one of these trips, in 1841, accompanied by the (Baltic German) Russian paleontologist Alexander von Keyserling and French paleontologist Edouard de Verneuil, that Murchison first met Pander: "Leaving our carriage at Neuermähler to go on to the next station, we went in a troika with von Keyserling to visit the naturalist and geologist Pander, son of the rich banker of Riga, who, according to Baron Casimir de Meyendorff, is the Barabbas of Livonia. Passing among hillocks of blown sand . . . and small lakes, through fir forests, and open tracts, we found the author at his chateau, surrounded by his seven fine children, and an agreeable, good-humored lady, a Petersburgian. The residence consists of a great chateau, with a Greek *façade,* which is only inhabited in three summer months, filled with casts of statues, and having inlaid floors. The flags under the peristyle are the same dark blue stone which we observed last year in the floor of the Citadel at St Petersburg, with long *Orthoceratites,* derived from Öland, etc. We were received in the little or winter villa adjoining, and breakfasted and dined there. We were loaded with kindness, and saturated with fossils and good cheer. The following notes were made in the highly heated room of Pander, amid myriads of fossils." One can understand Murchison's great appreciation of the "very accurate and painstaking Russian naturalist";[6] Pander handed him a large slab of geological territory on a plate. Murchison repaid the debt by becoming a great publicist for Pander, mentioning him in his widely read *The Silurian System* (1839), *The Geology of Russia and the Ural Mountains* (1845–46), and *Siluria* (1854). With Murchison's assistance, Pander's name and discoveries were made available to the English-speaking world and beyond, even before Pander had discovered his enigmatic teeth.

His father having died two years earlier, in 1844 Pander returned to the capital to pursue a career in the scientific section of the Russian Department of Mines. Now he began to prepare his great paleontological work – eventually published in four volumes – *Monograph of the Fossil Fishes of the Silurian System of the Russian Baltic Provinces.* Meticulous and pioneering in its use of the microscope to reveal the histology – the microscopic anatomy – of these fossils, the first volume, published in

1856, gained international attention. It did so because it revealed something that had remained unseen by the thousands of eyes that had looked intently at rocks over the previous half century. It was here that Pander first described those strange, tiny, tooth-like objects, the only evidence of "Lower Silurian" fishes. To come across such fossils, in rocks that had no right to possess them, was – to say the least – surprising. With intense curiosity and excitement, Pander peered down his microscope. But he did so for too long and contracted an eye infection that nearly blinded him. For almost two years he kept his microscope covered; the moment of revelation simply had to wait.

By the early 1850s, paleontology was a science vastly different from that Pander had known in the 1820s. It was now intellectually mature: rigorous in method and rich in theory. A vast network of European museums recorded the history of life in all its variety and with a great consistency of understanding. Naturalists had become connoisseurs of this variety, and so when sorting material under his microscope, Pander would have possessed a rich mental database of things previously seen that he could use to understand the new things he found. But this was no dry mechanistic exercise. It could so easily become a journey of exploration producing feelings of excitement and puzzlement, bizarre imaginings, and hopeful ambitions. As Stephen Jay Gould once noted, in typically picturesque fashion, "The legends of fieldwork locate all important sites deep in inaccessible jungles inhabited by fierce beasts and restless natives, and surrounded by miasmas of putrefaction and swarms of tsetse flies. (Alternative models include the hundredth dune after the death of all camels, or the thousandth crevasse following the demise of all sled dogs.) But in fact, many of the finest discoveries . . . are made in museum drawers." Gould would undoubtedly have known, however, that the camel's death and thousandth crevasse had their equivalents for those who made their discoveries in museum drawers or looking down a microscope tube. Pander's exile from St. Petersburg, his eye disease, and his extraordinary difficulties finding an illustrator to draw his fish discoveries show that he too had moments when he felt the sled dogs had deserted him. He did, however, overcome these difficulties, and in Trutnev[7] he found a gifted artist willing to painstakingly draw the microscopic details of his fish just as he wanted.

Panders' new fossils were quite remarkable: objects that were translucent, colorful, and beautiful, that were unexpected yet evocative.[8] Naked and mute, these things needed to be clothed in the arcane and specific language of science so that others could appreciate the discovery and discuss it using the same terms. By these means, science as a whole could then seek the objects' meaning and truth. Here Pander had to make decisions about what descriptions to use to distinguish these fossils from others. He believed, from his connoisseurship of nature's objects, that he understood their language; making sense of them was thus a kind of mental conversation. The objects may have remained silent, but to him they seemed to speak. At the time, he was studying fossil fishes, and it was inevitable that he would look at these objects through eyes accustomed to looking at fish: They looked like fish teeth and he would see, name, investigate, conceive, and understand these objects as he would fish teeth.

He tested his ideas by looking at every facet of each object. If all the different aspects of a tooth said the same thing, he could be assured that his initial interpretation was probably correct. He soon found, however, that there was much to contradict the idea that these fossils were teeth. Chemical investigation suggested that they were made of lime (calcium carbonate) and not, like every other vertebrate tooth, apatite (calcium phosphate). A hollow internal "pulp cavity" seemed to support the tooth theory, but, unusually, it did not mirror the external form of the tooth. Indeed, unlike vertebrate teeth, which grow by the addition of layers to the surface, these microscopic teeth appeared to have grown by the addition of layers to the inside of the pulp cavity. Cone shaped in form, they grew cone *in* cone. Grinding the teeth down to make translucent slithers or thin sections that could be viewed under the microscope, he searched the fossils' interior structures for further clues. Color seemed to be important. In the "yellowish, transparent, flexible, hornlike teeth," he found layers, or lamellae, which showed this cone-in-cone growth. The "snow white, opaque" teeth seemed to be immature forms of the yellow kind. He noted every recurring feature from small cells or bubbles arranged side-by-side, which lay between the lamellae, to large cavities randomly placed. Some pinkish teeth, made up of multiple "points" or cusps – "compound teeth" – showed a puzzling structure of "alternating light and dark cross-striped areas," the dark stripes apparently formed of small cells. He thought these pink ones "different in every respect" from the yellow and white teeth.

Pander also detected what would become mysteriously known as "white matter" – parts of the conodont fossil lacking laminations that appear white in reflected light and an opaque black when light is transmitted through the fossil in thin section. White matter formed a distinctive feature of these new fossils.

These strange tooth-like objects were not like any known fish teeth, but Pander still considered them teeth: "Not only are we completely ignorant of the animals that possessed the lower Silurian teeth but also there are no descendants or living animals that have a similar type of tooth structure. We do not know what kind of teeth we are investigating, whether they belong to a jaw-bone, the palate or the tongue . . . or whether the various forms originated from the same animal." He concluded that they could not have been attached to a jaw and fell upon the only possible modern analogy: "We can maintain, however, with reasonable certainty, that these teeth were inserted into the mucous membrane of the throat, similar to the teeth of the cyclostomes and the squalids." Cyclostomes include the jawless hagfish and lamprey, while the squalids are a family of sharks that includes the humble dogfish.

As no one had ever seen teeth like these before, it was logical for Pander to imagine they came from a previously unseen type of fish. In his lifetime he had seen many inexplicably strange fossil fishes come to light, so why not stretch the boundaries a little further? Strange teeth demanded strange fish. Living cyclostomes were strange enough; they lacked jaws, true teeth, scales, bones, and a bony spine. Who could say how much stranger vertebrate life had been in the distant past? Pander's Conodonta, named after their cone-shaped teeth, were simply a new group of fishes no less exotic than others that appeared at this early stage of vertebrate life.

In everyday language, the name of these fishes became simplified to "conodonts," but in the process the name lost the precision of its original meaning. Soon, if one found one of these fossil teeth, one found a "conodont"; the fossil became a conodont rather than the animal. In time scientists came to forget Pander's original usage, and so ingrained did this thinking become that much later still it became necessary to refer to the "conodont animal." When this happened, some conodont workers complained that this was like talking of the "lion animal" or the "giraffe animal." The word "conodont" is thus always a little ambiguous, and one has to ask, is this referring to a tooth-like fossil or to an animal? As much as possible I shall use the terms the scientists themselves used.

Having described the teeth in extraordinary detail, Pander then had to divide them into species so that each type could be given a scientific name. He admitted that he did not know whether the different types of teeth came from one mouth and thus were all the same species or whether each belonged to a different species of fish. This deficiency in his knowledge was not as important as it might at first appear, as paleontologists were used to naming parts of things knowing that they could correct any errors later on. This permitted Pander to opt for a pragmatic solution to the problem, and so he divided them into fifty-six species on the basis of their shape. He knew he was not creating true species but attempting to precisely define the different kinds of these objects. Armed with this knowledge and these names, others could also find conodont fossils and build upon what Pander had discovered.

What Pander had not foreseen was that by claiming his fossils to be the remains of the earliest vertebrates, he would unknowingly create an El Dorado myth. For like the lost city of gold, his fish relied on evocative yet contestable evidence. Just as many found in the lost city an irresistible draw, so science would not rest until it had finally resolved the truth of these anomalously old fishes. It was this controversial aspect that brought Pander's book international attention, and this in turn began to write the animal's mythology. Some would now go in search of the animal, while others would attempt to impose their more conservative views and deny Pander his fish. Indeed, the doubters began to take the high ground even before Pander's book hit the presses.

There has been much speculation about when Pander discovered his fossils. It may have been as early as 1848 or as late as 1850. News of the discovery first escaped Russia in January 1851, in a letter written to Murchison by Pander's friend and colleague, Gregor von Helmersen.[9] That letter said little other than that Pander had found a new fauna in the older rocks. Murchison told his French disciple, Joachim Barrande, and he in turn wrote to Pander, who responded immediately, enclosing sketches of the rock succession and even examples of the fossils themselves. Pander's ideas about the conodont fishes were already well formed and it was with Barrande's reading of his letter to the Sociéte Géologique de France that Pander announced his discovery of "an immense quantity of teeth of fishes, the first traces of vertebrates" to the world. "These teeth, though similar in form to those of certain Placoids, Ganoides, and even fish now alive, are distinguished completely by their microscopic structure," he told its members. "If one wanted absolutely to classify these fish among the forms nowadays, they would show analogy only with Cyclostomes."[10]

Complimenting the Russian on the remarkably attentive and meticulous work that must have been required to find fossils no one else had seen, Barrande refrained from expressing his own opinion about them. Rather, he saw them as a welcome and timely arrival, for in Britain an argument had recently raged as to where in Murchison's Silurian rocks the oldest fish occur. That argument ended with the denial of fish as ancient as those Pander now believed he possessed. But if Barrande was only too willing to bow to Pander's superior knowledge of these fossils, he was somewhat disturbed by Pander's claim that this discovery provided evidence for Russian rocks older than the Silurian. Barrande thought Pander wanted to use his tiny fishes to insert a new geological system beneath the Silurian, and on this point he predictably objected.

Pander soon discovered that he had been a little naïve in permitting examples of his fossils to circulate in the jealous scientific communities of Paris and London before his detailed analysis was published. It left him exposed to a preemptive strike – one his friend, Murchison, was already primed to make. Murchison was then preparing his *Siluria: A History of the Oldest Known Rocks Containing Organic Remains,* a book he aspired to make definitive. Pander's teeth were something of a problem, and particularly so if they might be used to claim a fossil-bearing system older than the Silurian. Were this to succeed, Murchison's book and his Silurian would be seriously challenged. So he called upon the expertise of Barrande and William Carpenter, professor of physiology at the Royal

Institution in London, and the superior technology of their "powerful microscopes," in the hopes of causing a little tooth decay. He was in luck. With no evidence of bone, Pander's fish teeth, "not larger than pin heads," could be dismissed as mere fragments of trilobites.[11]

It was Pander's reading of these objections, two years before his own book was published, that spurred him on to develop a fully argued case for the fish. But Pander was perhaps too open in even this analysis as his unsympathetic colleague in St. Petersburg, Karl Eichwald, turned Pander's own doubts about the conodont's peculiar construction against him and suggested that the conodont fossils came from the skin covering of that sausage-shaped relative of the sea urchin, the sea cucumber.[12]

The debate concerning the identity of the fossils took a new turn, however, when, in the late 1850s, some English analysts discovered that conodont fossils contained traces of calcium phosphate, the material which makes up vertebrate bone and teeth. Pander now felt even more confident about his conclusions. Murchison, the self-proclaimed historian of the oldest fossiliferous rocks, was once again on the back foot and now called upon the distinguished British anatomist Richard Owen to provide a definitive diagnosis. Owen, who had taken charge of the natural history departments of the British Museum in the year Pander's book was published, was a human encyclopedia of animal physiognomy, expert in both vertebrates and invertebrates, living and fossil, and a student of teeth. It was reasonable to believe that he was the man in all Europe most capable of solving the riddle.[13] Pander himself had used Owen's work to interpret these tooth-like fossils.

Owen approached the problem with an open mind. He knew that recent decades had brought to light "every type of invertebrate animal" in the older rocks and that the development of life appeared to follow a non-evolutionary progressive and branching course. At first this growing list of very ancient animals had caused surprise, but now Owen was willing to consider the possibility of vertebrates also being present in these older rocks. At the time, the earliest fishes came from the Upper Silurian. Pander's discovery pushed them back into the Lower Silurian, or what we would now know today as the Lower Ordovician (a term introduced later in the century to resolve the disputed overlap between Sedgwick's Cambrian and Murchison's Silurian).

Owen examined Pander's "minute, glistening, slender, conical bodies" under his microscope and published essentially the same report three times between 1858 and 1861, making only minor alterations with each new revision.[14] In 1860, for example, and perhaps in deference to Murchison, Owen described Pander as an "accomplished naturalist and acute observer" and his book as "an important work." But in 1861 he found no place for these compliments. In 1860, he raised the possibility that a few of Pander's fossils might have claim to "vertebrate rank" but then admitted "that the parts referred to jaws and teeth may be but remains of the dentated claws of *Crustacea*." The recent recognition of calcium phosphate caused Owen to give greater consideration to the vertebrate origin of conodonts in that year. Here, Owen alighted upon a new analogue, the whale shark (*Rhinodon*), which although typically nine meters in length has rows of thousands of two-millimeter-long teeth. But Owen agreed with Pander that these teeth were never attached to bone. So he considered Pander's hagfish theory but here felt the conodonts were "much smaller, slenderer and far more varied" than the teeth of that group of fishes. Then he asked, if they did need to be attached to a "soft substance," then why not a "soft Invertebrate genus"? Certainly they seemed not to belong to an animal with a shell (because no shells had been found), and by a process of deduction, and not finding any perfect solution himself, he concluded that "they have most analogy with the spines, or hooklets, or denticles" of the slug-like naked mollusks or worms. Owen offered these as the least problematic and contradictory possibilities. He certainly saw no grounds for Pander's more narrow interpretation: "The formal publication of these minute ambiguous bodies of the oldest fossiliferous rocks, as proved evidences of fishes, is much to be deprecated."

For some, Owen's verdict was definitive, his scientific indecision concealed behind his clear preference for an unproblematic invertebrate. Beyond the inner circles of metropolitan science, Owen's self-styled greatness was received wholesale, even if many in the city, such as the young Thomas Huxley, "feared and hated" him: "He can only work in the concrete from bone to bone, in abstract reasoning he becomes lost."[15] Pander's fish had relied upon Pander's imaginative capabilities, and in Huxley's view this was a facility Owen lacked.

✷ ✷ ✷

In June 1861, when Owen's most recent opinion was published, Huxley was engaged in the Panderian act of studying the embryology of the chick's skull. He, too, would soon be called upon to ponder the conodonts. As secretary of the Geological Society of London, he presented a paper on these fossils written by his associate, John Harley of Kings College, a Ludlow fossil collector who had earlier provided fossils for Huxley's investigation of giant fossil "sea scorpions," or eurypterids. Believing Murchison's Silurian to be "the Crustacean Age," Harley had gone in search of crustaceans in the Ludlow Bone-Bed, a deposit made famous by Murchison, who described it as "a matted mass of bony fragments, some of which are of a mahogany hue, but others of so brilliant a black, that, they conveyed the impression that the bed was a heap of broken beetles."[16] Murchison interpreted these fossils as the remains of primitive fish and thus possibly the earliest record of vertebrate life, but he could not get any good specimens of these fish and so could say little about them.

Harley's solution was to break and wash pieces of the bed. It was by a similar technique that Pander found his first conodonts. Now Harley found "minute bodies," too, and looking at them through his microscope, he thought they were like those described by Pander, a fact that seemed to be confirmed when Huxley gave Harley two of Pander's own specimens. But Harley, who looked with crustacean eyes, could see nothing piscine in his specimens or in Pander's, only the spiny protection of the modern horseshoe (or king) crab (*Limulus*). He concluded that all his finds, including the conodonts, came from crustaceans and in all probability from the shrimp *Ceratiocaris,* on which the conodonts performed as "minute spines which were attached to the tail-spines." He convinced himself that this was the case by examining specimens of mantis shrimp in the collections of the Museum of Comparative Anatomy at King's College in London. Sure in his conclusions, Harley set up a new genus to accept all his material: "*Astacoderma* is a name I would give this genus, in which I would also include the whole of the so-called Conodonts, and thus give them at once a natural association, and a more appropriate name."

By coincidence Dr. Alexander von Volborth of St. Petersburg was making a short visit to London at the time. In conversation with Murchison and Huxley at the Museum of Practical Geology, he told them that he was convinced that the conodont was no limited phenomenon but was to be found wherever the Silurian occurred. Huxley told Volborth of Harley's discovery, and this sent Volborth in search of the author. On viewing Harley's collection, Volborth was convinced that a small tooth was indeed one of Pander's conodonts. Making the long trip up to Ludlow, he too found conodonts in the rock, hundreds of them, he said. But these were quite unlike those he knew in St. Petersburg: They were white, opaque, dull, and brittle. Nevertheless, when Pander saw them, he felt he could confirm the conodont's peculiar internal structure. Volborth also found conodonts on the Swedish Baltic island of Öland, though these were of the familiar kind.[17] Whether Harley's rather odd fossils really were conodonts was, however, not beyond doubt. Certainly those he figured looked nothing like them, but then no one was quite sure how varied these new fossils could be.

By the end of 1861, the conodont had at least six possible identities: fish, sea slug, marine worm, sea cucumber, trilobite, and now crustacean. Each was the product of a rather different outlook. Pander had unquestioningly been rather narrow in the kinds of animal he considered but had been very open and imaginative about the anatomical possibilities of vertebrate life. One might expect a fish paleontologist who had in his lifetime seen the discovery of the most remarkable and unexpected fossil fishes to have this open mind. Owen was by comparison an encyclopedist. If this permitted him to admit a huge diversity of animals into his analysis, it by implication made him rather more the generalist and content to work within the known categories of life. He was not one to think very far beyond the apparent natural order of things. Harley was an entirely different kind of naturalist: an enthusiast, rather less informed and as a result having a tendency to naïve assumptions and conclusions. To these we might add Carpenter and Eichwald. With the mythologizing of the animal, the varying capabilities, scientific status, and outlook of these authors would be overlooked; all views would be considered equally valid. By 1861, the extraordinary riddle of the animal's identity was becoming the fossil's most important feature. The conodont was acquiring the mantle of perpetual uncertainty, which others would recognize and develop. As its mythological status developed, so it acquired a long tail of speculation. It became the very portrait of puzzlement.

Nevertheless, each actor who confronted the problem thought he possessed the solution. In the 1867 edition of *Siluria,* for example, Murchison published Owen's 1858 opinion and stated with satisfaction that "the question [of the conodont] is completely set at rest."[18] It wasn't that Murchison believed the issue of affinity settled – he knew Owen was tentative on that point – but that science could now be sure that these were not the remains of vertebrates. This was the critical point for both Owen and Murchison. It was a rather secondary concern to know what the conodont actually was.

However, even if paleontologists like Owen had grown accustomed to the strangeness of the fossil past, the conodont could not easily be pushed into the shadows, for whenever these evocative little things came to light, usually as unexpected and chance finds, they astonished the finder. Throughout the nineteenth century conodont fossils were sufficiently rare to produce in collectors those same feelings of wonder and amazement that had once affected Pander. These were precisely what Bath amateur geologist and fossil collector Charles Moore felt when he found them.

Moore had long collected from the Carboniferous Limestone and had, by washing debris found in fissures in the rock and sorting them under his microscope, made the sensational discovery of Triassic mammalian teeth.[19] Nevertheless, he was not prepared for the sight of conodonts when this same technique turned them up in the late 1860s. These were quite unlike those Pander had seen: "Their forms are about as eccentric as can be imagined."[20] Rather than illustrate these fossils, he painted a word picture that began with teeth that were remarkable for their scale and imitation of a larger form. It ended with forms that to this seasoned collector seemed utterly bizarre: "a minute conical fish-tooth"; another "not unlike the central cusp of a hyboid tooth"; another with "teeth arranged on an irregular or waved platform . . . too eccentric for description"; "nine curved teeth, arranged somewhat symmetrically, graduating in height from the centre, but with a much larger fang at one end"; "a long curved tooth at one end, throwing off a semicircular spur, which passes under a base-line, on the top of which follow numerous small depressed regular teeth"; "some instances with serrations as close-set and minute as are the bristles on an insect's limbs"; like "an old-fashioned rat-trap."

Moore had discovered an Aladdin's cave of organic wonders, and it was to Murchison's *Siluria,* and particularly to Owen's diagnosis of Pander's "curious fossils," that Moore now turned in an attempt to explain his own finds. He believed that he was the first to find these fossils above the "Lower Silurian." Certainly the 1867 edition of Murchison's *Siluria* recorded that Pander and others had found them from the Upper Silurian, Devonian, and Carboniferous Limestone. As to what the fossils were, Moore agreed with Owen but added, "One objection to it is the variety of forms they present, and that we have no existing analogues." Moore's bizarre specimens seemed to remake the imagined animal. It now appeared even less certain.

As the controversy entered the mid-1870s, so its center moved from Britain to North America. As it did so, it could be said to have progressed in a limited way. First, Charles Moore was certain that conodonts were restricted to the Paleozoic era – a period of geological time stretching from the Cambrian to the Permian and typified by particular forms of life. Second, Pander's fish was dead. The animal that had left behind these peculiar fossils was an invertebrate. Moore might also have recognized that while many possibilities could be eliminated, no prime contender had been found. Owen's two favorites were attractive as much for what they lacked – a hard shell, skeleton, or backbone that might become fossilized and associated with the teeth – as for any morphologically similar thing they possessed. This is how things looked in Britain; this is what science could be said to have deduced from the evidence. However, no sooner did the conodont arrive in America than it found itself, once again, a fish.

The North American awakening owed much to the booming city of Cincinnati in Ohio, which drew in immigrants from across America and Europe. Here, in response to the local abundance of fossils, there emerged a group of avid collectors later known as the Cincinnati School of Paleontology. The Cincinnati Society of Natural History became the hub of the school's activities and spawned many eminent geologists, important papers, and fine collections.[21]

The conodonts, when they arrived, entered the American consciousness with something of a jolt. As rising Cincinnati geologist Edward Oscar Ulrich recalled in 1878, "Somewhat more than a year ago, the paleontologists in the vicinity of Cincinnati were considerably disturbed by the announcement then made, that fish jaws had been discovered in large numbers in rocks of the Cincinnati group. Two of the collectors here sent specimens of the supposed fish jaws to Dr. Newberry, and in a letter to me, he stated that he considered them to be identical with Pander's Conodonts."[22]

John Strong Newberry was director of the second Ohio Geological Survey and professor of geology and paleontology at Columbia University in New York. An Ohio man from Cuyahoga Falls, south of Cleveland, he had studied in Paris under the great French paleontologist Adolphe Brongniart and explored the famous Eocene fish excavations at Monte Bolca in Italy. He could claim the pioneering American geologist James Hall among his early influences and had shown considerable talent on expeditions into the West, most notably in establishing initial stratigraphic sections of the Grand Canyon. At the Ohio Survey, he was working on a nine-volume series giving the first comprehensive account of the state's abundant fossils, and it was in one of these volumes, published in 1875, that he described the first American conodonts, which came from the Cleveland Shale at Bedford. To him they seemed so like, and yet so unlike, fish teeth. Like Murchison before him, Newberry needed to resolve the matter of their identity if he was to achieve his goal of producing a definitive account of the state's fossils. If not the most important fossils he would describe, these were certainly the most controversial, as they had profound implications for determining the history of life on Earth. So he sought the assurances of America's greatest experts in those fields that encompassed the debate in Britain. Surely they would recognize whether these things came from a vertebrate or invertebrate animal? If only the problem were that simple. Harvard professor Louis Agassiz, a native of Switzerland and a leading figure in American science, and possibly the greatest living fish paleontologist, thought the fossils were the remains of sharks and rays. In contrast, Edward Morse, an authority on the structure of invertebrates, thought them very like mollusk teeth and could imagine them belonging to the ancestors of living species. William Stimpson, America's foremost – and soon to be deceased – expert on crustaceans, could at least confirm that conodonts were not part of the armament of his group of animals. He, too, thought they might be the lingual teeth of mollusks.[23]

Newberry's consultation had done nothing to resolve matters. American opinion was now divided along exactly the same lines as in Europe. Was it a fish? Was it a worm? No one could tell him. He had to make his own call, and being rather more an expert in fish than an encyclopedist, he tended to see the problem as Pander had. He could see that conodonts were quite unlike shark's teeth: They had no enamel, or dentine with radiating and ramifying canals, nor a distinctive base (though he knew this latter was often lost in fossilization), and could not have been set in a jaw. But his point-by-point analysis of the possibilities fell short of Owen's overview, and before long he began to speculate – "I take the liberty of offering, as a possible and plausible explanation of the enigma" – that the animal might indeed be a relative of the hagfish. While he admitted there were problems with this view, not least the "horny or chitinous" chemistry of the hagfish's teeth, he challenged anyone to compare the teeth of these two animals and not see "a very close and remarkable similarity between them." He became captivated by the thought that he possessed in his hands, in great variety, "the first fishes that existed on the globe." Although his final diagnosis relied rather more on his own gut instinct than on the opinions of the great men he had consulted, nevertheless, their names certainly gave his conclusions a sense of weight.

Ulrich was not the only one yanked into action by Newberry's controversial claim. In 1877, Professor Albert Gallatin Wetherby, a mollusk expert and Cincinnati Society member, sent George Bird Grinnell at the Peabody Museum of Natural History at Yale a large number of varied specimens for identification. This museum had been established in 1866 from a $150,000 gift made by George Peabody, uncle of Othniel C. Marsh, who became Yale's (and the United States') first professor of paleontology in the same year. Grinnell had been a student member of Marsh's first Scientific Expedition of 1870, undertaken in the company of

William F. "Buffalo Bill" Cody, which with a military escort made some remarkable fossil discoveries. Marsh, who became wealthy as a result of his uncle's generosity, built extraordinary fossil collections and famously became an antagonist in the "Bone Wars," which surrounded the discovery and naming of the American dinosaurs. Grinnell became one of Marsh's collectors and had also joined George Armstrong Custer's military expedition in 1874, which confirmed the discovery of gold in the Black Hill's of Dakota and then set off a gold rush. As white settlers encroached upon the Indians' sacred lands, the Sioux and Cheyenne left their reservations. In June 1876, Custer and the Seventh U.S. Cavalry entered South Dakota in the hopes of controlling the "Indian problem" only to meet with their own annihilation at Little Bighorn. Fortunately, Grinnell had declined to accompany Custer on that fateful trip and instead found himself alive and well, and examining Wetherby's tiny fossils. Aware of their similarity to conodonts, Grinnell asked Marsh to obtain examples of these problematic fossils from Pander himself. These confirmed that Wetherby's fossils were not conodonts. They differed in color, structure, and chemistry. The Cincinnati fossils were, Grinnell concluded, the chitinous hooks of annelids (worms). Showing close affinity to the living worm *Nereis,* he gave these new fossils the name *Nereidavus.*[24]

For Grinnell, this was an unremarkable discovery – fossil worm trails had long been known; he had simply found evidence of the worms themselves. By implication, this meant conodonts could not be worms, or at least not the ancestors of modern forms as these were now known to be little different from those still living. Grinnell was emphatic about the distinctions. But by then Grinnell's focus was already beginning to change. The 1874 expedition had given him a great interest in local Indian tribes, and he would soon become an anthropologist and in time a fervent campaigner for the protection of the American wilderness.

Ulrich, who in 1878 had just obtained his first paid job in paleontology as curator at the Cincinnati Society of Natural History, thought the conodont question unresolved but favored the views of Owen and Morse. Paradoxically, having acquired specimens of several species of the modern *Nereis* from Wetherby for comparison, Ulrich took Grinnell's conclusions to mean that conodonts might indeed be worms. Ulrich was delighted to find "a striking resemblance between their jaws or hooks, and the little Conodonts that are so common in our rocks."[25] However, the young Ulrich hesitated, aware that the greatest authorities had not been so certain and that the fossils "present so little from which accurate conclusions can be drawn, and for that reason all the theories that have been advanced to solve the enigma are based on some points, of which they give a possible, and in some cases altogether probable explanation." Without wishing to further mull over the different theories, he fell upon the annelid as the most probable answer to the puzzle and then speculated upon an ocean that "at times swarmed with innumerable worms" that left the slightest traces but "which the palaeontological collector has as yet not unearthed, but some of which he will undoubtedly bring to light in the future." Ulrich would claim, nearly fifty years later, that he had been mistaken. It seems that he had not seen any real conodonts at this time and had been looking at worm jaws. Of course, no one knew this and so Ulrich's name was added to the list of those who had expressed yet another opinion on the identity of the animal.

By the mid 1870s, then, Pander's fish still had little support, despite Newberry's attempted resurrection. Newberry may have been the senior investigator, but behind Ulrich's conclusions stood Owen's tentative yet authoritative opinion, which continued to hold sway. But now perhaps the most astute mind to engage with the problem during this initial North American phase entered the field. It belonged not to one of those elevated authorities who seemed so versed in thinking within their own boxes, but to George Jennings Hinde, an amateur British geologist then in his early thirties, the son of a paramatta manufacturer who had taken up farming and had acquired a taste for geology around the age of sixteen, inspired by the writings of that popular Scottish literary "Robinson Crusoe," Hugh Miller. Perhaps he had been motivated to pick up Miller on reading the eulogies that followed the sensation of Miller's death – he shot himself. A captivating writer, Miller had famously found extraordinary fossil fishes in the Devonian Old Red Sandstone: "Creatures whose very type is lost, – fantastic and uncouth, and which puzzle the naturalist to assign them even their class; – boat-like animals, furnished with oars and a rudder; – fish plated over, like the tortoise, above and below, with a strong armor of bone, and furnished with but one solitary rudder-like fin; – other fish, less equivocal in their form, but with the membranes of their fins thickly covered with scales; – creatures bristling over with thorns; others glistening in an enameled coat, as if beautifully japanned. . . . All the forms testify to a remote antiquity, – of a period whose 'fashions have passed away.'"[26] Miller's discovery of these fish in the 1830s was so well known that it was certain to have affected Pander's outlook, permitting him to think beyond known forms and see the conodonts as fish.

Hinde also attended the lectures of cave explorer and peripatetic popularizer William Pengelly, which he gave in Norwich in 1862. A man of solitary habits and a "silent and retiring disposition," Hinde arrived in North America to study geology under Professor Henry Nicholson at the University of Toronto – which he did for seven years. He traveled widely, and as he traveled, so his interests moved to the "well-nigh invisible contents of the rocks"; "where other paleontologists went into the field armed with hammer and chisel . . . he took with him only a magnifying lens." While at Toronto, Hinde published his first papers on a diverse range of geological topics, many of which reflected Nicholson's own interests. Nicholson, who had arrived in Toronto in 1871, had been asked by the government of Ontario to investigate the province's Silurian and Devonian fossils. Newberry also had invited him to describe Ohio's fossil corals and bryozoans for the volumes he was producing.

Hinde might have read of the debate then taking place in Cincinnati, because he decided to take up the cause of the conodont. Traveling and collecting around the Great Lakes, he knew the fossils themselves might not resolve the matter but that their associations with other fossils, and their distribution in time and space, might help formulate a more informed interpretation. On his return to Britain, Hinde also benefited from information from two contemporary British workers: Moore, who could now extend the conodont's range up to the Permian, and the Scottish fossil collector and local geologist John Smith. Smith, who was manager of the Eglington Ironworks on the Ayrshire coast of Scotland, had found conodonts in the Carboniferous in 1876, extracting them from the rotten limestone in a fashion that must have owed much to Moore: "Having collected a number of specimens, I sent them to this Society [the Natural History Society of Glasgow] for exhibition, and, if possible, to procure some information about them. As no one seemed to know what they were, I left them in the Hunterian Museum that they might be shown to visitors. The first caller who knew anything of them was Dr Hinde, who, on their being shown to him by Dr Young, at once pronounced them to be Conodonts."[27] Completely freed from the rock matrix, Hinde recognized that these conodonts were far superior to those he himself had collected in America.

Hinde's identification of the first Scottish conodonts prompted the society's vice president, John Young, to reflect on the animal's growing mythology: "Although these curious tooth-like organisms have now been known to Palaeontologists for more than twenty years, great doubts still exist as to what group of animals they belong." Smith would later find conodonts from many English localities and even noted, rather prophetically but without effect, "They are found chiefly in the powder of rotted limestone, and, but rarely, in shale."[28]

Hinde reported the results of his Great Lakes investigations in two papers read to the Geological Society of London in March 1879. His search revealed that between their earliest occurrence and the Lower Carboniferous, the conodonts had no regular association with other fossils, and thus the tooth-like fossils really were the only surviving parts of the mysterious animal.[29] He had found conodont species very similar to those described by Pander that provided the first indication that these new and tiny things were distributed over immense distances and might, as a result, be used to correlate rocks. In contrast to Charles Moore, who had been overwhelmed by their eccentric wonder, Hinde could show that simple teeth, which Pander had found in such profusion, were restricted to the older rocks while the compound forms, composed of multiple points or cusps, had a far more extensive range. The Ordovician (as we now know it) and Middle and Upper Devonian were rich in these fossils. By contrast, in the intervening Silurian and Lower Devonian he could find none.

He had found that particular strata had particular conodont distributions. One layer, less than three inches thick, exposed at Eighteen-mile Creek near North Evans on the southern shore of Lake Erie, was so crammed full of these fossils that he named it the "Conodont-bed." This rock belonged to the Middle Devonian, and in this and the overlying Upper Devonian he found the variety of forms unparalleled. And then, at North Evans in the Upper Devonian Genesee Shale, in one lucky hit he split open a slab to reveal a cluster of twenty-four conodont fossils and associated "plates," all squashed into an area no larger than a quarter of an inch in diameter. Hinde could not believe these fossils had been swept together by water currents, as the fossils themselves were too delicate and would have been destroyed. An alternative explanation was to believe that they represented a natural association, reflecting their presence in the soft tissues of the animal. They had simply been "crushed together into a shapeless mass" during fossilization. His conclusion was striking, for it suggested that different kinds of the conodont teeth occurred within the same mouth. He called the animal that possessed them *Polygnathus dubius,* a name which signaled not doubts about the truth of the association but rather its zoological meaning, as the new find did nothing to resolve the affinity of the animal or the zoological structure and function of the animal's jaw apparatus. It was a revolutionary interpretation that Hinde saw as a step closer to the truth. Unfortunately, the rest of his specimens were isolated finds and he had no way of artificially constructing associations between them. This meant that each of these fossils – if they were unlike those making up *Polygnathus dubius* – had to be given a species name of its own. In other words, for these fossils, he was forced to continue that practice Pander had reluctantly adopted. Like Pander, Hinde knew that twenty of his species names had no biological meaning; he reckoned that in all he possessed only two or three true species. In time he thought these true species would be recognized, but for the moment he had to accept a mixed system that, while it sought to establish a biological truth, sacrificed clarity and simplicity. This sacrifice was small, however, if all one wanted to do was understand the animal.

Huxley told Hinde that his *Polygnathus dubius* closely resembled the teeth of the hagfish such "that it would be difficult to prove that they did not belong to fishes of this order." But he added that no living fish held such an assemblage. Hinde had found annelid worm jaws in the same strata as those in which he had found his conodont fossils. Like Grinnell before him, he could show that the worm jaws were distinct

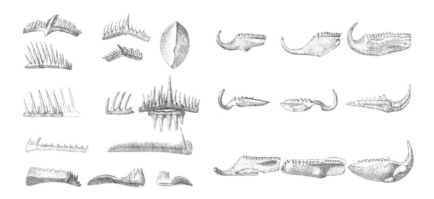

1.1. Hinde's proof. Hinde claimed that these different kinds of conodont fossil (*left*) came from a single animal, which he called *Polygnathus dubius*. He showed that these were in every way different from worm jaws (*right*). From G. J. Hinde, *Quarterly Journal of the Geological Society* 35 (1879).

and possessed the morphology and chemistry of modern forms (figure 1.1).[30] Hinde was emphatic: The animal that possessed the conodont fossils was not a worm. All the evidence pointed to conodonts being relatives of the hagfish as Pander had proposed, and he told his audience not to prejudge the past diversity of these fish on the basis of the "pauperized descendents of the present day." While Hinde admitted that his evidence permitted only tentative conclusions, in two complementary papers, he had masterfully broken the thread of thinking that linked Owen to Ulrich. Now the dissidents – Pander and Newberry – appeared prophetic. The worm was dead, the fish resurrected. And in a few years, as American geologists began to investigate the same rocks and localities, so Hinde's ideas began to be consolidated. For the first time a complex biological species of animal – *Polygnathus dubius* – began to inhabit the American scientific mind.[31]

While the French, British, and Americans had each in turn gotten a little excited, even perturbed, by the arrival of the conodont, the Germans had remained surprisingly quiet on a subject that had first been published in their language. It was the evangelizing Darwinian Friedrich Rolle who first broke this silence. Reviewing opinion, he concluded that if conodonts were not relatives of the hagfish, they might be related to the small eel-like amphioxus (or lancelet)[32] or to the sea-squirts (which have remarkable free swimming larvae). Both these latter animals possess a notochord, a structure in place of the spine, seen in the embryos of vertebrates. It suggested that these invertebrate animals were related to vertebrates and that they might be brought together with the vertebrates in a group known as the chordates. It was a suggestion with connections to the pioneering embryological work of Baer and Pander and seemed to place the conodont in the realm of those animals from which vertebrate life first evolved.

German-speaking scientists could now no longer ignore the conodont. For Germany's Owen-like encyclopedist, Karl von Zittel, who had risen to fame during the 1870s, the conodont became a problem requiring resolution. Zittel was now working on his five-volume *Handbuch der Palaeontologie*, which was to become the definitive late-nineteenth-century guide to the study of fossils. Zittel's interest in the conodont may have arisen from the necessities of the book or have been triggered by the arrival of Hinde in Munich. Hinde had traveled there to begin a doctoral dissertation on fossil sponges under Zittel – an expert in the group – not long after his conodont paper had been read in London. To tackle this new conodont problem, Zittel teamed up with established Baltic fish zoologist Josef Victor Rohon, who over the previous decade had very usefully published on the anatomy of all those vertebrate and near-vertebrate animals considered to be relatives of the conodont: sharks and rays, lampreys and amphioxus. Rohon was also no stranger to the microscope.

From a review of the literature on the enigma, Zittel and Rohon noted that the conodont's "most faithful companion" was the annelid jaw. Perhaps this was important given that the conodont fossil itself was rather poor evidence for such anomalously early vertebrate life. Believing that they had the advantage of superior technologies and new knowledge resulting from recent discoveries in other fields, the two men felt a solution to the "conodont question" was in sight. What they lacked, however, were conodonts – a deficiency soon rectified with material from St. Petersburg and from Hinde's American collections.

Their investigation revealed the presence of organic material in the conodont fossils, which tacitly signaled a link to the chitinous teeth of worms. The sheer variety of form was to them even more conclusive: "Against an interpretation as the toothed jaws of vertebrates; it points rather to seizing, catching or supporting organs." They could confirm the presence of interior structures described by Pander, but then denied most of them any significance. The lamellae survived their assault and they introduced fine radial canals, but the rest of Pander's carefully detected structures were now to be considered mere artifacts of fossilization. Their final and most effective blow, however, relied upon the kind of visual argument Hinde had used to crush the worm, only here they were to destroy the fish. These pictures conclusively showed that the horny teeth of hagfish and their relatives "do not have anything in common with the conodonts." With the fish destroyed, Pander's figures were compared with Ernst Ehler's published illustrations of bristle worms and with another kind of worm, the gephyrean, *Halicryptus spinulosus*.[33] The similarity of external form was remarkable, and stripped of the materiality of the fossil and a history of interpretation that had long doubted that simple morphological comparisons were useful, Zittel and Rohon's argument was made complete. One did not need to be able to read German; one could simply look at these illustrations and believe.

The two men were in no doubt that these animals were worms of some kind and that "from the large diversification of form it can be concluded that the Conodonts represent numerous kinds and that thus in the Palaeozoic age the coasts of the seas were populated by a substantial number of the most different worms." Only a few years earlier, Newberry had imagined something similar, only for him the seas were filled with small fish. Now that fish was dead, its lifeblood apparently an illusion brought on by erroneous interpretations. However, a more open-minded reader might have seen in Zittel and Rohon's arguments a conspiracy to murder an unsettling – if tiny – monster.

Rohon confirmed the truth of their new worm four years later when searching through hundreds of conodont fossils collected from the same rocks in which Pander had found his specimens. Rohon was looking for evidence of fish teeth and found ten examples. These specimens were confusingly similar to conodont fossils but were nevertheless different. It made him even more certain that conodonts were worms, not fish. This discovery was picked up by Britain's leading geologist, Sir Archibald Geikie, who wrote in his popular textbook, "According to Dr. Rohon, however, all 'Conodonts' are not annelidian, but include undoubted teeth of fishes with recognizable dentine, enamel, and pulp-cavity."[34] Something had been lost in translation, but fortunately for Rohon, no conodont worker ever noticed Geikie's interpretation.

In truth, in the nineteenth century, few people looked at the conodont and puzzled over its meaning. The question that troubled science was whether the fossil signaled the early arrival of vertebrates into the history of life. Progress became a matter of assertion and counter-assertion based on evidence that was evocative yet ambiguous, contradictory, or simply misleading. Owen, the encyclopedist, felt he could delimit and weigh the possibilities using the concepts of "analogue" and "homologue." In zoology, an analogue refers to a part of an animal that has the same function in different animals. The classic example is the dorsal fin of the shark, dolphin, and ichthyosaur – a broadly similar structure existing in unrelated fish, mammals, and reptiles. In contrast, a homologue is a structurally similar part that bears the same relations to other organs. The human forearm is homologous with the front leg of a cat by position, detailed structure, and embryological development. To our eyes there is an evolutionary relationship, but Owen was no overt evolutionist. Nevertheless, he understood that the detection of homologues was key to making sense of the animal world and the different categories of life that compose it. He knew that homologues provided reliable data about the zoological relationships between animals in a way that analogues did not.

Owen's overview of the natural world permitted him to see both analogues and homologues, but it also convinced him that he understood the structure of the natural world well and thus, like Zittel, was predisposed to believe that such an early fish was unlikely. In contrast, specialist workers like Agassiz had a more focused outlook; they were more inclined to think within their own specialist boxes. Faced with an ambiguous fossil whose morphology said nothing particularly certain but spoke evocatively with its shape and color, these specialists were inclined to see the object as homologous with other objects in their understanding. They were more inclined to admit them into their territories because they did not fully comprehend the full range of possibilities. Specialization permitted the ambiguous to mislead; the analogue to be perceived as homologue.

One other cornerstone of paleontology was illustration, and in Hinde's hands this was turned into a powerful pictorial argument. The technique was, however, turned against him by Zittel and Rohon. All these latter two men had to do was find the right image to convince the reader (and the non-German speaker). It was a more contrived argument than Hinde had produced, but readers are easily seduced by such things. Rarely do they consider the constructed and political nature of "logical" arguments.

The resulting seesawing of opinion, some of it carefully judged but much of it ill founded, half-baked, or simply speculative, gave the object a growing mythology. Progress had been made, not least in Hinde's complex *Polygnathus,* and it seemed likely that it would continue into the next century. But the future was not so certain. American paleontology was about to undergo a revolution that would make it central to the nation's economic future. The conodont would be torn from this arcane debate over its biological affinity and made into something useful. The science of the animal was about to get a whole lot messier.

In a few days the Eldorado Expedition went into the patient
wilderness, that closed upon it as the sea closes over a diver. . . .
I looked around, and I don't know why, but I assure you that
never, never before, did this land, this river, this jungle, the very
arch of this blazing sky, appear to me so hopeless and so dark, so
impenetrable to human thought, so pitiless to human weakness.

JOSEPH CONRAD,
Heart of Darkness (1902)

TWO

A Beacon in the Blackness

JUST THREE YEARS AFTER THE PUBLICATION OF PANDER'S BOOK on the conodont fishes, oil was discovered in the United States. A new black liquid flowed out of the ground and into American minds, altering them forever. (The conodont played no part in this discovery, but it too was altered). Notorious wastefulness followed. Successive wells ran dry. But calls for conservation fell on deaf ears, as America developed its obsession with the automobile. In 1921 there were 10.5 million motor vehicles on the road. By the end of the decade there were 26.5 million. Demand for oil grew exponentially, but without the predicted oil shortage as discovery continued to outpace demand. A plague of oil derricks advanced across the American landscape, from Pennsylvania, New York, West Virginia, Ohio and Indiana into California, the mid-continent (Kansas and Oklahoma), the Texas and Louisiana Gulf Coast, and Illinois.[1] In the unregulated American economy, oil was soon in overproduction and prices plummeted, falling below that of water in some states during the drought years of the early 1930s. By then the once buoyant economy was in freefall and oil had played no small part in that collapse.

As each state had located its own easy resources, so it profited from an oil boom, but as those wells ran dry, oil producers were forced to drill deeper and at greater expense. Ignorance of geology had led some wells to be drilled a thousand feet below the deepest productive horizon, while others had been abandoned before the oil-bearing strata had been reached. Such imprecision had accompanied attempts for coal in England more than a century before, but there a remedy had been found. The surveyor and engineer William Smith showed that rocks had a consistent order and that each contained its own peculiar fossils. This information was then used to correlate one stratum with another across country and to assign to each a relative age. This branch of geology – "English geology," Sedgwick called it – became better known as stratigraphy, and it dominated the science in the nineteenth century. It permitted ores, particularly coal, to be located intelligently and consistently rather than through luck and brawn. Oil is a rather different kind of mineral resource, but this same knowledge could help find it. There really was no excuse for committing these old errors in the New World, except, of course, for that same race for riches that had once fueled an English coal prospecting fever.[2]

The science of geology eventually began to slowly enter the thinking of American oil companies around the turn of the century when it was understood that oil often accumulated under anticlines (rocks folded into arches). However useful such knowledge, it often proved insufficient. By 1913, for example, all the known anticlines in Illinois had been drilled. Oil production in that state fell and continued to do so until 1936. It only rose again when new technologies were introduced into oilfield exploration.

Perhaps surprisingly, no one was consistently examining the cuttings brought to the surface in drilling operations. Ordinary fossils – the things Smith had used to order his rocks – were often destroyed by the drill. After World War I, however, attention turned to microscopic fossils, and particularly to those simple single-celled animals known as foraminifera. By 1925, the seven largest oil companies on the West Coast possessed laboratories that employed some twenty-three people in geological investigations. Soon nearly every wildcat was sampled and studied, and hundreds of geologists were engaged in this work. And as the paleontological community grew and diversified, so it professionalized itself. Fossil-focused organizations burgeoned. The Paleontological Society was established in Baltimore, Maryland, in 1908, when "the immensely rich oil fields of the Mid-Continent were being discovered." The American Association of Petroleum Geologists (AAPG) was founded in Tulsa in 1917. Its *Bulletin* grew from 159 to 1,319 pages per annum in just eight years. By then it was issued monthly and was under the control of editor extraordinaire Raymond Moore of the University of Kansas. Its physical weight spoke of how geology in America had changed: The science was now dominated by a deeply utilitarian outlook that sought only to serve the needs of business. The Society of Economic Paleontologists and Mineralogists (SEPM) was established about a decade later, at a meeting of the AAPG, again in Tulsa. SEPM's focus was to be devoted almost entirely to the utilitarian science of microscopic fossils, or microfossils. Its *Journal of Paleontology* first appeared in July 1927.[3]

With this changing professional demographic, paleontology more generally underwent a shift of emphasis: Universities began to introduce courses in "micropaleontology." At the University of Chicago, Carey Croneis set up a masters program in the subject, converting the top floor of the Walker Museum into a laboratory. Paleontological doctorates and masters dissertations were soon dominated by this new specialism. Increasingly, paleontology took on a business aspect and began sweeping up every possible tiny fossil, turning fossil curiosities into objects of great utilitarian potential. Laurence Sloss, a student at Chicago in the 1930s, later recalled how the Chicago department's "graduate body of mature men and women" swelled at this time "because their jobs in oil and mining had evaporated with the Depression."[4] University departments were reinvigorated by this influx and there was an explosion of interest in these newly fashionable microfossils.

It was one thing to know that microscopic fossils might be useful, but in the 1920s it was difficult to know which ones. Many early studies considered them all. U.S. Geological Survey (USGS) worker Paul Roundy, a recent convert to this microscopic world, now understood that some foraminifera were useful but that many were not. Sponge spicules and tiny snails were sometimes numerous but hardly studied. Worm jaws were too rare. Ostracodes, or "seed shrimps," on the other hand, were abundant and most important. Plant seeds and spore cases held potential but again had attracted little attention. On the conodonts, however, Roundy was emphatic: "These I believe will prove to be of considerable importance in stratigraphic work when they have been more fully studied."[5]

Roundy was at the time investigating Mississippian (Lower Carboniferous) rocks in Texas. Rocks of this age and type had, since 1912, fueled a feud between government geologists. They were part of a sequence of black shales that occur throughout much of the interior of the United States and part of Canada and range "from a featheredge to several thousand feet in thickness."[6] At some point in this shale sequence the Devonian passes up into the Mississippian, but geologists could not agree precisely where. Knowing the position of this boundary was critically important and not something that could be ignored or worked around. But the fossil evidence, which in other parts of the world had decided the matter, remained elusive or contested in North America. The problem proved persistent and threatened to make a nonsense of the burgeoning science. Even in 1931, it was still possible to claim that rocks of this age "have almost as many names as there have been sections studied": Chattanooga, Hardin, Berea, Bedford, Sunbury, Saverton, Grassy Creek, New Albany, Sweetland Creek, Arkansas, Woodford, Cleveland, and Ohio.[7] Different names for essentially the same thing, and the same name for different things. What had started as the "Ohio Shale problem" soon became a more pervasive "black shale problem."

At the heart of this controversy was E. O. Ulrich, who by the beginning of the twentieth century had risen to become a USGS geologist of some distinction. Known for his outspoken views, one friend half-jokingly remarked, "The trouble with you, Ulrich, is that you think you are God." Ulrich had been at his most godlike in 1911, when he published a huge and radical revision of the older fossiliferous rocks (the Paleozoic). Shaped by his own peculiar outlook and his dislike of the "paleontological autocrat" who privileged fossil evidence above all else, he refashioned the geology of America. So all-encompassing was the resulting publication that all who now studied these older rocks had to pay attention to his opinions, and this was particularly true of those wishing to resolve the problem of the black shales. For here, Ulrich had been at his most contentious. With a few taps of his typewriter keys, he had moved great swathes of black shale into the Mississippian. Providing little evidence to support his geological choreography, opinions became inflamed. His only follower was his protégé, Ray Bassler, who published a supportive paper the same year. A fellow Cincinnati geologist, Bassler as a boy had adopted Ulrich as "Uncle Happy" and followed him to Washington, where he would establish a career for himself as a paleontologist at Columbian University (now George Washington University) and the U.S. National Museum (the Smithsonian).[8]

In February 1912, Ulrich's colleague at the Survey, Edward Kindle, pressed Ulrich to publish his data. Kindle did so in print, demonstrating his superior knowledge of the black shale problem and implicitly suggesting that Ulrich and Bassler lived in a hypothetical world far removed from reality: "This opinion concerning the age of the Chattanooga shale is comparable to some which have followed it in the poverty of evidence on which it rests and the positive phrasing which might mislead one unfamiliar with the subject to suppose that it represents an established fact." It was now that the conodont was introduced for the first time as a key actor in the debate. Kindle continued, "Although generally considered to be nearly barren of organic remains, the writer has found the carbonaceous beds of the Chattanooga shale to carry a conodont fauna which is quite as abundant in the lower or Huron shale of Ohio and Kentucky as it is in the upper or Cleveland shale. These minute but beautifully preserved fossils may be obtained at any locality and at any horizon in the black shales from Lake Erie to Alabama. These fossils have long been known in the Ohio Shale, but with the exception of a very few species have remained undetermined and undescribed. When they have been described and the species which are confined to the upper and lower horizons of the shale distinguished, they will prove to be an invaluable aid in correlating the different parts of the Ohio Shale with their equivalents in the Chattanooga Shale in Kentucky and further south. Until this has been done, however, any attempt to make use of these fossils in correlating subdivisions of the Ohio and Chattanooga shale must be considered premature and futile." Bassler had also observed these fossils in the shales but had seen no particular use for them. Now, with Kindle's words, the conodont was thrust into the limelight.[9]

Kindle believed the black shales were mainly Devonian. Indeed, this was the orthodox view until overturned by Ulrich. Kindle knew Roundy already agreed with him and he wrote to other authorities to get their opinion on the matter. Among these was E. B. or "Ted" Branson, a fossil fish specialist who had collected from the black shales. Branson, who was soon to switch his focus to conodonts, was only too happy to agree with Kindle. The Survey's specialist in Carboniferous fossils,

George Girty, on seeing Kindle take his stand against Ulrich, asked him for his opinion on the Bedford Shale. Girty admitted that for years he had been "playing the pendulum" on the subject, repeatedly changing his mind. He now supplied Kindle with evidence to convince him of its Devonian age, telling him that on this matter they should stand together. Kindle was delighted to have such an experienced authority on his side. Now the USGS was split into opposing factions, and Ulrich led the minority.

Ulrich acted quickly to rebuff Kindle's attack, publishing his data in the August issue of the *American Journal of Science*. Probably as a result of the editor's sleight of hand, Kindle's counterview appeared in that same issue, just a few pages later. The detail of their arguments need not concern us: Ulrich and Kindle were now head to head, the tenor of their papers scarcely concealing their personal animosity. Ulrich's arrogant and lofty pugnacity was met with Kindle's blows below belt. To the falsetto-voiced Kindle, Ulrich was a descendant of the German geological pioneer Abraham Werner, a man long stereotyped as the overly theoretical Antichrist of modern geology. Behind the scenes, Survey geologists discussed the assault on Ulrich's position as many of his key strata began to fall to the other side.[10] And when the appointment of a new Devonian specialist came under discussion, there was a campaign for it not to be an Ulrichian.

Matters came to a head when a paper that had been submitted for publication was rejected by a referee. Neither Kindle nor Ulrich had written the paper, but Kindle had supplied the fossil identifications and Ulrich had been the referee. Kindle became enraged when he heard the news and wrote to the USGS's chief geologist, David White, telling him, "You will find it increasingly difficult to get any self-respecting palaeontologist to take a place in your organization, in which a man with Ulrich's peculiar views is invited to dominate everything relative to Paleozoic palaeontology. . . . You know as well as I do that Ulrich delights in scrapping on any pretext whatsoever."[11] He sent the letter to Girty, who checked into the matter. He found that Kindle was inventing demons. Ulrich had behaved impeccably. But by then Kindle had had enough and he left to join the Canadian Survey, there to rise to an elevated position (not least metaphorically so, for Mount Kindle was named in his honor).

✳   ✳   ✳

The excess heat in the black shales dispute dissipated with Kindle's departure. In what was still a tiny geological community, where everyone knew everyone else, individuals were inclined to takes sides, and the science and its personalities became inextricably entangled. All, however, agreed on one thing: The conodonts might provide a solution to the black shales problem. By 1914, Ulrich believed he possessed distinctive conodont faunas that supported his views.[12] The fossil as a result moved closer to becoming a utilitarian object, valued only for what it could tell the field geologist. "The study of fossils apart from their stratigraphic relations is pure biology," Ulrich wrote.[13] And those who thought otherwise, in oil-obsessed America, were increasingly finding themselves in the minority.

William Bryant, Devonian fish specialist and director of the Buffalo Museum of Science, did not think as Ulrich did, but he too valued the conodont's utilitarian possibilities. A supporter of Kindle, Bryant washed and boiled lumps of George Jennings Hinde's "Conodont Bed" from Eighteen-mile Creek and discovered that conodonts made up 50 percent of the mass left behind.[14] He found the fossils easy to identify once freed from the surrounding rock and could show that Hinde, who had examined his fossils on the surfaces of the rock, had made a number of misidentifications. He also thought Hinde's anatomically complex *Polygnathus dubius* another mistake. Both he and Roundy believed the association was merely coprolitic (fossilized feces) or "the ejecta of some fish." More than anything they wanted the conodont fossils to be effective "horizon markers," which meant that each kind needed to be narrowly defined. In pursuit of this goal, Bryant disassembled Hinde's *Polygnathus,* restricting that name to the distinctive plate-like forms and renaming all the other fossils.

Ulrich and Bassler were also pushing ahead with these "difficult toothlike organisms," hoping to locate a definitive answer to the black shales problem that supported their view. By the early 1920s, Bassler had began to nurture an interest in conodonts in Ray Hibbard, an optician, World War I veteran, and fossil enthusiast living in Buffalo. Over the coming decades, Hibbard would maintain regular correspondence with

Bassler and contribute significant collections to the National Museum long after Bassler's own interest in these fossils had waned.[15] In time, Hibbard would develop his own private museum and library, swap collections with distinguished foreign paleontologists, experiment with acids and photography, and even install a microfossil washer machine. Hibbard was that vital collecting cog that had long powered the geological engine. And for those who drew upon his services, he had the added advantage of possessing no great ambition to publish. He was simply a servant to others.

Ulrich and Bassler now drew upon collections belonging to Hibbard and to the National Museum – probably the best collections of conodonts in the country – to work on "a reasonable and, it is hoped, natural classification." Bassler presented the preliminary results of this work at a meeting of the Paleontological Society in Ithaca, New York, in December 1924. Two years later, their jointly authored paper appeared.[16] In it, and with relatively little fanfare, the two men claimed that "conodonts" – the fossils – were the remains of fish and that each type of conodont belonged to a distinctive species. The simple pointed ones belonged to fish species related to hagfish and their kind, the more complex arched and pick-shaped ones possibly belonged to relatives of the sharks and rays, while those they called "dermal plates" may have belonged to a distinct group but were not to be considered true conodonts. The two men also rejected Hinde's complex *Polygnathus*, claiming it was contradicted by fish biology. But in that subject they had no real expertise. The arguments they made for their simple fishes were superficial. Ulrich now admitted that his earlier preference for a worm had been a mistake – he simply had not then seen "true conodonts," and when he did, he soon changed his mind.

Armed with a comprehensive history and bibliography of the subject, which had been prepared by the museum's Grace Holmes, Ulrich and Bassler grouped the conodont fossils into four families according to their shape: simple conical teeth, pick-shaped teeth, arched bars, and plates.[17] In doing so, they increased the fossils' usability by making them easily recognizable. They then sought to demonstrate that conodonts could resolve the black shales dispute. They said it was possible to prove that none of the conodonts recorded in the Devonian Genesee and Portage rocks of New York state could be found in the Ohio black shale, the upper New Albany in Kentucky, or the Chattanooga in Tennessee and Alabama. These Genesee and Portage fossils could, however, be found in the rocks below these horizons. They concluded: "In so far then as the evidence of conodonts is concerned, the post-Devonian age of the Chattanooga and Ohio shales, as long advocated by the senior author, seems conclusively established." But how could they have concluded anything else?

It was this apparently unequivocal solution to the black shales problem that thrust Ulrich and Bassler's paper into the limelight. Its readers now took the conodont seriously. And by simplifying the animal so that each kind of fossil could be said to represent a single species, the conodont became as easy to use in stratigraphy as single-celled foraminifera.

By the time of Ulrich's retirement, in 1932, the paper had been widely publicized in the United States and abroad, and their conodont classification was said to be in "extensive use." By then, microfossils were so fashionable that Ulrich's brief excursion into conodonts was considered a career high point. It had been a groundbreaking piece of work that Bassler thought had performed a little revolution.[18] No one saw it as an attempt to take a resource Kindle had identified and turn it against him. But if Ulrich really had manipulated the fossils to support his outspoken views, or simply been blind to objective reason, surely the conodont would in time expose him, for those who now examined the black shales were as attentive to the conodont as they were to the words of Ulrich.

By 1932, Clinton Stauffer was a seasoned professor at the University of Minnesota. A specialist in Devonian stratigraphy, with experience of the contentious black shales in Ohio and Ontario, he had worked for the Canadian Geological Survey and was another of Kindle's associates, agreeing with him on the matter of the black shales. By happy accident, Stauffer was now to become a national expert in conodonts as a result of taking an interest in building work on the campus. To his delight he discovered a thin lens of sediment in the Ordovician Decorah Shale exposed in the excavations that contained finely preserved conodonts.

Recognizing the novelty of this find, he described and published the conodont fauna knowing that all his species were entirely new to science. No one had previously looked for conodonts in this rock. He imagined Ulrich and Bassler's fishes "may have migrated over a much wider area than that covered by the Decorah shale" and were likely to be found in other rocks of similar age that might then be correlated with the Decorah. Sure enough, a short while later, Fanny C. Edson of the Gypsy Oil Company in Tulsa, Oklahoma, got in contact with him having found conodonts in well cuttings that penetrated the Decorah Shale in distant Reno County, Kansas. Edson, who was one of many women involved in oilfield geology at the time, needed to be sure that her local Decorah really was the same as that named after the town in northern Iowa. Stauffer told her that the conodonts proved it. News of Stauffer's newfound expertise spread and soon he found himself collaborating with foraminifera worker Helen Plummer, one of the instigators of the Society of Economic Paleontologists and Mineralogists and half of a noted paleontological marriage. Together, Stauffer and Plummer pioneered an investigation of the conodonts of a wholly different group of rocks belonging to the Pennsylvanian. By chance, then, Stauffer had stumbled into a paleontological lacuna and at once found himself the expert. As a result, and seemingly with no particular plan in mind, he then began to populate the American landscape with these mysterious fishes. At each location, he put in place a time marker to which others could correlate their own local rocks using conodonts.[19]

Another in this advance guard taking up the conodont was Frank Gunnell, of the University of Missouri. He began by undertaking reconnaissance collecting throughout Missouri, Kansas, and Oklahoma, finding abundant conodonts in the Devonian, Mississippian, Pennsylvanian, and Permian.[20] He discovered that they occurred mainly in shales and decreased in abundance in sandstones, conglomerates, and limestones. This idea – that conodonts mainly occurred in shales – soon became embedded, and it was on these rocks that conodont-seeking geologists would concentrate their efforts.

Gunnell realized that the conodonts' size and occurrence in otherwise "unfossiliferous" strata had made them invisible. It now made them peculiarly useful. Indeed, it seemed quite magical that these fossils should come to light in rocks that so desperately needed them. Like Stauffer, he too thought Ulrich and Bassler's notion of free-swimming fish added greatly to the fossils' utilitarian potential. That potential was heightened still further by Gunnell's belief that many species had a very limited stratigraphic range. It meant that a single species might signify a narrow period of geological time yet be found across wide geographical areas. Buoyed up by a new optimism, which came from finding these fossils in unexpected places, he felt that lithology and depositional environment had little impact on their distribution. If this was the case, then the conodont was the ideal utilitarian fossil, provided it could be found in sufficient numbers. Gunnell now turned his attention to the conodonts of the Pennsylvanian, wishing to document the stratigraphic range of every species he found. Noting and valuing every ridge and node as he identified them, he soon possessed more than one hundred new forms.[21]

A third pioneer was Chalmer Cooper, a West Virginian who had returned from fighting in World War I to complete a degree in geology and engineering at the University of Oklahoma. It was in this state that he found his first job in the science, as chief geologist at the Oklahoma Geological Survey. Inspired by Ulrich and Bassler's assurances of a distinction between the conodont fossils of the Devonian and Mississippian, he began his studies by gathering reference material from Ohio in the late 1920s, then used this to demonstrate the Mississippian age of several local rocks. As a disciple of Ulrich and Bassler, though unconvinced by their fish, Cooper soon found himself embattled with Roundy, Girty, and others at the National Museum.[22]

Fired up by Gunnell's discovery of conodonts in the neighborhood of Columbia, his colleagues, Ted Branson and "Doc" Mehl, hatched a plan to reveal the stratigraphic utility of the conodont completely. They were to do what Gunnell had began to do but on a grand scale. Both were former students of Samuel Wendell Williston, who himself was a former assistant of dinosaur addict Marsh. Williston had risen to a position of distinction as a vertebrate paleontologist, and Branson had followed him from Kansas to Chicago in order to continue his studies. The first to arrive in Missouri, Branson had orchestrated a renaissance in the Geology Department's fortunes. Ambitious, and still known as a fish paleontologist, the conodont fish must have come to him as something of a revelation. No other fish fossils held this kind of stratigraphic potential. Branson's collaborator, Maurice Mehl, had as a young man teaching in Oklahoma been active in the founding of the A A P G. They made a good team. Branson was a man of ambition and strategy, while Mehl was regarded as an exemplary scientist.[23]

Their extraordinary conodont odyssey started in 1930, when the two men, with the help of their students, began an assault on the Missouri strata. They published as they went, mapping a geological landscape strewn with the teeth of thousands of corpses and fundamentally altering the animal and its utility. "We discover new conodont localities and horizons nearly every week," they reported in 1933.[24] Their data began to accumulate so rapidly that they, rather remarkably, acquired permission to fill all four issues of the university's research journal, *University of Missouri Studies,* with conodonts. At 349 pages in length and published between June 1933 and October 1934 under the title *Conodont Studies,* this was the first book-length treatment of these fossils.

Like Ulrich and Bassler, Branson and Mehl believed conodonts were the teeth of more than one group of primitive fish. It was the finding of bony material attached to some conodonts around this time that convinced them that this was the case, not the weak arguments that had come from the National Museum. They also differed from the Washington men on one other important point: They believed "the teeth in some species were arranged shark-like with two or more kinds in the same mouth." But, ever the utilitarians, they recognized that the "likelihood of finding the teeth in their original associations is remote." They used this to justify giving each kind of tooth its own species name, and in doing so they included the leaf-like forms Ulrich and Bassler had considered dermal plates. These were now thought to be "crushing teeth" and were to become particularly important. While praising the "epoch-making contributions" of Ulrich and Bassler, they thought the Washington classification premature and refrained from using it. Branson and Mehl were their own men and were determined to forge their own way. They saw conodont studies as a blank slate and made it their mission to fill it. They wished to create a new research field and place themselves at the head of it.

They realized that this would involve both evangelism and education and so began *Conodont Studies* with a "how to do it" guide. It gave assurances to those who doubted they would be able to find what were still largely unknown fossils. Branson and Mehl encouraged these novices to exploit plastic and sandy clays, suggesting that they were often more productive than many shales. They also implied that limestones were often devoid of conodonts, so making their investigation unattractive. Instead, readers were advised to adopt mass-processing techniques that involved boiling samples of sediment, sieving the results, and then separating fossils from the remaining sediment on the basis of their differing densities using heavy liquids, before finally picking over the resulting residue to extract the fossils. This, they claimed, was far more reliable than picking individual conodonts off bedding planes by hand. "Starting with a meagre knowledge of general micropaleontological technique" and unable to draw upon the technical expertise of "industrial organizations" (oil companies), who saw such information as commercially sensitive, they pulled together an arsenal of methods from their own innovations and borrowings. Among these was the innovative use of stereoscopic photography to study the fossils. They even went so far as to publish stereoscopic images, thus enabling their readers to see the fossils in 3D and better understand their complexity.

Branson and Mehl's optimism for these new fossils was, however, overshadowed by one great fear: contamination. They worried that their samples might be corrupted by splashes from boiling pots and residues left on sieves. But with care they knew they could avoid these problems. They were rather more concerned about natural contamination resulting from older rocks and fossils being eroded and redeposited as younger rocks, and so mixing together conodonts belonging to different periods. They also imagined younger deposits containing conodonts penetrating cavities in older rocks, much as Moore's Triassic mammals had fallen into fissures in the older Carboniferous Limestone. What they did not consider was that their minds could also become contaminated as a result of this obsession. It affected their ability to see. An unexpected conodont was, for them, always an error and never simply a rarity. But perhaps they did not have the luxury to think otherwise in a science so new.

Branson and Mehl's "how to do it" guide was followed by papers discussing rocks ranging in age from the Lower Ordovician to the Lower Mississippian. In these, the two men demonstrated how conodonts could be used to solve existing stratigraphic problems. The Ordovician Harding Sandstone, for example, was problematic because it held Ordovician invertebrate fossils and fish fossils that looked rather Devonian. The conodonts alone proved that this sandstone could be correlated with the Middle Ordovician Joachim of Missouri. The fish were merely deceptive. The conodonts were, in contrast, quite distinctive: dull and amber colored, "a very primitive group" that possessed what is "best described as 'fibrous' structure" that splits lengthwise. They seemed to be fused to parts of the animal's jaw.

In the Lower Ordovician Jefferson Formation of Missouri, they found their oldest conodonts. They were simple, diverse, and distinctive, and they offered the possibility of correlating this rock for the first time. An examination of the Silurian Bainbridge Formation of Missouri produced rather different results. Here they were pushed to find any conodonts at all, but when they did they were a predictable mix of "typical Ordovician genera" and "obvious forerunners of typical Devonian and Mississippian genera." Although these rocks were positioned between the Ordovician and Devonian, they could not connect these two groups of conodont fossils. Indeed, their difficulties in finding conodonts suggested to them that the Silurian was a low point in their history.

As they worked their way through the stratigraphic column, a more comprehensive picture of changing conodont faunas began to emerge in their minds and it became possible to guess the relative age of the strata on this basis. Locating rocks of equivalent age was not, however, as easy as might be imagined. The geological map of America, which can so easily be painted in large blocks of uniform color, conceals a complex, three-dimensional patchwork of rocks. It was from the pieces of this jigsaw that Branson and Mehl extracted their fossils and thus began to solve the puzzle of their age relationships. It was to their advantage that they treated their subject abstractly, as by removing their samples to the laboratory, they also removed themselves from the problems and complexities of the field. It simply became a matter of matching fossils at different sites. This also helped them to imagine the conodont fossils as an evolving continuum of forms and to guess the form of those conodonts that occupied gaps in their knowledge. They became connoisseurs of their fossils, and this gave them an ability to predict.

※　　※　　※

Before long, Branson and Mehl's climb up the stratigraphic ladder led them to consider those conodonts that marked the division between the Devonian and Mississippian – those contentious rocks Ulrich and Bassler had claimed to have tamed using conodonts. To aid their interpretations, Branson visited Hinde's type specimens at the British Museum in London. With a lack of empathy for a science still in its early days, Branson was rather dismissive of Hinde's achievement: "He spent twenty days in making the study and it was far from thorough in many respects." Hinde's material was embedded in matrix, and for this reason Branson considered it imperfect, but it had been supplemented by specimens from Bryant. Branson also had some five hundred specimens of his own from Eighteen-mile Creek, which his son Carl had collected. Armed with this material, Branson and Mehl now entered this contentious terrain.

Their study, and the whole of the third issue of *Conodont Studies,* published in June 1934, concerned the Grassy Creek Shale of Missouri. Branson and Mehl's conodonts correlated it with the Chattanooga, the low Huron of Ohio, the Portage, the Hardin, and the Woodford; across them all there was a fairly consistent and distinctive fauna very much like that Bryant had described for the Genesee, which they took to mean that all these rocks were Upper Devonian. Ulrich and Bassler had claimed the Chattanooga and Hardin as undoubted Lower Mississippian precisely because they had not found Genesee or Portage conodonts. Branson and Mehl, however, drew upon their superior connoisseurship to claim that the faunas found in these different rocks were more closely related to each other than to those of undoubted Middle Devonian or Mississippian rocks. Indeed, they felt that distinctive faunas marked the Devonian-Mississippian boundary and these were not those claimed by Ulrich and Bassler. They also pointed out that Ulrich had foolishly dismissed the evidence of Devonian fish fossils. Ulrich's blinkers were now fully exposed, his godlike command of nature a matter of self-delusion.

Branson and Mehl's new faunal criteria boiled down to just six seemingly diagnostic groups of conodonts: The "*Polygnathus*-like genera, *Polylophodonta, Ancyrodella, Ancyrognathus,* and *Palmatolepis*" were restricted to the Upper Devonian and were only ever found in the Mississippian as contamination. They were replaced in these latter rocks by *Pseudopolygnathus* and *Siphonognathus.* All of these kinds of conodont would become known as "platforms." Ironically, Ulrich and Bassler had dismissed them as being "dermal plates," not conodonts. Now they were the most important conodonts of them all.

Branson and Mehl were in awe of the Grassy Creek conodonts, which were "in a highly plastic stage of their development." The distinctive forms of major groups seemed to merge seamlessly one into another. Some genera, such as *Polygnathus,* were so variable that species could not be distinguished. Small forms could not be related to larger ones. The sheer variety and blending of form was perplexing and yet marvelous. And alongside these remarkable forms were other conodonts that seemed to belong to an earlier era. Branson and Mehl thought they must be contamination and they established a principle to guard against it: "It is the new elements that determine the age."[25] Old ones, they presumed, may result from erosion and redeposition.

As the final issue of *Conodont Studies* was packaged up for the presses, Branson and Mehl must have realized the enormity of what they had achieved. They, rather than Ulrich and Bassler, had laid the bedrock for a new science, and across the American landscape they had left markers others could now use to locate the age of their own rocks using conodonts. Ulrich and Bassler may have asserted the utility of these new fossils, but in doing so they merely echoed Kindle's call. The proof of the method had not been theirs; they had merely stated what they thought they knew. The proof belonged to Branson and Mehl, who now said it was only necessary to identify a few kinds of conodonts in order to resolve the problem of the black shales. As a consequence of this work, in 1935, Stauffer felt it was possible to say that "Conodonts in Paleozoic sediments assume much of the importance of the Foraminifera in later sediments."[26]

*   *   *

But not everyone was enamored with the Missouri conodont factory. Ulrich and Bassler's disciple, Chalmer Cooper, must have felt his own work had been devalued by Branson and Mehl's proof that the Washington men had been wrong. In a review of *Conodont Studies,* Cooper found every possible excuse to complain: about their preference for mass processing, their re-identifications, their use of broken specimens so "valueless to other workers," and their splitting of fossils into narrowly defined species and genera "to a degree that makes it difficult to recognize them."[27]

At Indiana University in Bloomington, the young John Huddle also grew concerned about the new Missouri orthodoxy. While Branson and Mehl were wrestling with the Grassy Creek, Huddle had been completing his doctoral thesis on the New Albany Shale of Indiana.[28] Like the Grassy Creek, this was a disputed sequence. Most thought it entirely Devonian, but in 1911, Ulrich had suggested that it might be part Mississippian. Bassler had then supported this view.

Huddle had entered this disputed territory quite by accident, having been shown some Devonian brachiopods from near the top of the shale. This seemed to challenge Ulrich and Bassler's claim. It was while Huddle was investigating the source of these fossils that he came across some conodonts that he knew from Ulrich and Bassler's recent work might resolve matters. He began his fieldwork in the summer of 1932 and was soon corresponding with Bryant and Cooper. This "brownish to black in color, more or less massive, hard, brittle" shale becomes fissile when weathered and is easily split, giving it its local name of "black slate." But Huddle found it resisted bulk processing. It was a technique Huddle distrusted anyway, having found that it destroyed the most delicate species. Instead, he painstakingly extracted or revealed his fossils using needles, in the process producing one of the most thorough investigations of those early years. Nevertheless, Huddle felt disappointed with his results. Of the 158 species he described, only 37 had been found previously. Only they would offer any chance of correlating these rocks with others, thus proving their age. He had no contact with Branson and Mehl and knew nothing of their Grassy Creek work. Huddle tentatively concluded that the lower part of the shale seemed to correlate with the Devonian Genesee. Those conodonts from the middle were most like those of the Hardin and Chattanooga but were also associated with a few Portage and Genesee species. These were found well below the Devonian brachiopods and proved that Ulrich and Bassler's supposedly distinctive Chattanooga species were not Mississippian. Conodonts from the upper part of the rock included the most new species, and while he found Chattanooga and Hardin species among them he could not say that the Mississippian had been reached. He did, however, find a shark's tooth, which Bryant suggested might indicate a possible Mississippian age. Huddle had independently and simultaneously reached the same conclusions as Branson and Mehl: Ulrich and Bassler were wrong. Cooper, who was now at the Walker Museum in Chicago and working on his own doctorate degree, remained oblivious to these new discoveries and continued to work on the Oklahoma rocks in the belief that Ulrich and Bassler were right.[29]

Branson and Mehl soon consolidated the position of their "index genera."[30] The "index Devonian genera" were now *Icriodus, Palmatolepis, Ancyrognathus, Polylophodonta* and *Ancyrodella.* The "index Mississippian genera" were *Pseudopolygnathus, Siphonognathus, Pinacodus* and *Solenognathus.* Such clear presence-absence criteria were a stratigraphic ideal, and among them there was one particularly widespread Devonian form, *Icriodus,* the various species of which also permitted the division of the Middle and Upper Devonian. There was nothing like it in the Mississippian and it appeared to have no descendants. There was no ambiguity about this fossil, and to prove their point they revisited some disputed and imperfectly understood black shales, including those studied by Cooper and Huddle (they did not mention Cooper by name, perhaps wishing to avoid a personal dispute). They confirmed that the upper beds of the New Albany Shale were indeed Mississippian and moved all Cooper's Mississippian rocks into the Devonian. Now it seemed they were able to resolve the black shales problem on the basis of a single type of conodont.

Remarkably, this advanced state of knowledge was achieved in 1938. But by then even Branson and Mehl knew that progress had been at the

expense of a chaotic proliferation of names. Unrestrained by any biological paradigm, workers had attempted to locate as many distinctive
species as possible. Now a new rationalism broke out, led by the soft-
spoken and self-contained Will Hass. Hass had moved to the National
Museum in 1935 to work with the USGS. Here Girty trained him in the art
of micropaleontology and Roundy gently directed him toward the study
of conodonts. In 1938, Hass coauthored an account of some Montana
conodonts, calling on Cooper to help with his identifications. In all Hass
identified forty-six new species. Here, a little naïvely, he took a swipe at
Branson and Mehl's index genera – a division Stauffer had recently and
unknowingly confirmed. Hass claimed that Branson and Mehl's Mississippian *Siphonognathus* also occurred in the Devonian, while their Devonian *Palmatolepis* could be found with two of their Mississippian indicators. Branson and Mehl told Hass that he must have used contaminated
samples, but Hass thought the associations a "curious coincidence."[31]

In the following year, Cooper was bedazzled by 256 remarkably preserved conodont species in an eleven-inch-thick Mississippian (so he
believed) shale, "the most striking assemblage of these forms encoun-
tered in my experience," he noted. "Some species are so preserved that
they are almost transparent and appear as though constructed of cellophane, giving the appearance of artificiality." As we shall see in the
next chapter, this extraordinary material permitted Hass to undertake
internal investigations of conodont fossils and reveal the true extent of
the name proliferation. It fundamentally altered Hass's outlook, and in
1943 he returned to his Montana specimens and decided that he had not
found a single new species.[32]

Two years later, Branson and Mehl began to talk defensively about
the actions they had taken. Critics were all too ready to forget the wilderness out of which the two men had led the conodont. Branson and
Mehl argued that without names there could be no progress, even if the
full history of a form remained unknown at the time. Each named species then existed as a discrete entity in time and space. It possessed no
future, no past, and no geography. It was from such flags in the ground
that science built its knowledge, and inevitably it would then need to
make corrections. It was to be expected that, as knowledge matured, a
period of name inflation would be followed by deflation as they realized

that the same fossil found in different rocks had been given different names.[33]

The other source of error was contamination, which Branson and Mehl wrote about at length in 1941, using examples from Huddle, Cooper, and Stauffer. Detection, they said, in a rather circular argument, relied upon "knowledge of conodont successions rather than in field evidences." There were, they said, such self-evident reports of mixing as the discovery of unworn conodonts in the Triassic of Germany that the problem could not be denied. Branson's son, Carl, had been one of the first to find Permian conodonts in the United States and fancied he saw the "senile specializations of a rapidly declining group."[34] Branson and Mehl thought the German fossils looked like those found in the American Pennsylvanian and Permian and must therefore result from contamination of the overlying Triassic rocks. To have persisted into the Triassic, these animals would have survived the great extinction that took place at the end of the Permian and crossed from the Paleozoic world into the Mesozoic virtually unchanged. The idea seemed impossible.

At the heart of the contamination argument was a growing feeling, which had certainly taken hold by 1938, that the conodont fossils showed progressive advancement over time. One could guess the age of these fossils simply from how advanced they seemed. It was a powerful idea. One that would prove irresistible to later generations and that, in this era, caught the imagination of Bill Furnish, a young doctorate student at the University of Iowa. Here a hard-working cephalopod paleontologist, A. K. Miller, was pulling together the beginnings of one of the most influential conodont research schools of the mid-twentieth century. To achieve this, Miller, a former student of Ted Branson, drew upon the influence and encouragement of his Missouri and Minnesota neighbors. Furnish was the first product of this new school and was already examining some of the oldest faunas and giving serious consideration to classification. He concluded that the fossil record indicated "beyond reasonable doubt that compound teeth were derived by several distinct lines of evolution."[35] In other words, and contrary to what most believed, there was no single evolutionary progression; development had followed a number of parallel paths. This meant that Ulrich and Bassler's classification, which had grouped together "teeth" on the basis of superficial resemblance, was no more reliable than Pander's. These "lines of evolution" were, Furnish believed, open to interrogation and would permit a truer classification. Only by this means could the science move beyond the connoisseurship of abstract forms. But in order to achieve this, further action was needed to tackle the proliferation of names.

Resolution of this problem required a curatorial mind – someone who could collate, distinguish, and arrange. Fortunately such a mind was possessed by Mehl's doctorate student and protégé, Sam Ellison. Ellison sought to bring order to the Pennsylvanian.[36] Reviewing thousands of specimens, when his predecessors Gunnell, Stauffer, and Plummer possessed only a few hundred, he could confirm that minute differences of form were an irrelevance. With this he began a mass extermination of species. Whole lists of names were now vaporized as one name came to define what had previously required a dozen or more. This was in 1941.

Delighted with his results, Ellison now turned his attention to the whole of the Paleozoic. In 1944, his former mentors, Branson and Mehl, had summarized the state of knowledge in Shimer and Shrock's definitive *Index Fossils of North America*.[37] Frequently cited, the book nevertheless lacked the clarity and security of argument that Ellison, then at Stanolind Oil and Gas in Midland, Texas, would find just two years later. Extracting the stratigraphic ranges of some fifteen hundred species mentioned in more than 180 papers, Ellison mapped the ranges of those genera he considered valid on charts ranging from the Lower Ordovician to the Upper Permian. The result had a clarity and logic never before seen in conodont studies. In this respect it was much like earlier pictorial arguments. It spoke directly of the wonderful utility of these fossils and their seemingly distinctive forms. It seemed to offer an almost perfect portrayal of their progressive evolution: "The fibrous and non-fibrous conodonts develop in form from simple cones through the bladed and bar stages to the platform types. The simple cones give rise to blades and bars by addition of denticles to the basic cone. The basic cone becomes the apical or largest denticle. Bladed and bar forms give rise to platforms by a process of lateral thickening of the blade or bar. . . . Such excellent examples of evolution impress the idea that the conodonts are among the

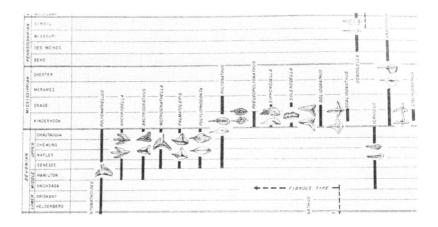

2.1. The solution to the black shales dispute. Ellison's mapping of the stratigraphic ranges of conodont fossils demonstrated the fossils' utility at a time when most paleontologists knew little about them. This part of the chart, illustrating Branson and Mehl's index genera, showed that conodonts permitted the detection of boundary between the Devonian and the Mississippian. From S. P. Ellison, *Bulletin of the American Association of Petroleum Geologists* 30 (1946). A A P G ©1946, reprinted by permission of the A A P G, whose permission is required for further use.

best fossils for family tree studies."[38] Here was the key to a new utilitarian future: abstract things on an evolutionary journey that left behind the footprints of their development.

This convincing summary of knowledge had implications for the black shales, for it embedded Branson and Mehl's index genera and made concrete the error of Ulrich and Bassler's interpretation (figure 2.1). Kindle's vision for the fossil and the black shales had been realized. This was certainly Hass's view as he attempted to zone rocks across the disputed boundary in Ohio in the mid-1940s. The results demonstrated the truth of Branson and Mehl's index genera – evidence he, Cooper, and Huddle had previously doubted.[39] In the late 1940s, Hass began to sample against a tape measure at "at the tenth of a foot scale."[40] This was a level of resolution above that achieved by other workers, and on Hass's early death in the following decade, Carl Branson wrote that Hass had made "the first real progress in solving the black-shale problem."[41] This was, of course, untrue; that honor surely belonged to Branson and Mehl.

✳ ✳ ✳

In little more than a decade of intense activity, helped on by Ulrich's out-spoken views, the conodont fossil had become a sophisticated utilitarian tool. But it remained in the hands of a cottage industry. By 1941, some five hundred people were engaged in the oil industry studying microfossils at an annual cost of $2.5 million. The pages of the ever-expanding *Journal of Paleontology* indicate that six to ten times more attention was being paid to foraminifera, and three to five times more to ostracodes, than was given to conodonts.[42] At that moment of peak conodont activity in 1933, Brooks Ellis, working on a definitive *Catalogue of Foraminifera*, had eigthy-two artists, photographers, geologists, typists, and librarians working for him. They were funded through the Civil Works Adminis-tration, which had been set up to relieve unemployment during the Great Depression. By 1935, his catalogue had twelve to fifteen thousand spe-cies descriptions, twenty-five thousand illustrations, and three hundred thousand references.[43]

By 1953, four out of five geology graduates became oil geologists, earning salaries of three hundred dollars a month or more. By then the oil and gas industry employed eight to ten thousand geologists, and many, such as Chalmer Cooper (who lived beyond his one-hundredth birthday), would be involved in encouraging the next generation: "You may have to be a nomad instead of settling down for life in one spot. You may have to 'sit on' a well all night and then drive a hundred miles to report on it. You may have to burn in India, freeze in Alaska, or do both in the Texas Panhandle."[44] Some were deeply alarmed by the transforma-tion that had taken place in this science: "In North America, particularly, we have tended to be concerned with the *uses* of invertebrate fossils in the solution of geologic problems rather than with the *meaning*, in the broad-est sense, of the fossils."[45] Oil had made paleontology the handmaiden of industry. The dour Arthur Cooper, at the National Museum, thought the well-paid micropaleontologists were splitting the science in two.[46] Nowhere could this split be better seen than in the science's treatment of the conodont. The utilitarians now felt they possessed it. But they did not. Elsewhere, others were finding evidence of the animal.

The figure that faces the principal entrance is the most remarkable in this excavation, and has given rise to numberless conjectures and theories. It is a gigantic bust, representing some three-headed being, or three heads of some being to whom the temple may be supposed to be dedicated.

CAPTAIN BASIL HALL,
Fragments of Voyages and Travels (1832)

THREE

The Animal with Three Heads

IN 1933, TED BRANSON AND MAURICE MEHL BELIEVED THE
conodont would remain forever silent on the question of its anatomy.
But they were wrong. Indeed, at the very moment they took possession
of the fossil and turned it into a geological abstraction, new discoveries
were being made that threatened to tear their utilitarian dream apart.
These discoveries did not do so, however, because Branson and Mehl's
bubbling pots of mud and practical science fit perfectly into a country
infatuated with oil. Who, by comparison, really cared about the biology
of a tiny, obscure creature? Who would willingly sacrifice the fossils'
usefulness for the sake of incorporating this new anatomical information? Carey Croneis, doyen of the new micropaleontology at the University of Chicago, certainly valued this practical turn, but he objected
to the willingness of oil company geologists to sacrifice science for the
sake of economic gain. He felt that the very integrity of the new science was at stake and called upon the industry to employ "men not only
of adequate scholastic attainments but even more important, men of a
high type of intellectual potentiality, which is, of course, a very different
thing." His was not a solitary voice, but the economic reality of the new
industrial paleontology was never going to be affected by the moraliz-
ing of paleontologists in universities and museums. Ted Branson's son,
Carl, for example, working for Shell in Texas in the late 1940s, revealed
how fundamentally different this utilitarian world was: "It has been five
years since I have seen many non-oil seekers; too long. . . . I'm mostly
tied to hunting for grease and get no time for reading or research."[1] As a
result, in the United States, two overlapping cultures developed around

microfossils. One was committed wholly to the economic project. For it, fossils were no more than abstract tools, and biological concepts, such as evolution, simply devices to be used to distinguish as many unique "species" (or time markers) as possible. The other community also valued the practical benefits of fossils, but it saw the fossils embedded more properly in sciences that sought to understand the past conditions of the earth and life upon it. One group, fed on its greasy diet, soon grew obese in participants, while the other remained small and, since it trained the new oil men and women, could never fully separate itself from the practical science. For many types of fossil this division of labor caused few problems because the fossils themselves were simple objects. The conodont, however, was a biological mystery and it was, as we shall see, about to acquire considerable complexity. This produced an animal with a schizophrenic identity.

The development of this divided world was further aided by Branson and Mehl's belief that they were the pioneers of a new science. This encouraged them to think that no one else was doing equivalent science. With their rapid conquest of the fossil and its distribution, their work assumed an intellectual independence and authority blind to what was going on further afield. But they were not alone in this. The geological community in American universities and state surveys in the 1920s and 1930s was small and incestuous. Through networks of friends, students, teachers, and other contacts it sought to distribute opportunities so that no one was really in competition and everyone had their own patch. In a country so vast, this encouraged particular outlooks to develop in geographically isolated research communities. At the University of Chicago and the Illinois State Geological Survey, for example, the science never became as utilitarian as it did in Missouri. At each of these two centers, Croneis and Branson, respectively, shaped minds and projects and defined what, for their students, should be considered the new science of microfossils.

We have already explored one side of this polarized world in Branson and Mehl's successful attempt to realize the fossils' practical worth. The other side of this science sought to reveal more about the animal itself and it did so over the same period. As we shall see, this division of labor resulted in two incompatible truths and three distinct and irreconcilable animals. To understand how this came about, we must return to the beginning of the twentieth century and think not about rocks and oil but about the animal.

In America, in the early twentieth century, so Clinton Stauffer tells us, Karl von Zittel and Josef Victor Rohon's worm held sway: "This conclusion, although perhaps favoring the minority, was rather generally accepted by workers throughout America, and conodonts were classified with the Vermes." However, this worm was destined to have an ephemeral existence. Was it not inevitable that a new generation would look at these fossils and see the fish others had seen? E. O. Ulrich and Ray Bassler believed the conodont a fish in 1915. A decade later, their comparison of conodont and fish fossils in the U.S. National Museum convinced them that each type of conodont belonged to a separate species of fish, as we have seen. This was, of course, the invention of a convenient truth that chimed with Ulrich's belief that geologists should concern themselves only with the usefulness of fossils, not with their biology.[2]

Others were also remaking the conodont fish, and doing so several years before Ulrich and Bassler published their classification. Among them was William Bryant. It was he who had taken Hinde's *Polygnathus* name and restricted its meaning to the conodont's distinctive "crested crushing plates," soon to be called "platforms." These tiny objects seemed to speak to him – though they did so silently – of a past that was alive: "In upper Devonian times a grinding dentition was introduced, bearing mute witness to the struggle for existence of these creatures as the conditions of life became more difficult." In Hinde's remarkable Conodont Bed at Eighteen-mile Creek, which Bryant considered "one of the richest and most interesting storehouses of ancient vertebrate life to be found anywhere in the world," he found evidence to show that the animal had later acquired dermal armor. By 1921, a beast was emerging, in his imagination at least, and it was not a worm: "On the whole, the longer I have studied these organisms, the more have I become convinced that the true Conodonts have hardly anything really diagnostic in common with Annelid jaws. If, as I shall hereinafter try to demonstrate, certain of the leaf-like forms are of the nature of pavement teeth, then the conclusion seems almost unavoidable that the Conodonts must be considered as the dentition of some primitive type of fishes." As he stared at these teeth, he noticed they were paired, left and right; they had come from a bilaterally symmetrical animal.[3]

Being a specialist in fossil fishes, it was perhaps inevitable that Bryant would see the fossils' piscine qualities. But he was not alone. At that same moment, John Muirhead Macfarlane, the recently retired professor of botany and modernizing director of the Botanic Garden at the University of Pennsylvania, was wandering some distance from his main subject. Following Hinde's lead, he also believed conodonts to be types of jawless fish related to the hagfish and lamprey (the cyclostomes) and, perhaps influenced by Zittel and Rohon's findings, thought the fossils to be part carbonate of lime and part "horny." To demonstrate this relationship, he copied Hinde's trick of constructing a striking visual argument, this time illustrating the "teeth" of cyclostomes and conodonts next to each other, without giving any sense of scale.[4]

Macfarlane then took Hinde's cyclostome and combined it with an old and unpopular idea first published by Ambrosius Hubrecht in 1883. Hubrecht had proposed that vertebrates originated from proboscis or ribbon worms (known today as nemerteans), marine animals capable of injecting a spiked proboscis into their victims' bodies. Hubrecht suggested that the proboscis, if retained within the body of the animal, could have evolved into the vertebrate backbone. By uniting these ideas – all of which had been sleeping for thirty to forty years – Macfarlane could contend that conodonts "may all represent evolved and complex derivatives from the horny teeth formed in the mid part of the proboscis of metanemerteans. For if we accept that freshwater metanemerteans may have swarmed in Cambrian lakes and swamps, these by progressive change may have given rise to Cyclostomes, some of which retained horny teeth as in existing types, while others may have advanced to a more complex calcareous type."[5]

This simple piece of interpretive addition seemed to him convincing, and he extended it still further. He noted that the flesh of cyclostomes is rich in oil. If conodonts were similarly oily and had existed in the huge numbers claimed by John Newberry and others, then, he wondered, could they be the source of petroleum? A number of recent biochemical studies encouraged him to think the idea reasonable. So he asked himself in what circumstances might these fish become oil? For this he needed to understand the geology of those rocks that produced oil. He found three well-known American examples that suggested to him an elaborate theory involving earthquakes and volcanoes. These latter, he said, produced rock dust, which fell into and jellified local water bodies, killing off fish populations wholesale. The corpses of these fish were then enveloped in this dust, with the chemical interaction between dust and dead body leading to the formation of oil. The theory had a poetic logic, but little more than that. One reviewer was distinctly unimpressed: "This is an ingenious philosophy, combining all the weaknesses of the organic and inorganic theories of oil origin, with few of the merits of either, and, moreover, an excellent example of the danger of generalizing on two or three particular occurrences. This is a book to look into, but difficult to take seriously."[6]

Throughout the 1920s, American paleontologists became increasingly convinced that conodonts were fish. This was aided by the ease with which fish teeth and conodont fossils could be confused. Rohon had wrestled with this problem more than thirty years earlier and it had not gone away: If fish workers were predisposed to see conodonts as fish, then might not conodont workers interpret fish remains as conodonts? In 1928, Chalmer Cooper found what he thought were giant conodonts and sent them to the National Museum for an expert opinion. Paul Roundy, who Cooper thought always too slow to produce results, was impressed but doubtful. He told Cooper, "As regards the 'jaw' which you sent, I am not yet prepared to make a definitive statement. Two of them are evidently long, single units with many sharp teeth, and resemble a gigantic specimen of *Lonchodus? lineatus*. They, however, have several points of difference and my present opinion is that they are not conodont but are true fish remains."[7] Girty then passed Cooper's specimens to fish paleontologist Louis Hussakof for study. Cooper, frustrated at not seeing his material for nearly three years, asked Croneis, who was then his doctorate supervisor, to seek its return. A year later, Ted Branson had somehow received these Oklahoma specimens. Over the preceding period they had remained a well-kept secret, for Branson had not heard about them: "Some material from Oklahoma received today has what seems to be conodonts 40mm long. Our largest heretofore are about 2mm."[8] Given the often frosty relations between Branson and Cooper, it is hard to believe that this material had been sent by the Oklahoma man. Cooper did, however, confer widely, eventually settling on the belief that this material came from fishes. His conclusion denied the existence of a giant conodont but nevertheless seemed to confirm that conodonts were indeed fish.[9]

What appeared to be final proof that this was the case appeared in 1929, when Stuart Raeburn Kirk of the University of Manitoba published his examination of the Harding Formation at Cañon City in Colorado. Here, nearly forty years earlier, Charles Walcott had famously found "the first definite Ordovician fish remains." Kirk could now show that the conodonts didn't just occur in the same beds as these fish but that "they show basal attachment to fragments of plates, identical in composition with the fish plates which are so abundantly scattered through the various beds of the Harding."[10] These fish plates belonged to ancestors of those strange, armor-plated jawless fish, known as the ostracoderms, that had flourished in the Devonian.

Kirk was convinced by this association, just as he had been convinced by Ulrich and Bassler's insubstantial evidence for the conodont fish. However, he knew that conodonts had been found in rocks formed long after these fish had become extinct. Stauffer and Helen Plummer thought this a fatal flaw in Kirk's argument and, having found similar conodont material themselves, remained unconvinced that the plates really were composed of bone. Branson and Mehl, however, who also found this bony material, thought it was "the only evidence of the fish nature of conodonts thus far discovered."[11] They certainly did not take Ulrich and Bassler's word for it.

The conodont fish that had been so often denied now began to swim freely in American minds. For example, when Frank Gunnell discovered that a Mesozoic hagfish fossil was intermediate in form between the modern hagfish and the ancient conodont, he could not help but apply a naïve evolutionary determinism then prevalent in U.S. paleontology to connect the dots. He thought it likely that conodonts would one day be found in quite recent rocks, and knowing that the hagfish was considered a primitive vertebrate, he was sure that conodonts would "aid in establishing many heretofore unknown evolutionary relations among the vertebrates."[12] To later scientists such views might seem prophetic, but like his restatement of the fish-petroleum theory, they simply reflected a contemporary moment when these ideas floated unauthored and available for application in that broad church of the new and practically minded micropaleontology.

In America at this time, what might be considered true or known could so easily get mixed up with mere conjecture. Thus when Stauffer and Plummer immersed bar conodonts in hydrochloric acid and discovered that the tooth-like points (denticles) fell out, they could not help but imagine them as tiny jaws with vertebrate-like teeth set in them. Stauffer had visualized such jaws even before he could induce his "teeth" to fall out. The fish that swam through American minds placed expectations into the way these scientists looked. They imagined the fish. Indeed, the enigmatic nature of the animal was once again permitting science to dream and believe, as Macfarlane did, that one day the animal would be found: "It may be hoped however, either that more definite light will be shed by further study of conodonts, or that some layer of subaquatic volcanic ash may yet be discovered, in which as with the medusae, the annelids, and the skates of the Solenhofen slates, fossilized cyclostomes may be discovered."[13]

These confident assertions about the conodont fish traveled across the Atlantic to Germany. There, in 1928, near Osterode in the Harz Mountains, work began on a dam, the Sösetalsperre, which was to produce the country's largest water reservoir. It was along the forest road built for the construction traffic that Wilhelm Eichenberg, of the Geological Institute at the nearby University of Göttingen, found great numbers of conodonts on the surfaces of Lower Carboniferous shales.[14] His first inclination was to use this material to emulate the detailed microscopic analysis of Pander, Zittel, and Rohon, but his conodonts were extraordinarily brittle and simply shattered when he attempted to expose them. He was familiar with the work of Ulrich and Bassler, Kirk, Hinde, and others, and accepted unquestioningly the American position; conodonts were the remains of fish.

Fortunately for Eichenberg, his university possessed an important expert in these animals; zoologist Franz Stadtmüller was then busy investigating the gill apparatus of living fishes. Stadtmüller gave Eichenberg access to the zoological collections but also advised him on his finds. Eichenberg could now see that conodonts were "skin teeth, plate teeth, plates," and – a new interpretation – "filter extensions of the gill-arches." In other words, each type of conodont performed a particular function in a single animal and in so doing formed complex respiratory and feeding apparatuses. Following the suggestion of another Göttingen colleague, Hermann Schmidt, Eichenberg introduced the collective term Conodontophorida, or "carriers of the Conodonts," to refer to this group of animals. By implication, then, "conodonts" were now the "elements" (a zoological term he borrowed from Stadtmüller) making up these apparatuses. For Pander, "conodonts" were fish, but now Eichenberg made the term apply to the individual fossils. In doing so, he adopted what was already becoming common practice in the United States. His final piece of rethinking involved the naming of the animal itself. Since the elements came from the same stratum and were components in a single animal, as he imagined it, he grouped them together and called the animal that possessed them *Prioniodus hercynicus*. This name came from just one of the elements, and Eichenberg's use of it in this way was simply an application of the rules of zoological nomenclature, which said that when once separately named fossils are found to belong to a single animal, the longest-established name in that group of fossils should become the name for the animal and thus for all the fossils of which it is composed.

Although based on the importation of American ideas, Eichenberg's analysis meant that Ulrich and Bassler's simple taxonomy, which had yet to take hold, was merely an abstraction, and he called for it to be abandoned. Optimistically, Eichenberg felt the answer to the long-contemplated question of the conodont animal's identity was almost in sight: "All that remains still to be wanting is into which order of fossil fish are the Conodontophorida . . . to be arranged."

The fossils making up Hinde's *Polygnathus* had been found in close association, but that animal proved unacceptable. What chance of survival had Eichenberg's complicated *Prioniodus*? His animal was even less certain and, being published in German, had other obstacles to overcome before it could penetrate American circles. Now two radically different fish swam in the minds of paleontologists – one separated from the other by the Atlantic. They could coexist because they were separated by numerous physical, linguistic, ideological, and cultural barriers.

Eichenberg's fish acquired more flesh four years later when forty-one-year-old paleontologist Hermann Schmidt acquired some quite remarkable fossils. Schmidt had actually found his first example of these new fossils a few years before, in a brickyard near Hemer in Westphalia. It showed conodont elements in close association, but the specimen was so poor that he thought no more about it. Then a collector showed up with a similar fossil. It had come from the same rock at the base of the Upper Carboniferous, but from Arnsberg, more than fifteen kilometers distant from Hemer. Perhaps these were not chance occurrences? Schmidt returned to the quarry in the spring of 1933, doubtless looking a strange sight as he systematically examined the bedding planes with a magnifying glass. The effort paid off, and in time he located nine specimens showing conodont elements in groups. In some of these the elements themselves were preserved, but in others all that remained were impressions of them in the rock. Most of these groups of fossils were surrounded by a dark patch of bitumen enrichment that he interpreted as decayed tissues of the animal. This meant, of course, that the association was natural; these elements had once occurred together in the body of a single animal.[15]

On closer examination, Schmidt discovered that there was a consistency to the grouping of the different element types; some groups were in complete disarray, but others showed a repeated arrangement, which gave a clue to their relative positions in the living animal. Schmidt attempted a reconstruction – the first time anyone had tried to do this – by interpreting the arrangement he saw on the bedding planes but always with Eichenberg's functional arrangement of the different parts in mind. His was not to be an original interpretation but an extension of the German model. This suggested that a pair of strong, bladelike conodonts acted as mandibles, each showing signs of having grown in close relationship with the other. Behind the mandibles, which "seized and cut,"

3.1. Schmidt's fish. Schmidt's reconstruction of the conodont apparatus closely mirrored the arrangement of the fossils in his fine natural assemblages. So perfectly did this seem to fit the anatomy of Eichenberg's conodont fish, he had no difficulty producing this three-dimensional model. In this drawing, the jaws are on the left and the gills on the right. From H. Schmidt, *Paläontologische Zeitschrift* 16 (1934). © Verlag von Gebruder Borntraeger.

sat a pair of "spiky" elements that pointed forward as if to cause mortal injury to the prey. Behind these were the most numerous group of elements – slender, comb-like forms Schmidt interpreted as components of the gill apparatus (figure 3.1). For Schmidt the conodont animal was a fish even before he began his reconstruction. Of this he was already sure. He even considered the possibility that a fish fossil found in the same bed might actually be the owner of this assemblage. The head of that fish, however, had not been preserved.

Schmidt reported his finds to a meeting of the German Palaeontological Society, the Paläontologische Gesellschaft, in September 1933, publishing a fully illustrated account the following year. Like Eichenberg, he gave his assemblage of conodont elements a single name drawn from one component, *Gnathodus integer*. In doing so, he claimed that there were far fewer true species of conodont than the Americans believed. His findings added weight to the German call for Americans to abandon their simplistic taxonomy and group together conodonts from the same stratigraphic horizon, giving them one name and recording the frequency of each type found. It was not simply Eichenberg's anatomical model that suggested this but also Schmidt's photographs, which spoke convincingly of the natural association of the different parts. Anyone picking up the paper would have seen this; it required no proficiency in German. The animal itself had left its own visual argument, and one just as powerful as those constructed by Hinde, Rohon and Zittel, and Macfarlane. Schmidt's associations preserved exactly the same mix of forms Eichenberg had found, and while they demonstrated that Hinde's *Polygnathus dubius* possessed too many conodont elements to be the remains of a single animal, that specimen did contain elements in the same proportions as Schmidt and Eichenberg had found. What none of these associations possessed, however, were those simple teeth Pander had discovered in such profusion in the older rocks. Now Schmidt wondered if these had come from an entirely different animal.

In a few short pages, and some remarkably convincing figures and plates, Schmidt proposed a revolution in the way conodont workers should consider their subject. To German eyes, at least, progress was being made: The American fish had now been given a German apparatus and had become rather more biological. The results were in many respects conclusive, but was it already too late to change the American way of doing things? In 1933–34, the American conodont machine – Branson and Mehl, Stauffer, Cooper, and Huddle – acquired a head of steam and was moving forward at a tremendous pace. Had Schmidt published in English in the United States, would it really have stopped this machine in its practical tracks?

On this possibility we can do rather more than speculate, for when Schmidt published his finds, twenty-eight-year-old Harold Scott of the University of Montana's School of Mines in Butte was making a similar discovery.[16] The remarkable coincidence of these events has fascinated conodont workers, but Schmidt and Scott were not the only workers finding these "natural assemblages" at this time. As we have seen already, the conodont fossils only came to light when people started studying fossils with microscopes. By the early 1930s, micropaleontology was the new fashion and an increasing number of graduates were engaged in looking at rock surfaces in search of tiny fossils. In a few years most would turn to those mass processing techniques Branson and Mehl so favored, and in doing so lose any hope of understanding how the different conodonts are associated in the rock. But of all the microfossils being used in the new science, only conodonts were affected in this way by this change of practice. In the early 1930s, however, workers were not set in their ways and there were still doubts about the boiling and sieving of sediments. So, for a few years, there was a much higher probability that associations would be found: More people were looking and fewer were boiling samples.

Scott was an Illinois man, born and educated in the state, and it was with that state's geological survey that he got his first geological summer job in 1927. He graduated two years later and got married on the same day. His career took him to Wisconsin and to oil consulting in Texas before the stock market crash forced him back to school. The Great Depression, rather ironically, turned out to be his big break. He won a fellowship at the University of Chicago and found himself in the fall of 1931 in what he considered one of the two greatest geological departments in the country. While there, the head of the Montana Geological Survey came calling, looking for a new staff member. Scott's supervisor, Carey Croneis, put him forward and he got the job. Scott realized how remarkably fortunate he was in this jobless country, but when he arrived in "bleak and dreary" Butte, he realized that his good fortune had come at a price.

Scott wrote his recollections of these years late in life and in doing so appears to have conflated two summers of fieldwork. In the summer of 1933, he "collected some of the black shales and took them to the office." In October of that year: "I decided to look at the shales under the microscope. To my astonishment, the first sample showed [an assemblage] of what are known as conodonts, complex teeth of an unknown animal. I immediately recognized their importance; they were the first [assemblages] ever found. I rushed to report upon them and have Elizabeth Lochrie draw them after my models."[17]

We shall come to this discovery shortly, but it is helpful to consider just how lucky Scott was at this time. It was probably in the summer of 1934, as he recalls, that this lucky streak continued. He had been asked to go out into the field to search for "pumpkin-seed gold" – whatever that was. He was given a car and an assistant but no salary, so he felt free to indulge his own interests a little too. He made several forays into the field that summer, but it was the last one that really turned his life around. On that occasion, he and his assistant entered the Big Snowy Mountains from the north, walked along the top of "the big Madison Limestone," and chanced upon some shales containing concretions. "They were unknown, unnamed and unmapped," he wrote. "I broke open a concretion in the shale and oil ran out of the cavity. I was amazed because lying above the shales was a sandstone bed capable of holding oil. They all dipped eastward into the basins of eastern Montana and the Dakotas." He rushed the discovery into print and almost immediately oil companies began to arrive in town. Scott later recalled, "The summer of 1934, without salary, had paid off. National attention on the oil potential of Montana and the Dakotas and international attention on conodont studies made the summer a land mark for me." He rounded off this period of discovery by writing up the Montana earthquakes of 1935, again attracting much attention. This hat trick of discoveries put him in *Who's Who* and made him one of the most highly paid academics in Montana. They also bought him his escape, for in 1937 he was headhunted for the University of Illinois.

It was in this context that Scott's associations of conodont fossils were found, and when he first looked at them he knew nothing of Schmidt's discoveries. Scott's fossils did, however, come from the Pennsylvanian (Upper Carboniferous) Quadrant Formation and were very broadly similar in age to Schmidt's. Scott had located seventy-five associations – or "assemblages" – of conodont fossils. This was eight times what Schmidt had found, but not one was as good as Schmidt's best (figure 3.2). Scott moved quickly to make his finds known and, being still in distant Montana, asked Croneis to present his preliminary findings at a meeting of the Paleontological Society in Chicago in 1933.

He made his case more fully in a short paper published in December 1934 which he later said "attracted world-wide attention by workers in the field." Here, Scott began by drawing upon a review of previous conodont studies recently published by Stauffer and Plummer. This enabled him to emphasize that dramatic seesawing of opinion that had fascinated commentators from the late nineteenth century onward. This little historical sketch aimed to do two things. First, it destabilized the existing fish; it demonstrated that the great and the good had never been able

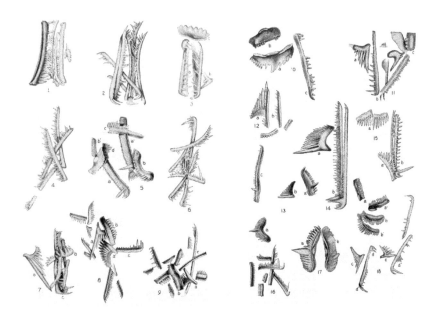

3.2. Scott's conodont assemblages showed repeated associations of fossils but lacked the explanatory power of Schmidt's finds. From H. W. Scott, *Journal of Paleontology* 8 (1934). SEPM (Society for Sedimentary Geology).

to agree on what the animal was and no new evidence had been presented to change matters, despite what some of his senior colleagues might think. Second, it acted like a Wagnerian prelude anticipating and reifying the significance of the drama that was about to unfold. With American opinion already firmly in favor of the fish, Scott knew he was about to introduce another sea change in thinking. He was sure the fish was simply an illusion and that he possessed the proof necessary to finally end its life. But just to make sure he was not making a grave error, he consulted Chicago professor A. S. Romer, author of the recently completed and groundbreaking *Vertebrate Paleontology*. Beyond their evocative – and yet puzzling – morphology, it was the fossils' phosphate composition that most encouraged opinion to favor the vertebrate. But Romer told Scott that such evidence could not be relied up, and looking at the conodonts he concluded quite emphatically, "I cannot accept them as the remains of vertebrates."[18]

In his paper, Scott presented his eighteen best, but still quite jumbled, specimens. Although conodont workers have claimed that Scott and Schmidt presented the same discovery, the quality of the evidence available to each man was remarkably different. While Schmidt's fossils were capable of making the argument for themselves, Scott's required a little more ingenuity if they were to have the desired impact. He did not possess a single specimen that gave a clear picture of the animal's apparatus, but he could reveal consistent associations of elements. Six specimens showed identical combinations. These, Scott believed, proved Ulrich and Bassler wrong; different elements did exist within the same animal. They did not represent separate species. The specimens that showed the best natural associations of parts, as judged by repeated symmetries, contained the long saw blade-like *Hindeodella*. These, Scott imagined, were probably grouped around the mouth orifice of the animal. That animal was, for Scott, most emphatically a worm. Scott ran through a series of numbered assertions, each of which he considered indisputable. Conodonts are paired and made up of right and left groups, thus not conforming to upper and lower jaws, a condition he felt essential for a vertebrate animal. The assemblages consist of no more than ten conodont fossils (Schmidt's reconstruction contained fourteen) – too small a number to be fish teeth. As many as four completely different types of conodont appeared in the same mouth. They lack a pulp cavity and appear to be attached to flesh in a manner incompatible with fish teeth. No other hard parts had been found, thus confirming the animal was soft-bodied. One can read in Scott's statements a predisposition to see a worm, but clearly he had written his paper after he had reached this conclusion. Thus, while he believed he gave an objective interpretation, this was undoubtedly an argument in favor of the worm: "It now seems impossible for conodonts to belong to any group of animals other than Vermes."

If the fish had been welcomed by those predisposed to such an animal, then we might wonder if Scott also looked at the conodont through similarly blinkered eyes. At a recent meeting of the Paleontological Society, Croneis and Scott had jointly presented a number of papers on fossil worms and even introduced the term "scolecodont," meaning "worm tooth," to cover a group of fossils they considered, like the conodont, of emergent paleontological and stratigraphic significance.[19] The term "scolecodont" is still used today to refer to fossil worm jaws.

Whether Scott's worm diagnosis affected the reception of his paper is unclear. He was certainly swimming against the tide. Some found his rather emphatic arguments convincing – one university lecturer even adopted the paper as a model of scientific argument with which to teach his students. But a fundamental corollary to Scott's discovery was that Ulrich, Bassler, Branson, and Mehl – indeed, everyone active in America at that time – was working on a false premise. Ulrich and Bassler's simple fishes were merely figments of their imaginations and certainly could not be used as justification for giving each type of conodont its own species name. Scott did not push the point. He was a practical paleontologist himself and valued fossils mainly for their utilitarian value. Also, as a young man, he simply could not afford to be outspoken, especially in a field where everyone knew everyone else, where jobs were scarce and enemies easily made. Scott had made his bold claim, but now he stepped back and recommended the continuation of current practices. Not really being a conodont worker, he soon returned to other things.

Unknown to both Schmidt and Scott (and to most conodont workers since), at this same moment, the young Dan Jones was also finding conodont elements in natural association. In March 1935, Jones had just completed his master's dissertation at the University of Oklahoma. In it, he had made his own report on conodont assemblages from the Pennsylvanian Nowata Shale, which occurs in the north east of that state. Like Schmidt and Scott, he had begun his field and laboratory work in 1933, though he was still in his teens, and his discovery was also the result of mere chance. Having found Branson and Mehl's washing methods useless, Jones had resorted to splitting the shale by hand then using needles to expose the fossils. The assemblages he found were far better than those Scott possessed. They showed conodont fossils arranged symmetrically, just as one would expect if a bilaterally symmetrical body had been compressed. But Jones was still very young and had just been beaten into press by the high-flying Scott. Perhaps because of this he made relatively little of his finds. He did, however, listen to his departmental colleague, R. L. Denham, who suggested that the conodont elements might have been used for grasping onto a mate during copulation, as he had seen such things in living worms, including nematodes. Jones thought this a possibility. The idea entered the animal's mythology, encouraging Denham to publish his own ideas on the "elusive little animal" a decade later.[20]

By the end of 1934, conodont studies had reached a new level of maturity. The conodont's place in stratigraphy was rapidly being established, a fish or worm affinity seemed very probable, and assemblages suggested biological complexity. Beyond those engaged in more thoughtful study, however, the enigma continued to attract armchair speculation. Thus within just two years of Scott's seemingly definitive and well-argued case, Frederic Loomis, a Massachusetts professor specializing in fossil mammals, muddied the waters by suggesting that conodonts might be the teeth of gastropods, thus reawakening an old idea which had been frequently dismissed. He felt sure conodonts would be found in deposits of much younger age "if sought." It was a piece of recreational dabbling, but it underlines the sense in which the conodont was continuing to develop as an object of mythology – an Arthurian sword in the stone by which all comers might test their intellectual strength. As yet, conodonts could claim few people who specialized in their study, and so there was little notion of the insider and outsider in the debate. Conodonts were there to be claimed by any specialist wishing to include them in his group of animals. It was assumed that they would inevitably be found to belong to one established group or another. Few could have been surprised, then, when in 1937, Henry Pilsbry, an American expert in land snails in his mid-seventies, looked at these fossils and also saw mollusk teeth.[21]

After years of economic and political turmoil in Germany, 1933 saw Hitler rise to become chancellor and the beginnings of anti-Semitism and book burning. Soon anti-German sentiments began to affect the reception of German science in the United States just as Schmidt's discovery was beginning to reach conodont workers there. In 1935, Stauffer was the first to record the arrival of Schmidt's innovations. He was convinced that the associations of conodont elements were valid and that "when good specimens or even good fragments of all these are found in close proximity, or as an assemblage of the remains of one individual, a long step towards the reconstruction of the animal can be made." Branson and Mehl felt the "long step" had not been taken. They looked at Scott's assemblages and dismissed them as chance inclusions in "excretionary matter." Their proof was that samples of isolated conodonts "fail to show proportional numbers of kinds supposedly found in one individual." They added that the different types of conodont making up these assemblages had different stratigraphic ranges – how could these different kinds belong to a single animal? Their objections were to become fair tests of the theory. If refuted, then assemblages would be proved anatomical rather than mere coincidences. Until these doubts were answered, Branson and Mehl would continue to class conodonts as the remains of fish and deny the validity of assemblages. As evangelists for conodont stratigraphy their interests lay elsewhere and they were actively attracting new recruits to their way of thinking.[22] As the most published and therefore the most authoritative workers in this field, their words had impact, particularly on those making practical use of microfossils. Increasingly, stratigraphers began to doubt the truth of Scott's claims.

In Continental Europe things were different. Eichenberg and Schmidt's fish was still being developed and enhanced. At the Royal Belgian Institute of Natural Sciences in Brussels, paleontologist and stratigrapher Félix Demanet had recently found his first conodonts in the Belgian Carboniferous. When, in 1937, Girty's successor at the USGS, Jim Steele Williams, called on Demanet on his way home from a Moscow trip with his former mentor and colleague, Ted Branson, he found the Belgian at a loss as to how to deal with his new fossils. Williams asked Branson to send Demanet some offprints. Demanet, however, soon became attached to the German way of thinking. The huge number of fossils he had collected confirmed the truth of Schmidt's repeated association of different types.[23] Schmidt's fish entered his mind. Conodonts were the remains of filtering "appendices" or "processes." He agreed with

Schmidt that there remained just one final question: To which group of fishes do conodonts belong?

Demanet knew that, in 1937, D. M. S. Watson of University College, London, had illustrated the gill rakers of the fossil fish, *Acanthodes*. He read over Watson's description and then went back to T. H. Huxley's much older account of another fossil fish where a similar arrangement was described. Demanet noted Huxley's comment: "Minute horny or osseous filaments seem to have been set at right angles to the branchial [gill] arches along their edges." This awakened a thought: Did this specimen preserve conodonts in situ? Demanet had Huxley's fish sent over from the Geological Survey Museum in the UK. It arrived as part and counterpart, one of which had formed the basis of Huxley's illustration. But to Demanet's initial disappointment he found that Huxley's illustration wasn't an entirely true reflection of the fossil but had been restored "at least with regard to the normal presence of "filaments" on both sides of the branchial arches." On the specimen itself there were mere scatterings of these filaments – the very thing Demanet wanted to study. Nevertheless, there was sufficient information in the fossil to confirm Huxley's interpretation and to relate it to Stadtmüller's more recent work. And then he found it, there, on the internal border of a gill arch, a 1.3-millimeter fragment of what he thought was a conodont: "It resembles a small toothed straight comb." This could be matched with a 4-millimeter-long impression in the counterpart that showed "the Conodont *in situ*," in contact with the gill arch. Although rather large for a conodont, he thought it proved Schmidt's interpretation was correct.

The Europeans' linear research program had in three important steps turned an idea into a material reality. Schmidt's combination of conodonts was also found in Scotland at this time, further confirming the truth of the European fish.[24]

Rather different evidence for conodont fish was found at this time by James S. Cullison of the Missouri School of Mines and Metallurgy in Rolla. He possessed two examples of "an outstanding specimen of a jaw" collected from two different localities. "These are of a bone-like substance on which cones like conodont teeth are set in a single row. These two specimens support the theory that at least some conodonts are the teeth of ancient fishes rather than the jaws of annelids."[25] Cullison imagined that these tiny and easily detached fish teeth might be taken for primitive conodonts. His strange jaws then became something of an enigma within the enigma. Some, like Ellison, were willing to let them stand as conodont jaws – at least until disproven. They remained objects of controversy.

By now Branson and Mehl's students were regularly writing master's dissertations that used conodonts in stratigraphic study. Almost every one of these introduced some new species, and one, by Gertrude Burnley, even reported finding conodont associations. She did not, however, take the next step and suggest these were reflections of the animal in life. Dan Jones continued his studies at the University of Chicago, writing a doctoral thesis on the Pennsylvanian Seminole formation under the supervision of Croneis. Using his fellow micropaleontology students of the class of 1937 to help with picking and mounting his finds, Jones was then coming around to Scott's way of thinking. He told him, "The evidence furnished by my assemblages tallies very closely with your conclusions that conodonts can be hardly anything else but the masticatory apparatus of annelid worms."[26] His finds, he said, repeated earlier ones and put the biological validity of natural assemblages beyond doubt; there were "definite types of associations, each containing its characteristic genera and species of conodonts." He did not give up the ideas of Denham and remained of the view that conodonts could possibly be grasping appendages in a buccal cavity of some sort, probably belonging to an annelid worm. Under Croneis's influence, Chicago was becoming another center for conodont studies. His micropaleontology course was an important draw, and before long pioneer conodont worker Chalmer Cooper, then with the Illinois State Survey, would arrive to register for a PhD.

By the end of the 1930s there were a number of conodont animals in the minds of European and American workers. In Europe the animal was a fish because the Americans had said so, but it had become a peculiarly European fish with a complex apparatus, which gave each conodont element a different functional role. It had been the result of a simple linear research project that revealed and extended the truth in logical steps. In the United States there were some, from Bryant onward, who had been building the fish from isolated parts who welcomed Schmidt's work. Yet another, far simpler fish swam in the minds of those who were devotees of Ulrich and Bassler, or who, like Branson and Mehl, wanted above all else a simple basis for their stratigraphic work. Another group, centered in the state of Illinois, particularly Chicago, had rejected the fish entirely: Here the animal was a worm. What precisely the animal was, it seemed, depended rather more on where one was than on the fossils themselves. But surely the science could not stay like this? More than anything science appeals to universal understanding, not intellectual segregation. It was inevitable that there would be a coming together of the various sides and when they then looked at what they had done, surely they would see that they had created a horribly complicated mess?

Well, here's another nice kettle of fish you've pickled me in.

OLLIE HARDY,
in Laurel and Hardy's
Thicker than Water (1935)

FOUR

Another Fine Mess

THERE WERE THREE WAYS TO SOLVE THE RIDDLE OF THE conodont. The first was to think differently about things known, but if anything too many people were thinking differently. The second was to find better material but this seemed only to deepen the problem. The third – taking advantage of the kind of technological change Zittel and Rohon thought empowering – was to journey into the object itself, and no one had attempted that since the late nineteenth century. An unexplored trail down which progress might be found, in the late 1930s it called to a number of those who had recently become fascinated by the fossil. Among them was Clinton Stauffer, who was perplexed by that simple paradox that now seemed to be at the heart of the problem: an animal with a wormlike arrangement of teeth composed of material indicative of a vertebrate. It prompted him to ask, was the phosphate truly part of the tooth or mere contamination? If simply contamination, then the mystery was solved: The animal was a worm. He asked his technically minded Minneapolis colleague, Duncan McConnell, to resolve the matter. McConnell examined the fossil's chemistry and crystallography and reported that the conodont was indeed composed of material structurally and chemically similar to that making up vertebrate teeth. Stauffer could only conclude, "It becomes evident that the only way to relate conodonts to the worms is to postulate an entirely new group of extinct forms with vertebrate-like teeth." He continued, "Which might be equivalent to suggesting that they are primitive vertebrates." Stauffer had been on this vertebrate track for some time, but just as McConnell seemed to give him the confirmation he needed, Bill Furnish, supported by Branson and Mehl, debunked his earlier suggestion that the bar-like conodonts were jaws with teeth inserted in them.[1]

Furnish, who was looking very closely at his fossils too, also believed he had examples of conodont fossils in his own collections showing breakage and repair but not abrasion or wear. This suggested to him that conodonts were perhaps used for grasping rather than mastication.

Stauffer's interpretation of the bar conodonts as jaws indicates how little the Americans knew of the interior structure of conodont fossils even in 1940. Up until then, every internal investigation had been undertaken in German-speaking Europe. This, however, was about to change. The young Wilbert Hass was convinced that too much faith had been put in the simple analogies that might be drawn from an examination of the fossils' external morphology. He was not the first of this new generation who aspired to look into the interior structure of fossil, but those before him had been prevented from doing so by the poor quality of the fossils they found. Situated in Washington, at the heart of the U.S. Geological Survey, Hass was undoubtedly in a privileged position as he could call upon some of the best conodont fossils in the country, and among these he found translucent specimens collected from the Mississippian by Roundy and Chalmer Cooper.[2] By grinding these down, he could produce thin sections for the microscope.

Recognizing that he must compare his findings with Pander's original descriptions, Hass had Pander's book translated. Armed with those descriptions in one hand and "beautiful sections" in the other, he began the most detailed examination of the structure of the conodont in nearly a century and in doing so called upon technologies far better than those available to Pander. Hass also had the advantage of large systematic collections, which even from a casual survey seemed to suggest that the animal's evolution had led to an increase in the surface areas of the elements, culminating in plate-like forms. He also concluded that conodonts grew cone *on* cone by the addition of material to the surface and not, as Pander had suggested, by internal secretion.

Hass's biggest contribution, however, concerned the more complex compound conodonts, which grew like the simple cones but did so from a number of points and in various directions (figure 4.1). Some of these

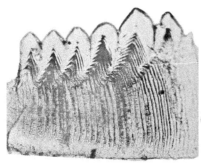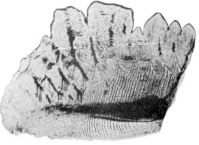

4.1. Hass's journey into the anatomy of the fossil. Hass's thin sections through conodont elements viewed in transmitted light. *Left,* a blade showing lamella growth. *Right,* an element showing 'aberrant effects due to suppression and rejuvenation of parts. From W. H. Hass, *Journal of Paleontology* 15 (1941). SEPM (Society for Sedimentary Geology).

conodonts became progressively simpler in form as the animal matured and as some growth points were suppressed. In other words, some of Pander's teeth – unlike vertebrate teeth – actually changed shape markedly as they grew. If these really were teeth, then this suggested that their function changed over time. This discovery itself seemed remarkable, yet it raised a rather more pressing issue: how to distinguish a mature specimen from a growth stage. There was simply no way to tell from the exterior morphology, as two quite different fossils might simply be growth stages of the same species. For Hass this too was a revelation because it meant one could not identify species on the basis of an additional bump or ridge, as had commonly been the practice. The conodont fossil had now acquired even greater ambiguity; it really was a master of illusion. Hass believed that given time it might be possible to document these growth stages, but in the meantime he advised workers to turn their attentions to the "pulp cavity." As this was the first formed part of the conodont and not altered, it was the best indication of the fossil's true identity and its evolutionary relationships. He then revisited all the new species he had invented in his earlier study of some Montana conodonts and deleted every one.[3] The conodont population was nowhere near as diverse as he had previously thought.

Hass's evidence looked disappointingly conclusive; conodonts were not teeth and the animal was probably not a vertebrate after all. He concluded, "Conodonts functioned as internal supports for tissues within or on the body of some marine organism at places subject to stresses." He could add, confirming Furnish's observation, that conodonts did not show tooth-like wear but that, in some cases, breakages had been followed by renewed growth. Some would dispute his "internal supports" theory, but there was widespread recognition that Hass had made the first major step in understanding the internal morphology of the conodont since Pander. Nevertheless, Hass remained uncertain about what the animal might be and refused to speculate. He was a believer in Scott's assemblages but also shared Scott's desire to maintain a utilitarian classification for the sake of the stratigraphic advances that had been made. His was a solid rather than flamboyant contribution.

The curatorially minded Sam Ellison was already making massive strides in introducing a simple yet comprehensive logic to conodont stratigraphy. He was clearly riding a wave of optimism, for he too believed that the answer to the biological conundrum was within reach. In 1944, he began to gather and curate facts. For him the riddle was merely a jigsaw to be solved by finding, sorting, and placing all the pieces.[4] With the aid of his colleagues, he compiled a comparative summary of the physical and chemical properties of conodonts, fossil and recent bones and teeth, and various naturally occurring phosphate minerals. Like many papers published on conodonts, it was short, consisting of just four pages of text, one figure, and three tables. His conclusion from this data was equally succinct: "The composition of conodonts is the same as the mineral matter in fossil and modern vertebrate hard parts." With his first critical piece in place, he now began to assemble the jigsaw around it in hopes that a picture would take shape. To do so, he needed to carefully select the parts: Cullison's jaws, Scott's assemblages, and the new structural and chemical data. Anyone assembling a jigsaw puzzle knows that reference to the box lid simplifies matters, and it seems that the box lid in Ellison's mind carried the image of a vertebrate. He had no more data than the cautious Hass, but he drew his conclusions with a conviction that sometimes comes with youth: "It is evident that conodonts may be considered as vertebrates on the basis of composition, size, shape, associated bone material, and assemblage associations. They are further restricted to the fish or lower vertebrates on the basis of internal structure and stratigraphic occurrence. To consider conodonts as belonging to any other group is to disagree with the evidence afforded by the composition." More seasoned workers might have seen in Ellison's assertiveness something of the young Scott of a decade before. Each was convinced of his rightness, though both were convinced of quite different things.

Harold Scott joined the University of Illinois in Urbana in 1937, where he would stay for the next thirty years. He had remained silent on conodonts since the publication of his first, potentially groundbreaking, paper in 1934. Fueled by Branson and Mehl's skepticism, that paper had received a muted reception – many simply did not believe it – but Scott seemed in no hurry to answer their tests. He eventually did in May 1942.[5] Now a high-flyer with a number of noteworthy discoveries under his belt, this was no longer the Scott who had once decided not to rock Branson and Mehl's boat. He had 180 assemblages from Montana, each composed of two or more conodonts in close association, as well as a further 3,000 isolated conodonts showing relative abundances in the proportions expected if complex apparatuses really did exist. All this evidence proved Branson and Mehl were wrong. Strengthened by the doubt that had greeted Branson and Mehl's methods, Scott asked why their lists of conodonts did not match the distribution he was able to demonstrate. Plainly the weakness lay not in the fossil apparatus but in their increasingly criticized lists of names. Scott further suggested that older rocks might contain rather different kinds of apparatus, meaning that one should not expect to find the same relative abundances of the different types anyway. But by then even Branson and Mehl understood that they had contributed to a proliferation of names, many of which were now considered invalid or unnecessary.

Scott had seen Schmidt's paper and he was not unconvinced by his arguments for the fish, though he found Demanet's discoveries very doubtful. Indeed, he was delighted by Schmidt's excellent specimens and the proof they provided that assemblages were real. Conodonts had by then been found in similar arrangements in Illinois, Kentucky, Oklahoma, and Montana, which to Scott made the objections from Missouri ludicrous: "It would be strange indeed to find a group of animals that had such a perfectly balanced diet that the excretal material would consist time after time of one pair of prioniods, one pair of spathodgnaths, one pair of prioniodells, and approximately four pairs of hindeodells." Branson and Mehl's test, which had so effectively undermined Scott's discovery, was now turned into a crushing blow against them.

Having emphatically destroyed the opposition, Scott set about reconstructing his assemblages, though he remained uncertain about the position of all the components: "They probably operated as rights and lefts, or possibly they were placed in a circular position around an esophageal tract." He included about twice as many elements in his reconstructions as had Schmidt. Indeed, Scott's reconstruction looked nothing like Schmidt's. In Scott's animal, each conodont element was set in soft tissue: "Such an apparatus would not only form an excellent screen to prevent undesirable objects from entering, but would also present a formidable barrier for the escape of desirable food once it had passed beyond the battery of teeth. . . . It could operate with equal ease either as the jaw apparatus of an annelid or as gill rakers of a fish."

Scott had sufficient material to define two genera. One – *Lochriea* – was composed of two species, and he remarked that it was possible to distinguish these on the basis of an individual "tooth," as he still called them. He also described another genus, which he named *Lewistownella*. These were new names, and as such they broke the zoological rules, for his named assemblages were composed of conodonts that themselves already possessed names. He had decided not to take – for the name of the animal – the oldest established name, as Schmidt and Eichenberg had done. Rather, he argued that he could not be certain that that name belonged to precisely the same element and thus animal. Perhaps that name was associated with a slowly evolving "tooth," while the animal itself, as understood from its apparatus, showed much more rapid change. How could that unadventurous tooth really help distinguish the different animals? It was clear to Scott that to follow the rules would be to throw conodont studies into "utter confusion." Instead, he introduced a dual system. The names of isolated conodont fossils were now to be understood as artificial "form genera" and "form species." It was these that the stratigraphic community would continue to use, but he could not use them as names for the elements within his assemblages. Instead, he took a form name, such as *Hindeodella,* and turned it into a noun (hindeodell) or adjective (hindeodellid) to describe the component elements. Although this approach was a radical step, it seemed logical to Scott, who had established a similar system for the spicules (the tiny skeletal components) of sponges back in 1936. It was a compromise that permitted both camps – the utilitarians and the biologists – to "have their cake and eat it." It also showed that Scott possessed a very sophisticated understanding of the problem.

Over at the Illinois State Geology Survey in Urbana, another disciple of Croneis, Ernest Paul Du Bois, was developing his own views on "the riddle of conodont origin," which he had first presented back in 1941 at the Illinois Academy of Science.[6] It may have been this that finally pushed Scott into press, as Du Bois was an able worker who possessed some extraordinary Illinois material. Du Bois had collected, split, and sorted some three hundred pounds of shale, locating five hundred individual conodont fossils and seventy-five assemblages, including one much like Schmidt's best specimen. He agreed in part with the German interpretation of their function: "The polygnathids perform the preliminary mastication, and the hindeodellids the final comminution or straining." This did not, however, mean that Du Bois looked at these fossils and saw a German fish. In one of his specimens there were "impressions" associated both with the conodont fossils and with a "brown carbonaceous film," which Du Bois interpreted as "a fossilized portion of the cuticle of some worm-like creature." He continued, "It is believed that the conical structures pictured here represent parapodia or cirri. Not shown in the figure is a hindeodellid which was 'sandwiched' in between two layers of the membrane and which may indicate in a more positive manner the origin of conodonts." Parapodia are foot-like projections seen on the sides of some worms which assist in locomotion. The common association of particular types of conodont pairings in some of the finds, where other elements were missing, further suggested to Du Bois that they must have been connected by muscle.

Reflecting on the affinity of the animal, he concluded that there were three possibilities: an unrecognized group of vertebrates, an unrecognized group of invertebrates, or annelids. He thought the latter most likely and felt that his finds suggested two different species, but he did not name them. The big stumbling block with this conclusion was, of course, the conodont's chemistry. But here Du Bois offered a new explanation. It drew upon information published by Frank Clarke, chief chemist to the U.S. Geological Survey, which stated with some surprise that some annelids, particularly tube-dwelling forms, were rich in phosphorus.[7] Du Bois postulated that an outer phosphorus-rich sheath or tube might, during evolution, have been turned in upon itself to form the mouth cavity of the animal and thus also the conodont elements.

In Du Bois's mind the animal achieved more solid form: "If it is assumed that conodonts are associated with both the problematic parapodia and the worm trails, it is possible to erect a picture which may represent the appearance in life of the animal which bore the teeth. The adult was an elongate worm, seldom more than three millimeters in width, with a length of at least three centimeters, and probably five or more. It probably possessed a ventral nerve cord and resembled modern annelids in many other internal structures. Metamerism [i.e., serial segmentation] may have been indicated by the serial development of the jaws, in which each type of tooth was restricted to a separate metamere, and by the presence of regularly arranged parapodia." He continued, "The anterior part of the digestive tract was divided into buccal and pharyngeal regions. The buccal cavity ["mouth"] had a single (but perhaps more in some cases) polygnathid on either side, with the blade directed anteriorly. These jaws were probably covered with hypodermis and cuticle so that only the actual cusps were visible. Protractor and retractor muscles supported and moved the teeth. Anterior to the polygnathids there may have been one or two teeth of the symmetrical type illustrated in Scott's figure 3c (1935 [1934]). The pharyngeal region ["throat"] supported the hindeodellids which probably functioned in the final straining or comminution of the food."

This was new. Based on the evidence and some good zoological reasoning, Du Bois had taken the argument to a new level of anatomical sophistication.

By 1944, Branson and Mehl were making accommodations. In their important summary of the stratigraphic significance of conodonts published in that year, they indicated that the jury was still out on affinity: "Most investigators *assume* that they represent jaw armor of an extinct order of primitive fishes." This was hardly an emphatic assertion of their own beliefs and appeared in a paper that was balanced and tentative on those many aspects in dispute. In this paper, they confusingly placed the conodonts in the Order Conodontophoridia, which differs from Eichenberg's grouping only by the addition of a single letter. It has been assumed that Branson and Mehl used this term to refer to a group of fish, but nowhere do they make this explicit. They seemed to be approaching the point of being able to accept assemblages but were still cautious: "Granting, for the purposes of illustration, that more than one so-called genus [of fossil] represents one individual, the stratigraphic range of any distinctive part of this composite, the so-called species, loses none of its value. The only offense of such usage is against biological veracity." Their classification was based on grouping similar forms together "in an attempt to indicate genetic relationships," but they admitted to rising taxonomic uncertainty, noting that "specific values are not agreed upon." As to the animal itself and its mode of life, they had yet to read Du Bois and could only draw information from the lithologies in which the fossils were found: "It is reasonably *certain* that the *normal* conodont environment was near shore in moderately shallow marine waters containing some clay, possibly somewhat modified by fresh waters around debouchures of streams." It was an ecological interpretation the ambitious Sam Ellison sought to imitate. But this simply revealed that virtually nothing was known; Ellison could write just seven sentences on the subject.[8] This did, however, suggest to him that the animal was not as restricted to particular environments as Branson and Mehl believed. In particular, Ellison was unconvinced by their assertion that conodonts were rare in limestones. His own literature review suggested that the fossils were found in almost every kind of sedimentary rock and that the animals were probably free swimmers.

The animal continued to develop in American minds during the war, but in Europe the conodont had been sleeping, locked up in unseen books or buried beneath the rubble of German cities. For fifteen years from 1934, conodont studies in Germany had ceased. Then, in 1949, an emergency committee, the Notgemeinschaft der Deutschen Wissenschaft, was established – as one had been after World War I – to mastermind science's recovery. The war had taken its toll on the universities, though sometimes that toll was more imagined than real. Schmidt's important specimens, for example, were believed by some to have been lost when really they remained untouched in Schmidt's office in a Göttingen, which had avoided destruction.[9] As the German paleontologists woke up from their self-induced nightmare, they believed the world was as it had been when they had fallen asleep. They had no idea of the progress the Americans had made. Thus in his *Outline of Historical Geology,* published in 1948, Roland Brinkmann adopted H. A. Pilsbry's never-fashionable molluskan teeth as an explanation for the conodont.[10]

The German recovery in conodont science was more properly led by Heinz Beckmann, a former student of oil geology at Cologne, who returned from the war to begin a doctorate on conodonts at Marburg. Beckmann first found conodonts accidentally, when working on Devonian brachiopods, and "immediately fell under the spell of these puzzling fossils." Unable to access the modern literature, and largely unsupervised but possessing exceptionally preserved local conodonts, "shining green, grey-blue with whitish, and often reddish, decomposition," he began a study of the inner structure of the conodont unaware of what Hass had achieved. During the war the zoological affinity of the conodont had become less certain, but Beckmann continued the prewar belief that he was looking at Schmidt's fish, so much so that he grouped his fossils anatomically: mandibles, hyaline teeth, and gill apparatus.[11] It was a wise move, as Schmidt, who was regarded as "the only conodont connoisseur in Germany," became Beckmann's examiner. In many respects, Beckmann discovered what Hass had already demonstrated, such as the importance of the pulp cavity. But he also found clusters of "fine hair tubes" in the white matter – also seen by Zittel and Rohon – which suggested that material might be delivered to the surface of the tooth from the pulp cavity and permit surface growth like the dentine in vertebrate teeth. However, he noted that continued growth resulted in the pulp cavity being closed off, and thus: "The arrangement of the cavities and lamellas forces us to accept that the outside lamella is not substantially older than the inside and proves the dentine structure of the conodont." Despite his intellectual isolation, or perhaps because of it, Beckmann's work threw up new ideas that both confirmed and contrasted with Hass's conclusions. It also demonstrated that Pander's anatomical descriptions remained, for the most part, robust. Yet Beckmann knew there was more to be understood, if only the technology was more advanced.

Schmidt again published on conodonts in 1950, following up his groundbreaking paper of 1934.[12] But like Beckmann, he found himself without the modern American literature and thus arguing with the past and with a Scott of sixteen years earlier. In his ignorance of Scott's more recent reconstruction, he made his own from Scott's published figures, in effect inserting another pair of conodonts into his own apparatus. In doing so, Scott's material was assured of a supporting role. The only recent paper to which Schmidt had access was Beckmann's, with its case for dentine tubes that simply strengthened Schmidt's conviction that the animal was a fish. Of course, Beckmann had begun by believing in Schmidt's fish, and the interpretation of his dentine tubes relied upon it. There was, then, a certain reinforcing circularity to German interpretations, which had probably accompanied the German fish from the moment it left Eichenberg's mind.

However, this particular fish did not have long to live. A new wave of conodont workers was waiting in the wings, and they would take possession of the fossil completely and remake it for modern science. The conodont assemblage might have been on the verge of acceptance, even among the most conservative of American stratigraphers, but it would take this new wave to deliver the final knockout punch. That punch came from an Englishman, one of the first Fulbright Scholars, who arrived in Urbana to work with Harold Scott. A precocious, young postdoctoral student, Frank Rhodes began his time in Illinois with an unexpected education; having arrived skeptical of conodont assemblages, he was soon converted and then became a great evangelist for them. Rhodes had recently completed his PhD at the University of Birmingham in the UK, the topic of conodonts having been suggested to him by Harry Whittington, then of Harvard University and later portrayed heroically in Stephen Jay Gould's *Wonderful Life*.[13]

In the early 1950s, Rhodes published a flurry of papers reviewing many aspects of conodont science. While most paleontologists only made excursions into conodonts (Branson and Mehl were exceptions), Rhodes was the first high-flying paleontologist to give them focused attention. Scott's mentoring hand can be seen as Rhodes launched himself into an attempt to "build up a detailed natural classification treating such assemblages as biological units" based on existing finds and some of Rhodes's own from Illinois and Kentucky. In doing so, Rhodes followed Scott in believing in the necessity of maintaining form species, and like Du Bois and Scott before him, he hammered home those proofs that demonstrated Branson and Mehl's skepticism had been ill founded.[14]

Following the precedent established by Scott, Rhodes introduced three new genera, and as he did, he shifted authority away from Schmidt, who, unknown to Rhodes, had recently put the Americans in a supporting role. Now Rhodes usurped Schmidt's original invention and made it American. He had not read Schmidt's recent paper and perhaps did not know whether Schmidt had survived the war; to Rhodes, Schmidt was a historical figure and perhaps as a result Schmidt did rather badly in Rhodes's review. Rhodes named *Scottella,* for example, in honor of his mentor, but it was essentially a new name for Schmidt's *Gnathodus,* which Schmidt had named according to the zoological rules. Rhodes believed Schmidt had wrongly identified the *Gnathodus* elements making up the assemblage, but this was perhaps a small point. Rhodes wanted an idealized and consistent natural classification; he wanted to make scientific sense and he saw that sense in the Scott's rationale for a separate system for assemblages. But soon Rhodes had to backtrack: Schmidt was alive and had not made a mistake. In any case, the name *Scottella* was already in use for another animal. Rhodes came up with a replacement, *Scottognathus,* which performed the same feat, but it too would be no permanent memorial to Scott. And if Schmidt had reason to be upset, Du Bois also had no reason to celebrate, for nearly all of his assemblages, which he had not named, were also accommodated within *Scottella/ Scottognathus.* The great irony here is that both Du Bois and Schmidt had described essentially the same assemblage – which was closely similar to the one Rhodes now recognized. Rhodes's interpretation, however, maintained the overall architecture but reversed the direction of each element.[15] In contrast, Scott's assemblage was strikingly different and largely speculative. Du Bois was, however, honored, in *Duboisella,* a name used to refer to the assemblages Du Bois had considered atypical. A third genus honored the state: *Illinella.*

Rhodes had thus manufactured a redistribution of honors, which echoed the spoils of war, but his rewriting of history would not stop there. He found the conodonts making up the assemblages graded from one form species into another, their distinctive names reflecting stratig- raphers' desire to highlight difference rather than any biological truth. To Rhodes, these differences were merely a reflection of natural varia- tion. In other words, they were all synonyms, and in a single stroke the taxonomic efforts of Stauffer, Plummer, Gunnell, Branson, Ellison, and others could be subsumed into a few form species.

On the matter of the animal's zoological affinity, Rhodes's open mind was also affected by the Illinois air: "Until more definite evidence is forthcoming no final opinion can be given . . . although the writer is of the opinion that evidence afforded by the study of assemblages tends to support their association with an extinct group of annelids."[16]

Rhodes's contribution formed a point of closure in the debate over whether to accept the reality of assemblages. In one short paper, he had traveled from skeptic to self-made authority. Now others would follow. By admitting to his skepticism and then conversion, he built a bridge be- tween the two opposing camps. For twenty years, conodonts had appar- ently existed within numerous overlapping and interwoven paradigms, which shaped the objects into different associations and configurations and dressed them in rather different flesh. For much of that time, the sheer momentum of microfossil-based stratigraphy permitted an un-complicated view to dominate, and, in the United States at least, even those possessing contradictory evidence did not attempt to counter it. The new *Journal of Paleontology* provided the vehicle for this American debate and a yardstick against which to measure progress, but as it was driven by the burgeoning oil industry it was also the voice of rational utilitarianism. Consequently, the question of biological affinity became secondary. It was a fascinating and attractive conundrum, which drew in its own workers, but it was also theoretical and contentious, and of little practical value.

In Europe, the situation was entirely different. The biological aspects of paleontology dominated and the affinity of the conodont animal was not in dispute. Conodont stratigraphy was yet to begin and there was no crisis – other than that determined by actions on the other side of the Atlantic. The arrival in the United States of a new generation of conodont specialist, epitomized here in the form of Frank Rhodes, meant that the two worlds would be forced into some kind of reconciliation. But this moment of reconciliation also added to a sense that the science was mired in a taxonomic mess of epic proportions.

In his detailed bibliographic review of 1952, Robert Fay was natu-rally buoyant about the American contribution: "No paleontologists in the rest of the world are doing work on conodonts comparable to that of the American workers." The Germans would doubtless agree, but perhaps for different reasons: They had been absolutely opposed to the simplistic artificiality of the American system. But the landscape was changing. Many, like Scott, who had contributed in their youth, were now senior workers, but few of these specialized in conodonts. Older workers were disappearing from the scene and with them ideas that had shaped the science. Ulrich's most important contribution took place late in life, not long before he retired from the Survey in 1932, then in his mid-seventies. He died in 1944, a geological colossus. His former assistant and protégé, Bassler, died in 1961 but had long been retired. Both he and Ulrich had been "eight-hour-a-day, seven-day-a-week" workers. Branson died in 1950, and although Mehl did not die until 1966, he only published one conodont paper outside their partnership. Croneis's influence on this field would also diminish as he followed a career path into university administration, eventually becoming president of Rice University in Houston and toastmaster at John F. Kennedy's last supper. He, along with many others, gained credit for bringing NASA to the city and was awarded nine honorary doctorates during his career. He also received, among many other awards, the Sidney Powers Memorial Medal of the American Association of Petroleum Geologists, the profession's highest honor.[17]

What Fay could not realize in 1952 was that American control of the field was about to end. Now northern Europeans would superimpose conodont stratigraphy on the better-studied rock sequences of Germany and Sweden. It was not that America lost its place as the senior nation in this enterprise, but it was only now, with the widespread emergence of new talent, that conodont research became properly internationalized. Beckmann and Rhodes had begun to contribute to the creation of a new field, and others would follow. Looking at what they had inherited, both the chaos and the knowledge, this new generation thought the conodont was there for the taking. However, before this new wave could take control of the field there was the matter of the mess that had been handed to them. Rhodes had started to make inroads, but soon his efforts would ignite a battle of a different kind.

Hence a chaos of false tendencies, wasted efforts, impotent conclusions, works which ought never to have been undertaken. Anyone who can introduce a little order into this chaos by establishing in any quarter a single sound rule of criticism, a single rule which clearly marks what is right as right, and what is wrong as wrong, does a good deed; and his deed is so much the better the greater the force he counteracts of learning and ability applied to thicken the chaos.

MATTHEW ARNOLD,
On Translating Homer (1862)

FIVE

Outlaws

THE ANIMAL THAT ARRIVED IN THE 1950S, IN AMERICA AT least, had been disassembled into its component parts, cannibalized to build mythological fishes of much simpler form. These parts carried names suggesting that *they* were the animal, but nearly everyone knew these animals – Ulrich and Bassler's aquarium of different fishes – were mere impostors. A growing number of Americans were starting to see the fossils with Wilhelm Eichenberg's eyes – no longer as discrete things but as disaggregated skeletons demanding reconstruction. A way of seeing had changed, and with it the very idea of what might be legitimately considered an animal. Now the science needed to take control of a language that had grown absurd and return the names to the animals themselves. The proper way to do this would be to follow Eichenberg's lead and name fossils and assemblages according to the rules of zoological nomenclature. This course of action, however, was considered ill advised as it would undermine the linguistic foundations of a utilitarian science that had only just found its feet. Every word would need to be redefined, every object renamed, as terms that once referred to single things came to define sets or assemblages. A quarter century of effort would surely collapse into chaos. This, at least, was a widespread fear.

The specter of turmoil and the politics of a close-knit field had led American conodont workers to compromise to the point of illegality. Scott's and Rhodes's invented names for fossil animals that technically already possessed names was at very least unorthodox. But now, rather than go backward and undo their earlier compromises, the conodont workers decided to stand their ground and see their actions legalized.

The solution was simply to establish two parallel systems: one for naming "mere 'nuts and bolts'"[1] and one for whole things. It was not such a preposterous idea. We can speak the language of cars using makes and models – the machine's own binomial system – and yet also talk of the carburetors and sparkplugs that compose them in equivalent terms. What the conodont workers suggested was something similar: a language for discrete conodont fossils distinct from that for whole conodont animals. It was not a new idea. Carey Croneis had been developing it throughout the 1930s. It was, in fact, his personal crusade: "If any of these orphaned micropaleontologic objects . . . are to have correlative value (and if, paradoxically enough, they are ultimately to find parents), they must first be classed as so many different bolts, screws, and nails regardless of whether they were made in the same factory or out of the same metal."[2]

One can imagine Croneis's garage littered with pots of bolts, screws, and nails of every kind, gathered from dozens of decrepit things. By putting the screws together he gained a special knowledge of them by simple comparison; by this means and no other it was possible to become a screw connoisseur. At the University of Chicago, trays of fossil bits and pieces were similarly arranged, and with the same purpose in mind. Since the birth of micropaleontology, tiny things of all kinds had become useful to science, but their accurate identification remained problematic. Croneis's solution was to give them names of their own. Something serviceable and utilitarian – a classification of useful animal parts that would permit them to be used in correlating rocks.

Croneis's radical solution, first published in 1938, was to remove these parts of animals from zoological nomenclature and place them in a newly constructed utilitarian classification, a military ordering (*Ordo militaris*) based upon the structure of the Roman army but paralleling the hierarchical ordering of systematic biology. He perhaps believed that this appeal to classical concepts added a sense of legitimacy:[3]

Linnean	*Ordo militaris*
Class	Exercitus
Order	Legio
Family	Cohors
Genus	Manipulus
Species	Centuria
Individual	Miles

Croneis and his Illinois disciples now campaigned to see this change implemented. They became optimistic. "It seems . . . obvious that such a system must eventually be adopted," Croneis wrote in 1941. But while Croneis attracted his supporters, few really wanted to move these objects away from the jurisdiction of the International Commission on Zoological Nomenclature (ICZN). It was the commission that established and policed the rules that ensured that each species of fossil or living animal had a unique and universally understood identity in the form of a Latinized binomial name. There was nothing more fundamental or more important in the natural sciences. Croneis was ambitious for his military scheme to be afforded the same protection, and it was to the July 1948 meeting of the once-moribund commission that two French scientists, Georges Deflandre and Marthe Deflandre Rigaud, took Croneis's scheme. In keeping with the animal that swam through minds in Chicago and Illinois (rather than Missouri or Göttingen), the plan addressed fragmentary *invertebrate* fossils. Any conodont workers not wishing to comply could, of course, simply claim to be students of fishes. But none of this mattered. The scheme met with complete opposition. It was rejected outright. The commission had no problem with the existence of such a plan, but only as a technical terminology outside of its control, which Croneis and his supporters knew could only result in a chaos of uncontrolled names.[4] Their protests won a slight reprieve; the commission would permit the subject to be discussed one more time.

So this problem of how to name and classify fossil components was "in the air" when Rhodes came to describe his conodont assemblages and in so doing redistribute the honors among his new Illinois friends. Rhodes had copied Scott's lead, but Scott had wisely referred to the elements composing his assemblages using his new adjectives. Rhodes's assemblages were composed of "form species" with Latin names of their own. Indeed, he never used the term "form species" without putting it in quotation marks, indicating that he understood its lack of status and questionable desirability. To outsiders he had created species composed of other species, and this was in a published paper. It prompted Scott

Warthin, the managing editor of the *Journal of Paleontology*, and Eugene Eller, a scolecodont (fossil worm) stratigrapher at the Carnegie Museum of Natural History in Pittsburgh, to write to Winston Sinclair, the chairman of the Joint Committee on Zoological Nomenclature for Paleontology in America. They questioned the legality of Rhodes's actions.

By coincidence, Sinclair was no stranger to this particular problem. A staff member at the Geological Survey of Canada, he had turned up some sponge spicules while "monkeying in things I know nothing about, like conodonts." Wondering what he should do with them, he fell upon Scott's sponge paper and considered designating them as form genera as Scott had done. Scott was only too happy to use form names in his stratigraphic work. Finding Scott's paper in a volume "constantly being sought by spider and snake men," he wrote to ask Scott for a copy in February 1952.[5] The two men at least knew of each other before this little storm blew up.

Sinclair's correspondence with Rhodes on the matter of form names began in the following year. It took place publicly in the pages of the *Journal of Paleontology*, where Rhodes had published his paper. Writing courteously, Sinclair hoped to circumvent the kind of polemics, "sarcasm and intemperate language" that often surrounded such matters, but he clearly intended to teach the Englishman "some principles which are often overlooked or forgotten."[6] Giving specific names to natural things is, of course, a scientific necessity, but it also confers a personal reward on the author by establishing him or her as an authority. Any assault on that name, therefore, is an assault on the author's integrity and intellectual offspring, which he is obliged to protect as he would his children. That this act could become so personalized was, however, beyond the ICZN's comprehension.[7] It saw only the virtuous aim of logical objectivity when in truth the names themselves resulted from complex and personal interpretations. The professional's façade could not conceal the fact that science had always been a political game.

Sinclair made a number of points that, he argued, demonstrated the illogic of Rhodes's actions. The first was that the names Rhodes had used to describe the elements making up the assemblages referred to animals and not to things. It was certainly true that, unlike biologists studying living animals, paleontologists were much more inclined to blur this understanding. But if an individual conodont element belonged to one species, Sinclair asked, how could it, by inclusion in an assemblage, then simultaneously belong to another? Using examples from better known and more straightforward fossils, he made Rhodes's actions seem a little ridiculous.

Sinclair and Rhodes were now engaged in a chivalrous duel, but Sinclair's initial lunge was rather ill-judged; Rhodes may have been young but he was no palaeontological backwoodsman. Sinclair sent a copy of his note to Rhodes prior to publication as a matter of courtesy. Rhodes responded immediately, and in equally diplomatic tone, but accused Sinclair of being "a little misleading."[8] This was not a problem that could be understood by extending the argument to other fossils, he said; it uniquely affected conodonts and scolecodonts, and no simple application of the rules could resolve it. The rules might result in the name *Hindeodella* being given to a whole assemblage, but, Rhodes argued, these and other elements were not restricted to single assemblage species or genera. Who is to say if this wouldn't be true regardless of the component chosen? In other cases, single elements in the same assemblage ranged from one form species to another, thus providing no reliable basis for naming the assemblage. The second problem arose from Sinclair's preference for Scott's use of descriptive adjectives formed from the element names. This, Rhodes felt, would produce insurmountable problems for stratigraphers and result in "the substitution of a clumsy and less satisfactory system of nomenclature."

Rhodes could also draw upon two decades of precedent to demonstrate that Sinclair's solution had been tried and had failed. The fact that Schmidt's 1934 application of the rules had been ignored by subsequent workers seemed to prove the point, though Rhodes made no attempt to elaborate why it had not succeeded. There were also other precedents for Rhodes's own methods, most notably Brazilian Frederico Waldermar Lange's 1947 paper on worm teeth. This paper was so "outstanding" that Lange was forced by popular demand to have it translated from Portuguese into English.[9] Lange was at the time aware of the conodont problem, not least because the discovery and study of scolecodonts had happened in parallel. Indeed, Pander had unknowingly found them, and Hinde had made them a personal interest and was considered a pioneer;

Stauffer was a modern master. Both fossil groups had been recognized as stratigraphically important and the fossil record of both seemed – in 1947 – to be restricted to the Paleozoic. In the case of scolecodonts this created a bizarre anomaly, for these very ancient fossils were just like modern-day forms. A huge lacuna of time separated the two in which, despite much searching, these fossils remained virtually unknown – though Lange admitted that until the eye becomes accustomed to seeing them, they do tend to remain invisible.

Lange's paper was of interest to Rhodes because Lange had discovered some unique scolecodont assemblages (until that time only discrete or isolated fossils had been found). Now Lange could attempt a proper zoological classification of these fossil worms, but he knew he could not do so by referring to the existing taxonomy based on discrete parts. Lange was aware that Hinde had been exasperated during his own attempts to make biological species from detached components: "On one hand very different jaws are met with in the same genus; on the other, identical jaws are frequent in different genera."[10] Rhodes was clearly right to pick up on Lange's paper as it exemplified the problem he was wrestling with, but in a group of very non-enigmatic fossils. Lange's solution was to completely ignore that very imperfect classification based on what he called "formgenera" and instead use comparisons with living forms. He had independently, but like Scott and Rhodes, introduced a completely separate biological nomenclature for these assemblages, and therefore the animals themselves, stating that the "wholly artificial and transitory" system used to describe discrete jaws should be retained for stratigraphic correlation. One needed no better precedent for the actions Rhodes had taken, and Rhodes concluded, "Dr Sinclair's suggested method of nomenclature, while perhaps theoretically preferable, [is] one which cannot satisfactorily be applied in practice."[11]

The argument between Rhodes and Sinclair attracted the attention of Peter Sylvester-Bradley, who, as a member of the ICZN, doubted that Rhodes had acted illegally under the rules.[12] He questioned whether the names of individual elements in an assemblage could ever be considered junior synonyms and thus disposed of once an assemblage was named after the earliest named element. The matter centered on the objectivity of the proposed relationship. Were the element species and assemblage species "objectively synonymous" (truly the same) or simply "subjectively synonymous" (merely appearing to be the same) and thus open to dispute? Since the type specimens (the actual fossil specimens described to produce a scientific name) of the element species were found separately from the assemblages, it seemed likely that the relationship would always remain subjective and beyond definite proof. It meant that Rhodes's actions were, in Sylvester-Bradley's view, legal. It was an argument grounded in the small print rather than in the intentions of the rules, a lawyer's reading, teasing out desirable interpretations from the inevitable ambiguities of a rule book.

The simple issue of legality, however, was not at the heart of the problem; it was whether any dual system of nomenclature was desirable. The very purpose of having a system of rules was to ensure that any zoological species would have one name and one name only. This principle had never been overtly stated, but it had been proposed for inclusion in the rules at a recent meeting in Copenhagen. Botanists, by contrast, had been only too happy to adopt an entirely different approach. The remains of fossilized trees, for example, were hardly ever found complete. These organisms were understood first of all as discrete parts that might eventually be united into a single biological species. Consequently botanists recognized and valued form genera and form species, giving names to each type of leaf, trunk, branch, root, and seed.

If the rules for zoological material were counterintuitive or simply unjustifiable, then Sylvester-Bradley was there to question them: "The Rules are not designed to trespass on the freedom of taxonomists to classify animals in any way they wish." With a long history of practice and some fifteen hundred conodont species names in use, this was a taxonomic problem of considerable scale. The rarity of assemblages meant that the majority of these names would remain in use, as he felt "there seems no likelihood that it will ever be possible to assign more than a very small minority of conodonts to an assemblage." However, that still left outstanding the conflicting interpretation of the rules by Schmidt and by Scott and Rhodes. One of these interpretations had to be outlawed, and a commission ruling would be required to decide which. Schmidt had acted legally, but his approach, it was argued, would cause grave problems for all engaged in stratigraphy. As one rising star,

Maurits Lindström, remarked, "To put it drastically, it will be in the interest of micropalaeontologists to find as few conodont assemblages as possible!"[13] Conodonts were in a terrible taxonomic tangle and the only solution seemed to be an appeal to the ICZN.

The contentious issue of conodont taxonomy had implications far beyond this still relatively obscure group of fossils. In the United States in 1953, Ray Moore had published the first volume of *Treatise on Invertebrate Paleontology*, which was to become one of the biggest publishing projects in the history of paleontology. Moore had acquired considerable financial support from the Penrose Fund of the Geological Society of America (GSA), having convinced Bassler to offer up an almost complete manuscript in 1948 as proof that the project could make quick headway. It would be the guinea pig.[14] Bassler's manuscript, on his favorite bryozoans, had been prepared for Otto Schindewolf's *Handbuch der Paläozoologie*, which Moore felt a casualty of the war (it certainly was now).

The *Treatise* was to be a massive catalogue of the invertebrate fossil world, with a summary of what was known about each fossil group. At its heart was a desire to clarify taxonomy and nomenclature, and in Moore's hands, taxonomic abstraction and regulation, and a scattered literature, were to be marshaled into a practical tool.

Wilbert Hass of the USGS had been selected to write the contribution on conodonts. His internal investigations were considered sufficiently groundbreaking – particularly for those wishing to correctly identity these objects. His willingness to destroy names he himself had invented and his closely measured stratigraphic sections demonstrated that he possessed the necessary disinterest and rigor. Moore resolutely believed that the series must obey the ICZN's *Règles*, which governed the naming of animals, and consequently the arcane debate taking place in the *Journal of Paleontology* was of great interest to him. It suggested that the conodonts might be difficult to tame, but this was not the only reason his attention was attracted. He saw in this argument a more fundamental principle and one about which he had long argued in the past.

A graduate of Denison University in Ohio, Moore was another who had gained his doctorate at the University of Chicago, though long before the era of Croneis. He had gone on to become, from 1916, a stalwart of the University of Kansas and the Kansas Geological Survey. By 1953, he was a man of distinction: "If there was a scientific office to be held, he held it. If there was a journal to be edited, he edited it. If there was a scientific idea to be argued, he argued it, and was nearly always remembered in the process." William Hambleton, his former student, painted a fine word-portrait of the man: "Ray Moore was a man of medium height, stocky, wore glasses and always seemed rumpled. His complexion was tinged with red, especially around the nose. In earlier days, he characteristically smoked a pipe, but later consumed uncounted cartons of Pall Malls which stained his index finger yellow. A kind of sly smile always lingered about his mouth, suggesting amusement or the contemplation that his next question to you might be unusually interesting. He was a person of great appetite – for food, drink, work, play, generosity and appreciation. He possessed a large ego or, perhaps more appropriately, was comfortable in his knowledge of his own worth. His gait was sturdy, suggesting a certain inevitability about reaching a destination. He drove an automobile, not as a mode of transportation, but as an instrument of retribution."[15] Moore could be forthright in his views, and as a consequence he created enemies and missed out on much-desired honors. To these people he was cold, demanding, and intolerant. He even seemed to treat his friends gruffly.[16] But in Sylvester-Bradley's desire to see the latest conodont controversy solved he saw his own views echoed. The two men formed an alliance. Both saw the ICZN's near-decision in 1948 to push fragments out into a lawless wilderness of technical language as a step toward scientific anarchy.

Moore had been among the first to write to Bassler to compliment him on his and Ulrich's 1926 paper and the significant turn it marked as the science faced up to the utilitarian needs of the oil industry. It was this work that made him rethink the possibilities of his own favorite fossils, the sea lilies, or crinoids, which although not microfossils did break down into small skeletal parts. In 1939, he suggested that if one could identify these complex animals from their isolated components, one had a fine tool for stratigraphy. However, as Moore admitted, it was rarely possible to allocate these parts to their true species – they were not sufficiently distinctive. But what if one constructed an artificial system, he thought, like that used for conodonts and fossil plants? The names need not reflect true species. Indeed, the practice of naming parts and grouping them, and then having to move them into proper species when a natural association was discovered, seemed to Moore to complicate the use of these fossils in stratigraphy. Better, he thought, to develop an independent taxonomy for these parts – perhaps a simplified version of Croneis's scheme. He set about the huge task of compiling such a scheme for his fossil sea lilies, but when Bassler, who shared an interest in these fossils, heard of Moore's plan, he killed it in its tracks. In a single sentence he dismissed Moore's herculean efforts as worthless.[17] This was not the first time these two men locked horns, nor would it be the last; each understood that one man's more radical project could completely undermine that of the other.

With Hass making rapid progress on the conodonts for the *Treatise,* which he would finish in March 1957, Moore and Sylvester-Bradley took up the cause and requested that the commission make a *Declaration* introducing the term "parataxon" as a refined notion for form taxa. They contended that this was a rather more straightforward system for paleontologists to understand than Croneis's Roman ranks of nuts and bolts. Had they succeeded, the idea would have been temporarily admitted into the code, to be ratified or rejected at the next international meeting. It would have been applied immediately and gained recognition from practitioners. The commission, however, felt it too radical a step and asked the authors to draft a proposal for the purposes of canvassing opinion prior to a full discussion at the next international meeting in London in 1958. So, in 1956, a draft proposal was circulated by the two men, who solicited support from the paleontological community. Under these proposals the names of parataxa would not be available for naming assemblages: "A wall should be conceived to separate the nomenclature of whole-animal taxa from nomenclature of fragments defined as parataxa."[18] The plan was exactly that which Rhodes had used. Although Moore and Sylvester-Brad-

ley could see wide application in other groups of fossils, workers in those
areas would need to make specific bids to the ICZN to permit parataxa
to be established for their animals. In June 1956, they supplemented this
proposal with the first two such applications. These concerned conodonts
and ammonite aptychi (the operculi, or flaps, that closed off the aperture
of an ammonite shell, which are usually preserved separately).[19] These
were to be test cases for the debate that would develop in the *Bulletin of
Zoological Nomenclature* in the run up to the conference.

The conodonts appeared to be the perfect case. By detailed examination of the named assemblages, they revealed complexities of synonymy
(two names for the same species), homonymy (one name for more than
one species), disputed identification, and unknown stratigraphic ranges.
The depth of subjectivity was considerable, but who could dispute the
usefulness of these fossils? Only two assemblage names stood in the way
of the parataxa locomotive, those of Eichenberg and Schmidt. Eichenberg's animal had been pieced together from discrete parts. It could be
considered entirely subjective and unproven. Schmidt's *Gnathodus* was
considerably more secure. It was the only animal that, according to the
rules, had a correct name. Schmidt, however, did not wish to stand in the
way of a new dual system and offered *Westfalicus* as a new name for his
assemblage if parataxa were adopted.

The proposal concerning aptychi arose entirely separately, and for
different reasons, but had been swept up into the parataxa debate simply
to remove a competing scheme. William Jocelyn Arkell of the Sedgwick Museum at the University of Cambridge, the world's most distinguished ammonite specialist and Jurassic geologist, had published his
own proposal for dealing with these fossils in 1954. Arkell did not want
the names given to these aptychi to undermine the beautiful, logical,
and long-established names of ammonites when both shell and operculi
were united. He knew that the laws of priority, which always required
the older name to prevail, would inevitably force such a calamity. When
Arkell read Moore and Sylvester-Bradley's proposal, he found himself
in complete agreement – "I think the authors have made their case completely" – and joined their cause.[20]

Opposition began to mount in some quarters before Sylvester-
Bradley and Moore commenced their consultations. Most vocal was

the American Society of Parasitologists. This particular problem arose because Sylvester-Bradley and Moore had discussed their plans with parasite worker J. Chester Bradley and then extended the concept of parataxa to admit the unidentified life stages of these animals. Perhaps the two men felt the support of medical scientists might win the day. Now these scientists were complaining as a system of parataxa would require rather more formality than they desired. The commission asked Moore to consult these parasite specialists, which he did in New York, but he thought the whole thing an unwelcome and unnecessary distraction. This was, however, one parasite the parataxa scheme could not shake off. It produced the first major blemish.

The ICZN sent out a succinct version of the parataxa plan to paleontologists and paleontological institutions in July 1957. Soon letters of support were flooding into the editor of the *Bulletin* from all areas of the paleontological community, but it was not all good news. Some saw it as a recipe for chaos, while others were clearly annoyed to see Moore championing this particular cause. While Moore could attract the world's best paleontologists to write for the *Treatise,* and could certainly write a courteous response if required to do so, he was also known for his outspokenness and had created powerful enemies. Ray Bassler was not one of these, although he and Moore were often at loggerheads. And in this new proposal Bassler saw Moore making a second attempt to have his sea lily fragments scheme accepted. Bassler was perceptive; Moore certainly did have these in mind. Bassler saw this as a subversive assault on his position that, if successful, would have personal consequences for him and he complained, "Moore acknowledged that the fragments had no value as genera and species in classification. These fragments probably gave rise to the later term PARATAXA." He dubbed Sylvester-Bradley "a European sponsor" and called for "all good naturalists to come to the aid of our taxonomy." He felt that "common good sense" could deal with the problem: "Any name based upon an aptychus can remain until the whole shell is known, whereupon the aptychus-name can go in parenthesis labelled as the operculum. Conodonts can be treated likewise until the entire animal is found, maybe centuries later, but the old unlocated names must be held for stratigraphic reasons." For the exasperated Bassler it simply wasn't worth wasting any further words on the subject.

Other objections came from groups in which similar problems had been dealt with using the existing rules. Only two conodont specialists objected. Both were at the Oklahoma Geological Survey. The first was Ted Branson's son, Carl, who was simply not convinced that natural assemblages were anything other than coprolitic: "Scott's assemblages are coprolitic associations. The validity of other assemblages is not demonstrated." He added, "The names of 'assemblages', said to be natural genera, should be suppressed as unnecessary, premature, and hypothetical." The second was Robert Fay, who used Branson's view to demonstrate how poorly conodonts were known: "At present there is no competent person to make a decision on the classification of conodonts" due to the unknowns. "We are proceeding along an odd path. . . . I vote that we dismiss this proposed insertion of parataxa into the *Règles,* because it is not sound, premature, and unnecessary, and highly subjective." Conodont workers in support of the plan included Ellison, Schmidt, Furnish, Hass, and new boys Klaus Müller and Rhodes. But the conodont community was simply not large enough to swing the argument in its favor without help.

In some cases groups decided to respond en masse. In Washington, Bassler's call for revolt had hit home. The city's "Nomenclature Discussion Group" received votes from 56 of its 113 members, all against the plan. Bassler was, however, not recorded among the votes cast. These specialists came from a wide range of zoological and paleontological fields and each had considered the possible impact of parataxa on their field; all were certain that confusion would result. Given that the proposal made no suggestion that it should be adopted by all fields, Sylvester-Bradley thought this a peculiar turn of events. Hass, who was then the only conodont worker in Washington, had spoken in favor of the plan but had preferred to write directly to the commission. Hass's colleagues seemed to believe that the simple solution was to recognize no synonymy between the assemblage and the discrete element. In other words, do what Scott had done and ignore the problem.

Sylvester-Bradley responded to the Washington scientists publicly and somewhat defensively, explaining his relationship to Moore in the proposal – he had an interest in resolving taxonomic issues, and not as a specialist in the fossils used as cases for change: "My present interest

... is therefore, not that of the parent of a fond child, but still that of one of the Commissioners who is attempting to find a solution to a difficult nomenclatural problem." Sylvester-Bradley no doubt understood some of the underlying issues that affected the Washington view. What Sylvester-Bradley wanted was constructive argument – not a vote. He encouraged the participants to contribute to the debate.

A meeting of the Institute of Zoology at the Polish Academy of Sciences also reached unanimity: The plan was rejected for use in zoology and accepted as a solution for a limited number of cases in paleontology. The meeting contained no paleontologists. Writing on behalf of the academy, Tadeusz Jaczewski saw parataxa as a kind of opiate that could lead to disfunction and laziness.

The British-based Palaeontological Association and Palaeontographical Society also voted and came out strongly in favor of a system that served the peculiar needs of fossils like conodonts and stabilized their nomenclatures. But they were split down the middle on whether the parataxa proposals were the best solution. They were aware, for example, that entomologist Curtis Sabrosky at the U.S. Department of Agriculture in Washington had proposed an alternative plan based on the system used in paleobotany in which form taxa were well established. Although he disagreed with Sabrosky's definitions, Sylvester-Bradley thought this a possible solution. For him, the terms "parataxa" and "form taxa" were interchangeable. By some clever and subtle maneuvering of terms invented to describe parts of fossil plants, he used this moment to distance the problems of fossils from those of parasites. The latter, one might infer, were no longer part of the parataxa debate.

When Sylvester-Bradley came to summarize the results of the consultation, it was clear that the vote in Washington had sunk the scheme, at least in numerical terms. He had received fifty-three votes in favor of the plan, of which thirty-six had come from individuals. However, those against amounted to seventy-nine, fifty-six of which were votes from the Washington census. Only nine individuals had written in opposition. Sylvester-Bradley mulled over the imperfectly formulated alternatives that had also been put forward. The consultation overall lacked a firm conclusiveness. His published summary was followed by a single-sentence letter from fossil fish specialist Errol White at the British Museum (Natural History): "I can answer your circular letter of the 8th July 1957, very briefly by saying that I am dead against special provision for Parataxa in the *Règles*."

The fate of the parataxa plan was decided at the Fifteenth International Congress on Zoology, which met at the British Museum (Natural History) in London in July 1958 to celebrate a centenary of Darwinian evolution. Just before the meeting, two letters arrived from the USSR, one from the Palaeontological Institute of the Academy of Sciences and the other from the Institute's Laboratory of Palaeoecology of Marine Faunas.[21] Both were translated from the Russian and revealed a remarkable conflict of view. The first letter, from the institute's president, Yuri Orlov, and secretary, Yanovsky, was succinct and fully in agreement with the parataxa plan. The second, from senior staff at the laboratory, which arrived a week later and had been drawn up independently, had not a single good word for Moore and Sylvester-Bradley's proposal. Their objections were considerable. They thought the plan had been proposed merely to preserve names – an act already possible under the rules – and not to extend knowledge. Indeed, they felt that the plan might actually circumvent scientific progress as workers could remain contented with the practicalities of an artificial system. Doubtless knowing precisely where American science had been heading, they said that even stratigraphers could not be excused the necessity of knowing their animals biologically. The argument gave strong indications of the chaos that might arise if two systems "separated, as by a wall" were permitted to exist. All paleontologists, they pointed out, deal with mere parts of animals.

The colloquium that preceded the meeting decided to defer the question, asking a committee, to which Moore was appointed, to report to the Washington Congress in 1963.[22] The conodont volume of the *Treatise* remained on hold a little longer, the outcome of the parataxa plan critical to its content.

On November 30, 1959, Hass died prematurely; Arkell had died the year before. Hass's loss was a tremendous blow to the *Treatise* project, so Moore brought in two young bloods who were already revolutioniz-

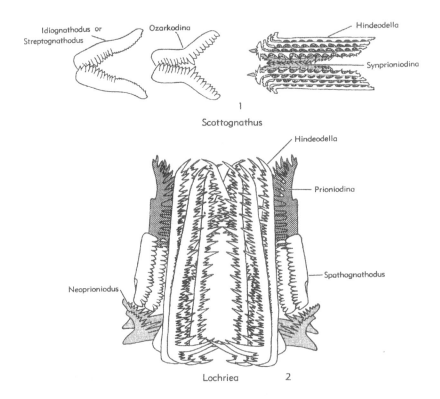

5.1. Hass's compromise. Scott's and Rhodes's conodont assemblages as published by Hass in the *Treatise of Invertebrate Paleontology* in 1962. Hass had died before the idea of parataxa had been rejected, and his dual classification of conodonts retained, at very least, the ideal of parataxa. Here "biologic" genera (assemblages) are composed of "utilitarian" genera (conodont elements). From W. H. Hass, *Treatise of Invertebrate Paleontology,* Part W Miscellanea (1962). Courtesy of and ©1962, The Geological Society of America and The University of Kansas.

ing the conodont in terms of its understanding and significance: Frank Rhodes and Klaus Müller. Both men had published reviews of conodont knowledge, and although neither were Americans, both had first met in Iowa, where Müller was on sabbatical in 1955. They immediately hit it off and had prepared a joint paper.[23]

By mid-1961, Moore recognized that the parataxa plan was dead – it was certainly delaying publication of the *Treatise* – and that the committee would recommend to the ICZN that it should give it no further consideration. Moore had, following the London meeting, consulted Hass, Müller, and Rhodes but remained dissatisfied. Hass had already retained Scott's dual system in the *Treatise* separating a "utilitarian" taxonomy for individual elements from an uncertain "biologic" group using the names Scott and Rhodes had introduced (figure 5.1). In an unprecedented move – Moore usually left the content of the *Treatise* volumes to the specialists themselves – he published his own statement. The concept of form species, he said, could not be used, and Rhodes's construction of species of assemblage composed of species of element was "unhelpful." He became an advocate for doing things according to the rules. His solution was to follow Sylvester-Bradley's observation and not to admit to an exact identity for the components making up an assemblage.[24] By that means all names might actually exist within a single system.

Rhodes, who had recently moved to the University of Wales in Swansea, felt Moore was entirely on the wrong tack. He told him that the fossils were not as ambiguous as Moore wished to pretend – one certainly could place discrete conodont fossils into assemblages. A fully paid-up supporter of the parataxa plan, Rhodes found Moore's plan more damaging than useful.[25] But Moore was emphatic, arguing that "dual nomenclature is not only unacceptable and illegal, but it is unnecessary." He warned: "Let us agree, then, on adopting a conservative, unassailable course which takes us around or away from conflict. . . . Bold workers who wish to proceed differently may do so, but then they are enjoined to tread carefully and follow through to ends that accord with the Rules."

In Moore's *Treatise,* all the important groups of fossils received the attention of one or more volumes of their own; the conodonts were, however, thrown into a ragbag of "Miscellanea": worms, small problematic "conoidal shells," and "trace fossils and other problematica." During the course of its production, Moore doubtless gained a new understanding of this latter term. In a book on *invertebrate* fossils, the conodont was for some only an honorary invertebrate – one of those rare vertebrates that had stratigraphic utility. It is interesting to note that Moore conceived of the *Treatise* in 1948, the year he also began his classic textbook *Invertebrate Fossils* with Cecil Lalicker and Alfred Fischer. That book formed the model for the *Treatise,* and its very last chapter was given over to conodonts.[26] It had been prepared by Lalicker, who personally believed a vertebrate origin likely, and indeed refers to these fossils as "teeth," yet there they were in a book on invertebrate fossils.

Although already out of date by the time it was published in 1962, the *Treatise* was undoubtedly a landmark publication, not least because of the debate that went on behind the scenes. In Moore's eyes the parataxa idea was well and truly dead, but as is clear from Rhodes's objections, it was unlikely that Moore's alternative plan would find universal support from a new, young, and ambitious generation that was then flooding into the field. Conodont science was at last breaking out from its American stronghold, and in doing so it would have to confront other ways of doing paleontology. Paleontology itself was undergoing a postwar renaissance; it was, once again, spring.

What a joy that was, what a boon to the eyes, after so much white!
But there was another green, surpassing in its tender softness
even the hue of new grass, and that was the green of young larch
buds. Hans Castorp could seldom refrain from caressing them
with his hand, or stroking his cheeks with them as he went on his
walks – their softness and freshness were irresistible. "It almost
tempts one to be a botanist," he said to his companion. "It's a fact,
I could almost wish to be a natural scientist, out of the sheer joy
at the reawakening of nature, after a winter like this up here."

THOMAS MANN,
The Magic Mountain (1928)

SIX

Spring

WITH A GENERATION LOST ON THE BATTLEFIELDS OF EUROPE and Asia, the 1950s felt like a new beginning. A sense of optimism and renewal altered the everyday. It was felt on the streets of London, New York, and Berlin, in cafes, offices, and even laboratories, and inevitably it affected the mindset of those who took an interest in fossils. A temporal rift seemed to separate this world from the prewar one, which now seemed old and remote. For the new conodont workers this sense of distance was aided by the death, retirement, or withdrawal from the field of Ulrich, Bassler, Branson, Mehl, and others. Now a new, young, idealistic, and ambitious generation took possession of the fossil as none had previously.

It was only now that the fossil truly entered Europe. With a land area roughly equivalent to that of the United States, this was a continent divided into geographically smaller but more densely populated nations. Each had its own geological ambitions, institutes, and cultures. Europe also possessed a far older and, in many respects, more sophisticated tradition of studying rocks and fossils. In the United States, the new postwar generation looked back at that earlier period of Stauffer, Branson, and Ulrich and saw something of the days of the pioneer. Science was then large in scale, low in resolution, and distinctly old-fashioned.[1] On both sides of the Atlantic, paleontology was to be reinvented as a project of international scope. The conodont was to be rediscovered as strange, evocative, and beautiful, and as possessing huge utilitarian potential. It became possible to imagine a scientific life with the conodont as its focus. In new minds the animal was to be reborn, repaired, and repeatedly reimagined. Winter was well and truly over.

✳   ✳   ✳

In the vanguard of this new generation was Frank Rhodes, who had been a schoolboy in Solihull, Birmingham, during the war. Rhodes discovered geology quite by accident while studying chemical engineering at his local university. It became a lifelong passion. After his sojourn to Illinois, he returned to Britain quite altered by the experience. He soon produced a flurry of papers on the conodont, including a comprehensive review of almost its every aspect, which he published in 1954. Here the century-old tale of mystery and illusion – "one of the most fascinating and perplexing problems in palaeozoology" – was remade for the modern audience. And as Rhodes reviewed, so he arranged. As he quoted, so he measured existing knowledge against his own beliefs and judgments. In doing so, he replaced the past with his own sense of the modern. Indeed, in this postwar world, it was rather easy to construct this sense of a new modernity and label earlier thoughts as obsolete. Branson and Mehl's anomalous fibrous conodonts from the Harding were an obvious target. These alone were attached to Kirk's bony material. Rhodes had not seen these fossils himself, but he knew that conodonts and fish teeth were easily confused. It seemed likely that these fossils were not conodonts at all, believing, a few years later, that these fibrous fossils were the remains of "primitive vertebrates." A decade later, however, Lindström would look at Branson and Mehl's original specimens and conclude that "all conodonts are lamellar *and* fibrous."[2]

Rhodes's analysis of the conodont problem led him to run down the long list of animal contenders. It was easy to strike a line through nearly all of them. The fish was floated – its plausibility reliant on the conodont's phosphate composition and those bony basal attachments. Neither were sufficient reason to imagine a fish, Rhodes said. The fish sank.

Wilbert Hass's mysterious animal suffered a similar fate. Rhodes had seen limited wear in the fossils he had studied – too little to suggest they were teeth but sufficient to know that they had not been concealed beneath flesh. He felt the animal most likely an annelid worm. Perhaps he had brought the worm back from Illinois? It was, after all, indigenous there.

In these arguments Rhodes transported the enigma and the science into the modern era and ensured that the conodont remained a most peculiar thing. Readers understood, implicitly, that the fossil was now in the possession of a youthful and precocious generation.[3] It could not be in better hands.

Rhodes remained actively involved in the conodont community from the 1940s to the 1980s, but his most important scientific work was undertaken at the beginning of his career. It, in part, earned him the moiety of the Lyell Medal of the Geological Society of London in 1957. Undoubtedly a high flyer, Rhodes was soon drawn into university administration, first in Britain and later in the United States. When he retired as president of Cornell University in June 1995, he was the longest-serving Ivy League president and among the most celebrated leaders in American higher education. Sanguine, unflappable, experienced, and just, Rhodes became wise council and diplomat for some in the conodont community, not least for his lifelong friend Harold Scott. Whenever conodont studies fell into turmoil or controversy, Rhodes always seemed to be on hand to calm the waters and find an amicable way forward.

Frank Rhodes rose to a position of some distinction while still a young man. Others grew up in less fortunate circumstances, and none more so than those who spent their youth in 1930s and 1940s Germany. Those who survived the political turmoil, witch hunts, extermination, and military campaigns returned to bombed-out cities. Defeated, and loathed by many in the international community, they also saw their country partitioned. To return to take up the study of tiny fossils might, in these circumstances, appear ludicrous. But in such things was salvation. These young scientists yearned for normality. Fossils and geology offered this as well as a distraction from the difficulties of everyday life. Rocks and fossils also became vehicles for ambition, and in those ambitions were the social motivations to build civil society. Such occupations were also a means to escape postwar drudgery. With good reason, then, these new workers entered the science with extraordinary motivation and energy.

They might have filled shoes left vacant by the war dead, emigrants, and the exterminated, but they wanted only to think of the future.

Klaus Müller's war had taken place on the eastern front. Encircled by Soviet forces, he had collapsed from typhoid in temperatures of minus twenty degrees Celsius. His body stacked in a train with the dead and living, he somehow escaped to medical care. He was – and would continue to be – a survivor. Repeatedly declining promotion, he had told his military superiors, "A Berliner will never take command of a company which is going broke!" He was always, spiritually at least, a proud Berliner. Like those around him, in the closing years of the war, he could see the hopelessness and the futility of it all. Allocated the unusual rank of "private first class,"[4] he found himself admired by his fellow soldiers and the confidante of various officers. It was in his conversations with these educated men that the seeds of his future career were sown; it was here that he acquired intellectual ambition.

The German advance eastward had been unmatched in its brutality. Fueled by hatred, the Russian counteroffensive in the closing year of the war was no less brutal. When Germany surrendered on May 8, 1945, its army was immediately dissolved. Müller returned to Germany a physical and mental wreck, entering Berlin on a beautiful day in early May. One of the first to arrive from the eastern front, Müller realized the city was still a dangerous place, and he had nothing, not even his papers. He had already decided that he wanted to study geology and headed for the city's Humboldt University. He was surprised by the welcome. "This was the first time people had been friendly to me, I would even say the first time in my life," he later recalled. He waited until the "chief" returned the next day. After an interview lasting more than two hours, Müller found himself inducted as the first student after the war. He was given the paid position of student assistant and set about helping his new boss rebuild the shattered department.

Situated in the east of the city, the Soviets soon began to exert their control over the university. Many years later, in August 1961, they would erect the Berlin Wall and divide the city in two. Remarkably, these important macro-historical events – the rise of Soviet ideology and the partition of Germany – played a fundamental role in the transformation of Müller into a conodont specialist. In what seems like a strange sociological experiment, these grand events interacted with Müller's knowledge of chemistry and his relationship with distinguished fish paleontologist Walter Gross, who had also returned to the university after the war. But for this experiment to take place and produce its result, the catalyst of Heinz Beckmann was required. To understand Beckmann's contribution, we must first, however, recover a little history from the United States.

In 1935, the U.S. National Museum's Arthur Cooper began using acids to extract brachiopods from Permian limestones collected from the Glass Mountains of Texas. Started merely as an experiment, it proved a sensational success; he separated some three million extraordinary fossils from sixty tons of rock.[5] The key to his success was the differing chemistry of the fossils and the rock. For most invertebrate fossils found in limestone there is essentially no difference and the use of acids will destroy both rock and fossil. Cooper's brachiopods, however, were preserved as silica and were thus relatively immune to the acids Cooper used to dissolve the limestone.

We cannot know precisely how this technique entered into the minds of conodont workers, but Cooper's work at the National Museum was well known, and certainly so to Ulrich and Bassler and those paleontologists at the museum. The museum was also well connected to paleontologists working in state surveys and universities. The phosphatic chemistry of the conodonts made them natural candidates for this kind of extraction and it was simply a matter of time before someone would put two and two together. All that prevented this was an assumption that limestones contained few conodonts.

It seems likely that the technique escaped into the conodont community at the hands of Bill Furnish, who was assisting A. K. Miller in a major study of the fossil nautilus collections at the National Museum. He began to use the technique with conodonts in the mid- to late 1930s, telling Mehl of his success in 1938. In Buffalo, Bassler's friend, Ray Hibbard, was obtaining wonderful results cleaning his worm jaws (scolecodonts) with hydrochloric acid in 1939. Branson, Mehl, and Ellison experimented with acids in the latter half of 1940, the method having been proven by master's degree student Freddie Strothmann earlier that year. Branson and Mehl published information about the technique in 1944.[6] By the late 1940s, acids were beginning to enter mainstream paleontological practice and Rhodes used them in his PhD work.

The growing use of acids in the 1950s revealed, to widespread surprise, that limestones could be extraordinarily rich in conodonts. The prewar belief that conodonts were preferentially associated with shales had been another illusion.[7] No-one had reflected upon, and few knew about, John Smith's earlier assertion that limestones might be a rich source of these fossils, nor had they considered that Smith's fossils had been leached out of these rocks as a result of the natural acidity of rainwater.

With the introduction of acid preparation, the conodont took on a new ubiquity: "With just a bit of preparation, almost any marine rock of Paleozoic or Triassic age, from almost anywhere on earth, will yield to the patient investigator an assortment of phosphatic microfossils termed conodonts." It was no longer necessary to trust in intermittent samples as complete temporal sequences could now be studied. This gave the conodonts a huge advantage over fossils which relied on the use of eye, hand, and hammer. It was now possible to collect conodonts in their thousands.[8]

Heinz Beckmann read Branson and Mehl's account of the use of acetic acid and improved upon the technique, thus making it possible to use less acid yet achieve more rapid digestion of samples. Published in 1952, this work had a huge impact on German paleontologists.[9] After reading it, Müller offered to use acetic acid to extract the Silurian fish scales Gross was studying. Gross, who had published several large monographs on these fossils, had always extracted these scales using needles. Gross agreed to a trial and gave Müller a piece of his limestone.[10] Müller returned with the residue, from which Gross picked out fish scales. But as he picked, he also found conodonts.

Like Pander, Gross came from that part of Europe we know today as Latvia; and like Pander, he too was an ancestral German. In 1925, when in his early twenties, he had moved to Marburg in Germany to pursue his career in the company of the slightly older Otto Schindewolf. Distinguishing himself in the 1930s in microscopic studies of the structure of the hard tissues of fish, Gross's work was very much in the Pander tradition. These studies naturally took him to consider the microscopic structure of those would-be-fish, the conodonts, for the first time in 1941. Unknown to Gross, Hass was at that time undertaking identical investigations in the United States and making discoveries similar to those Gross would make in Germany. Gross, however, was handicapped by the war, which made publication and wider communication impossible. His research was cut short, and from 1943 he found himself a soldier, then a prisoner, before finding relief from both in the countryside after the war. Only in 1949 did he join the university in Berlin, where he replaced his friend Schindewolf and found himself in the company of the recuperating Müller.[11]

The material Müller had extracted for him now provided the basis for a new paper on conodonts. Having now read Hass's study, and spurred on by the American's weakening of the American fish, Gross set about destroying the German one. He did so emphatically. He confirmed that the conodont fossil lacked a true pulp cavity and he refuted Beckmann's claims for a dentine-like structure. Nor did he find enamel. The conodont fossil was neither the tooth of a fish nor the grasping apparatus of a worm. It was not part of a gill apparatus or mandible, and it was not composed of bone or cartilage.[12] Indeed, it didn't appear to belong to an internal skeleton at all. If it reminded him of anything, it was the phosphatic external armor found in some Devonian fishes, though even here there were differences. The conodont elements must have grown, he thought, by the addition of layers at the surface. That surface, then, must have been beneath a protective secretive layer, certainly during growth, and perhaps always. In the most thorough zoological comparison to date, Gross could only conclude that the conodonts "perhaps belong to a special branch of the chordates or jawless vertebrate animals." He could certainly see structures that reminded him of other vertebrate animals, but so much was different about them. He could not place them with any known group of animals.

A few years later, Gross produced the first detailed description of the "basal structure" of the conodont, a component that had largely been ignored. He revealed that an extraordinary cone of material, composed of lamellae, sometimes extended well beneath the conodont element as commonly understood. It appeared to grow by additions inside the large cavity of the base such that the cavity might become completely filled. Unknown to Gross, the young Maurits Lindström, in Sweden, had published something similar two years earlier but had done so in a journal so obscure that no one had seen it.[13]

Gross took his work on the base a little further in a paper in 1960. By that time the argument was being conducted on points of extreme detail.[14] The base and the crown were then to be understood as a single continuous unit, with the crown – the conodont fossil of common understanding – being more significantly mineralized.

In a few short papers, Walter Gross completely altered the zoology of the conodont. Delivered with Gross's considerable scientific authority and building upon the doubts of Hass, it tore the conodont fossil out of the mouths of fish and worms. Now it was even more incomprehensible. So original, meticulous, and convincing was Gross's work that it cast a long shadow. In the 1960s, few would consider the nature of the animal and almost no one thought of these fossils as teeth. Those who might wish to dissent knew they would need better material and improved technologies. These did not arrive for a decade.

Few paleontologists shaped the fossil so fundamentally, but Gross was not to become a conodont worker. This was just an excursion by a specialist in fossil fishes who effectively swept the conodont out of his field of concerns. Gross remained in East Germany for a while but became an increasingly outspoken political dissident. Schindewolf orchestrated his "escape" to the West and to Tübingen in the early 1960s just as the Berlin Wall went up.

Müller was not particularly impressed by Schindewolf, who had risen to become the leader of German paleontology after the war. He favored other candidates. He had remained in the east of the city after the war, undertaking his doctorate research on the cephalopods of the Devonian of Thuringia in East Germany. But when he stood for examination, the university's Soviet-controlled authorities told him that he would also need to write on communism. The very idea was an anathema to him. Fortunately, just as the Soviets were inflicting their politics and ideologies on the old university, plans were being developed to erect the Free University in the west of the city. Now, with the aid of professors in both universities, Müller escaped, in what was still merely a bureaucratic exercise, to the west of the city.

At his new department at the Free University, Müller was given a small position, but he was already suffering from acute tuberculosis of the kidney. It was simply the next in a series of diseases that would affect his life. Müller was so stressed by the urgent need to get his PhD finished that his doctor found him unresponsive to treatment. "How long do you need to finish your thesis?" his doctor asked. Müller said, "Three days." When those three days were up, Müller returned to the hospital and spent the next year there, before leaving the "flatlands" to recuperate in the Alps, in the manner of Hans Castorp in Thomas Mann's famous prewar novel *The Magic Mountain*. At the insistence of his wife, Eva, who was always his voice of reason, Müller eventually left his retreat to return to the life of a geologist. He did so reluctantly.

Conodonts had still not entered his thinking, and back in West Berlin, he was a geologist trapped in a half-city surrounded by the communist "East." His research sites in Thuringia might just as well have been on another planet; he had no chance of collecting there now. Out of desperation, he took the remaining samples of limestone left over from his PhD work and dissolved them in acetic acid. In them he found hundreds of conodonts, and from these he published his first conodont paper in 1956. In the middle of writing it, he traveled to Iowa, the new U.S. capital of conodont studies. It was not his intention to stay there, for he had wanted to tour the country, but his active tuberculosis, the offer of free medical support, and the wise counsel of his wife, meant he remained in the one spot. There he spent two years in the company of Miller and Furnish and wrote his own summary of conodont studies.[15] It was a paper, like other contemporary accounts, that showed the conodont to be a vehicle for creative thinking. Of all the fossils, it seemed open to new views and new solutions. There were no senior workers to control or constrain thinking, or maintain orthodoxies.

In his review, Müller recognized that the rapid evolution of the platform elements – so useful to Branson and Mehl – provided the key for identifying and naming whole animals because it was these elements which distinguished them. One did not need whole assemblages to name whole animals, he thought. The approach had worked with fossil mammal teeth, so why not conodonts? Müller admitted to the subjective decisions paleontologists needed to make to document evolution but nevertheless believed it was possible and preferable to construct a natural rather than a utilitarian system, even if using isolated elements. However, even among the new generation there were those who disagreed. The young Lindström remained attached to the old ideals of the stratigrapher: "The object of the classification of isolated conodonts could not be to expound phylogenetic relationships between the different kinds of conodonts. Classification of isolated conodonts rather has to restrict its aims so as to provide the worker concerned with conodonts, especially in the field of stratigraphic geology, with clearcut morphological categories."[16]

In 1956, the physically weak Müller felt empowered as the conodont world was remade with acids and new thinking. Having written his thesis on Devonian cephalopods, he now felt he might be the person to tame the European Devonian using the conodont's newfound ubiquity. He was, however, unaware that many other German minds were thinking similar thoughts, for he was isolated from his compatriots by illness, politics, and travel, as well as his preference for lone working.

Müller had no knowledge of what had been happening at the ancient University of Marburg. Here, in 1949, Beckmann had demonstrated that Branson and Mehl's contentious but wonderfully useful Grassy Creek conodonts could be found in Germany. As his rocks were rather better studied, Beckmann could now confirm that the Grassy Creek was not only Upper Devonian in age but that it occurred in the lower part of that division. Beckmann's paper on acids followed, in 1952, and then, in the spring of 1953 at a meeting of the Deutscher Geologischer Gesellschaft

6.1. Four harbingers of spring, captured in later life. *Clockwise from top left:* Otto Walliser, Maurits Lindström, Willi Ziegler, and Klaus Müller Photos respectively: Dick Aldridge; Helje Pärnaste; Senckenberg Gesellschaft für Naturforschung; John Huddle, courtesy of John Repetski.

(German Geological Society) in Dillenburg, he revealed the true potential of the conodont for the high-resolution stratigraphic study in the German Devonian and Lower Carboniferous. The impact on his audience was immediate. A magical solution that overcame supposedly "unfossiliferous" strata and the patchy distribution of other fossils seemed to manifest itself before their eyes. The conodont was ubiquitous, could be processed with little effort, and opened up the possibility of studying rocks on a centimeter-by-centimeter scale. Exploiting fine and increasingly well studied German rock sequences, this audience imagined that a dedicated conodont worker might "influence biostratigraphic interpretations throughout the world."[17] Beckmann had fired a starting gun and a new generation rushed forward.

A school of conodont studies was established at Marburg in Beckmann's wake in which Willi Ziegler, Günther Bischoff, Ursula Tatge, Otto Walliser, Hans Bender, Dieter Stoppel, Reinhold Huckriede, Peter Bender, Hans Wittekindt, and others came to work on these fossils. It became a site of intensive research activity. Beckmann's influence spread to all corners of the soon-to-be-divided country. Dietrich Sannemann at Würzburg, for example, was the first to confirm Beckmann's results with his own pioneering study.[18] At Humboldt University, Joachim Helms, an assistant to Gross, began working on the conodonts of Müller's Thuringia around 1958.

In Marburg, Bischoff and Ziegler began as PhD students under the supervision of Professor Carl Walter Kockel, who, although an excellent tutor, actually knew little of conodonts. Bischoff took inspiration from Hermann Schmidt, then a Göttingen professor, and collected limestone samples from the Devonian-Carboniferous boundary under Schmidt's guidance during the Arnsberg conference of the German Geological Society in 1954. Bischoff was astonished by the wealth and variety of conodont material this produced, and at the clear faunal change that appeared to separate the Devonian from the Carboniferous.[19] Using this material and the security of the cephalopod fossils that already divided up these rocks, he could, unlike American workers only a few years before, be certain of this change.

While Bischoff worked at the top of the Devonian, Ziegler began an assault on the poorer faunas of the Lower and Middle Devonian. Before long the two men were working closely together and attempting to create their own parallel chronology to mirror that based on cephalopods. Throughout the mid- to late 1950s, the combined effort of these young Marburg scientists continued to reap results and the foundations were laid for major developments in the following decade.

In 1958, a position came up for a conodont specialist at the Geological Survey at Krefeld. Müller was approached, but hearing that the organization had no director and was moribund, he declined. Ziegler was less choosy, and already rather more attached to conodonts than Müller, and he accepted the post. Driven to succeed whatever the organization's circumstances, Ziegler's energy and single-mindedness, applied in a full-time position, left Müller standing. Ziegler probably understood that it would permit him to outcompete those who shared his ambitions. Müller now looked on helplessly as the door shut. The German, and with it possibly the world's, Devonian now belonged to Ziegler.

Müller's survival instincts kicked in. Better to enter an uncompetitive field than to fight; Müller was not much of a fighter. Perhaps he would be able to do in the Silurian what he had planned for the Devonian? Müller turned to the excellent Silurian sections in the Carnic Alps on the border between Austria and Italy. This field work also had the advantage of altitude, so beneficial to his ongoing health problems. But he had not been collecting long before he discovered he was not alone. Otto Walliser shared Müller's new ambitions.

Like Müller, although somewhat younger, Walliser was a survivor of the war. He had imagined a career in the realm of living plants – in his case forestry science – but the university in his home town of Tübingen offered no such course. However, geology seemed to fit with his nature-loving instincts, and he soon developed the ambition of becoming a museum curator. Tübingen also had the distinction of being home to Schindewolf, from whom Walliser would learn some of his geology. On completing his degree, Walliser moved to Marburg. He was already becoming a specialist in Lower Jurassic ammonites, but he was also aware that he was entering the new German center for conodont studies. And as he was to be senior to the younger Ziegler, Bischoff, and Tatge, he thought it judicious to take an interest in these fossils. With the Devonian and Triassic niches already filled by those he was joining, he fixed his attention on the Silurian. From his first trip to the Frankenwald, he returned with a single sample that proved by far the richest in conodonts in his whole career. This demonstrated the potential, and so he began working on the Silurian of the Carnic Alps.

Walliser had worked on these rocks for a year before Müller appeared in search of upland solitude. For both men this was a problem. Müller had funding from the national geological society but now found himself in the embarrassing position of possessing a grant for a project that lacked the scientific novelty he had claimed. Walliser was already considerably more advanced in his studies. Müller asserted his right to proceed on the basis of the support he had been given. Walliser claimed the advantage of being the first and of being self-funded; he was not going to give way. Someone suggested the compromise of collaboration, but Müller, who kept his tuberculosis hidden, knew his health would not take it and declined. He knew Walliser was strong and an exceptional field man. In the end a compromise was negotiated: Müller agreed to deal with the simple cones and Walliser with the rest. Walliser gave Müller all his simple cones, but Müller failed to keep his part of the bargain. Walliser then had no opportunity to do the work himself. The two men remained on bad terms throughout their careers. Only in retirement, decades later, did Müller write to Walliser to express his regrets.[20]

The outcome for science of this face-off was, however, hugely positive for both men. Müller did not, in the end, stand his ground and was forced to drop down even deeper in the stratigraphic column, into the Cambrian, which for most conodont workers was completely off the radar. Here he had space to make the most remarkable discoveries and achieve real distinction. In 1975, Müller, now perceived as a Cambrian specialist, found his life change again when he accidentally discovered the finely preserved "Orsten" fossils. Then fifty years old, he effectively gave up conodonts to pursue these new and extraordinary fossils, and a new community of researchers grew up around him. Fossils preserved in three dimensions and in exquisite detail, he was able to challenge the interpretation of one of paleontology's most iconic objects, the tiny and strikingly odd trilobite, *Agnostus*. Müller felt he proved, on the basis of remarkably preserved soft parts, that this famous fossil was not a trilobite at all.[21]

As for Walliser, he soon had the Silurian to himself and in a few years had singlehandedly tamed it. He in time became drawn to far bigger questions concerning the history of the planet. We shall return to both these men.

Before 1950, Sweden had played little role in the development of conodont studies, and in common with much of the rest of Europe, it barely knew these fossils. Among these earlier Swedish workers was Assar Hadding, who had described a dozen conodonts in his 1913 doctoral thesis on graptolites. These fossils had been found in Ordovician rocks not far from Lund, where Hadding was, from 1947, rector of the university and a prominent figure in Swedish geology. A stratigraphically isolated fauna, Hadding's finds had no real scientific impact, but Swedish geologists became aware that there was a group of fossils they had overlooked.

The Swedish workers who entered the science in the 1950s invariably possessed botanical roots. This interest in plants and their taxonomy owed everything to Sweden's great hero, Carl Linnaeus. So it was that Sweden's conodont pioneer of the new generation, Maurits Lindström, was required as a schoolboy to collect and curate a herbarium, along with every other twelve year old. He was expected to know and to be able to identify some twelve hundred plant species. Stig Bergström, who became established as a leading conodont worker in the early 1960s, began as a particularly outstanding botanist. "But when he came across the conodonts he found in them such potential and such beauty," Lindström later reflected, that Bergström decided "that this should be it!"[22] Lennart Jeppsson, a still later worker, also came to this science through botany. We shall come to these younger men in due course. For Lindström, plants were not a first love, but they were a means to gain rapid entry into university, and this he did.

As a schoolboy, Lindström had made a small study of the Middle Ordovician, not far from where Hadding had collected his material. He wrote this up and presented it to his biology teacher, who was completely amazed by it. Encouraged, he continued to work on it after 1950, publishing it in 1953 when nineteen or twenty years old. Lindström was a schoolboy paleontologist.

In 1949, and still in his mid-teens, he had the opportunity to study geology under the distinguished Swedish stratigrapher J. E. Hede, where he now found himself by far the youngest in a class of just three. His compatriots were, to his eyes, like uncles; he still considered himself merely a schoolboy. Hede was by this stage rather elderly and his method of teaching rather archaic; he would read out the essential content of a paper published perhaps twenty years before, showing illustrations and drawing beds and listing fossils, and expect the students to make copious notes. Lindström remembers clearly the fleeting moment when the conodont appeared mainly for its political message, as this was a university where Hadding ruled: Hede said, "Conodonts are probably worm jaws but one can generalise and say that they are essentially anything small and spiny described by an incompetent palaeontologist!" Lindström soaked up this mass of geological information undeterred by the formality of its delivery. Now he at least knew of the conodonts, but he was yet to *really* know them.

Around 1950, Lindström came across papers by Russian and Indian workers discussing the sensational discovery of very small plant fragments, including wood, in rocks of Cambrian age.[23] These were puzzling discoveries of things that had no right to occur in rocks so old. Lindström was encouraged to begin his own investigations. In his student room, he set up a small laboratory and began dissolving bituminous limestones with hydrochloric acid. The stench of acid and bitumen was everywhere. Among the fossils he found were some odd spiny fragments, which were clearly not plant remains. He took these to Sweden's leading micropaleontologist, Fritz Brotzen, in Stockholm. A fugitive Prussian Jew and German autocrat, Brotzen told Lindström, in his strange Swedish, that they were probably trilobite fragments and not very interesting. He advised Lindström to take up the study of fossil fishes instead – fossil fishes then being a Stockholm strength – and then sent him to see Erik Stensiö, the country's leading fish paleontologist. Stensiö presented Lindström to his assistant, Erik Jarvik, who later became Stensiö's successor and who was then working on the structure of the teeth of primitive fishes from Gotland. Jarvik had extracted these teeth from limestone using acetic acid, after which he thin-sectioned them and studied their microscopic structure. Lindström was impressed. Jarvik showed interest in Lindström's fossils but advised him, "Use 10% acetic acid and you will experience wonders."

Lindström returned to his experiments, hoping that acetic acid would transform his bituminous limestones into a fossil wonderland, but those rocks dissolved poorly. Undeterred, he located some purer Ordovician limestones, and in these he found his first conodonts. He now realized the potential of his discovery, for he could order and correlate rocks in ways his colleagues could not. He could imagine himself a solitary pioneer in a new field of Swedish study, and he worked day and night to produce his first paper, beginning his collecting for it in 1953. That landmark paper brought recognition, and before long, he saw his career stretching out before him, with the chair at Uppsala, then occupied by Per Axel Thorslund, already in his sights.

Like others, Lindström published his own summary reflections on the field early in his career. As his career had begun in a rather precocious manner, it is perhaps unsurprising that his summary took the form of a monograph – the first of its kind in the field – published in early 1964. The publisher had been sent in Lindström's direction by Anders Martinsson, a brilliant and universally admired micropaleontologist. Extremely competitive and temperamental, Martinsson was then at the University of Uppsala, which was Sweden's palaeontological center and Lindström's goal. Lindström later wondered, with a smile on his face, whether Martinsson had recommended him to the publisher in order to distract him with a book that would never be recognized by those who elected that university's chair of paleontology. The book certainly did not help Lindström's career, but the Uppsala dream effectively died, so Lindström thought, when he had the audacity to suggest that Sweden's Ordovician rocks were deposited in deep water. The Swedish long beards knew that this was not the case. In that country's small and claustrophobic geological world, you were either in or out; Lindström was now out. Luckily, Otto Walliser threw him a lifeline from Germany, having located a chair for his Swedish friend in Marburg. German science had no problems with Lindström's unorthodox ways. In 1966, Ziegler moved back to Marburg from the Geological Survey, when the opportunity arose, drawn not least by the chance to work with Lindström, whom he considered the "senior conodont scholar."[24]

Had any of Lindström's Swedish colleagues really understood conodonts, they might have been rather less dismissive of his book. Simply called *Conodonts,* it was rather more than a textbook. Unaware of what was in the still unpublished *Treatise,* Lindström used the book to reimagine conodont science. It brimmed with imaginative ideas, many of which had a real impact. It was positively reviewed in the British press (it was published in English), being deemed both timely and well executed and seeming to invent a science of "conodontology."[25] It was the grandest of the reviews of conodont science, which marked the rebirth of the subject. Like the others, it was infected with optimism.

In Germany, Sweden, and Britain, the return of peace saw the rise of entirely new communities of conodont workers, but in the United States change was more subtle and generational. Iowa then became important for producing conodont workers, and among these was Walt Sweet, whose education had benefited from the G.I. bill after the war. He had ended up at Iowa because his college mentor in Colorado had graduated from there; there was no more logic or ambition to it than that. What Sweet found there was a kind of paleontology that doubtless had something of Branson and Mehl's Missouri about it. Miller, who became Sweet's supervisor and friend, had been one of Branson's students and had inherited Branson's obsession with publication. "Miller published like crazy," Sweet later reflected. "He was big on illustrations but short on real innovations." It was Branson and Mehl's personal encouragement that had persuaded Furnish to work on Lower Ordovician conodonts in the late 1930s. "Bill was A. K.'s assistant on a large project involving nautiloids in the U.S. Nat. Museum's vast collections," Sweet remembers. "It is my recollection that A. K. kept Bill busy on this project night and day and that Bill undertook the conodont study as his own PhD project, which he conducted late at night and at other times when

A. K. did not demand his presence." Shortly afterward Furnish was off to Saudi Arabia to work for ARAMCO, not to return to Iowa City until 1953. When Sweet arrived in 1950, he was the only one interested in conodonts. However, he remained for the most part a cephalopod man at Iowa, though he spent his summers back in Colorado teaching on a field course and working on the Harding and Freemont – strata that, respectively, contained conodonts and cephalopods.

Sweet took up an academic position at Ohio State University in Columbus in June 1954 and remained there for his entire career. In 1956, he obtained a Fulbright Research Grant, which enabled him to spend a year studying nautiloids at the University of Oslo in Norway. While there, he traveled to Lund and Stockholm and became acquainted Lindström, Stensiö, Jarvik, and others. Lindström made a return visit in 1959 or 1960, spending six months in Columbus, and the two became close friends. It was in part a result of this that Stig Bergström ("small, officious, very good, very sharp and with a photographic memory")[26] arrived in Columbus in 1960 to undertake his doctoral degree under Sweet's supervision. This was before Sweet considered himself a conodont specialist. Bergström traveled to Sweden briefly in the hope of finding one of those rare academic positions there but then returned to stay in Columbus, where he was hired as an assistant professor and curator of the Orton Geological Museum in 1968. He became a professor of geology in 1972, and a few years later, director of the museum and a U.S. citizen. He and Sweet became close collaborators.

Sweet's full conversion to conodonts occurred rather later than others mentioned in this chapter. The turning point came when Miller and Furnish failed to complete their promised biological description for the cephalopod volume of the *Treatise*. Frustrated, Moore asked Sweet to take it up. He knew Sweet had no choice in the matter because his own work was in the volume waiting to be published. So that was how Sweet spent the summer of 1964. Never really a biologist, he cribbed whatever information he could: "By the time I got through with that volume I decided that there was never ever going to be a time when nautiloid cephalopods are worth a damn stratigraphically." It was with these thoughts in his head that he found himself converted unexpectedly into a conodont specialist. He had already set his students to do some self-contained

stratigraphic projects using conodonts and was delighted to see the results of their work. Although he had published on conodonts in the 1950s,
it was this work by his students that completely altered his research focus.
In the years ahead, he and Bergström would have a major impact on the
way others visualized the unknown animal. Near the end of his career,
Sweet would also write the only other solo English-language monograph
dedicated to these fossils. And like his mentors in Iowa, Sweet would turn
his department in Columbus, Ohio, into a conodont factory, pouring
forth conodont specialists.

Before World War II, the science of conodonts had laid down its roots.
It had not fully realized its potential, and thinking was constrained by
geographical isolation and the dominance of American utilitarianism.
The science that emerged after the war was quite different and little connected it to that earlier period. Now it branched out and blossomed, liberated by acids and in the possession of this new generation of actors who
were set on pursuing a better future. Geographical isolation vanished – at
least in the West. The fossil and its workers began to build international
connections almost immediately. There was to be one aspiring and sophisticated global science.

Evolution proceeds continuously, and all jumps
are deceptions caused by gaps in the record.

ROLAND BRINKMANN,
Monographie der Gattung Kosmoceras (1929)

SEVEN

Diary of a Fossil Fruit Fly

IN GERMANY, EVERY STUDENT OF PALEONTOLOGY LEARNED OF their compatriot Roland Brinkmann's 1920s centimeter-by-centimeter study of the English Oxford Clay. His three thousand beautifully preserved, ornate, and nacreous *Kosmoceras* ammonites recorded the reality of evolution with wonderful picture-book clarity, each twist and turn on their evolutionary journey permitting a moment in time to be defined and used to order and correlate rocks elsewhere.[1] Brinkmann's study traced the evolution of these fossils through just fifteen meters of the clay. Now Beckmann's disciples, armed with the gift of efficient acid preparation and the ubiquitous and rapidly evolving conodont, aimed to perform the same trick on a far grander scale – through the whole of the German Devonian. Quick to recognize the heroic possibilities of this new challenge, in the mid-1950s, this generation believed they possessed nothing less than the makings of a worldwide standard.

These new German workers were perhaps more attuned to evolutionary thinking than were their American forebears before the war. Otto Walliser, for example, was a devout Darwinian. For him, it was this that made paleontology attractive. While many in America were also now Darwinians, this had not always been the case. Before 1930, according to George Simpson, professor of vertebrate paleontology at Columbia University in New York, most paleontologists in North America believed in evolution but few placed any significance on "natural selection" – that element that distinguished Darwin's theory from the work of other evolutionary theorists. There, for the majority, "Darwin was dead." If the American paleontologists thought about evolution at all then, it was "evolution by random mutation" or "evolution by the inheritance of acquired characters." Simpson recalled that his own teacher, when lecturing on the subject, "gave equal billing to all the conflicting theories on the causes of evolution . . . but personally espoused none of them." Evolution was in open season, with relatively little to constrain personal interpretations.[2]

It was the American utilitarian mission for paleontology that had created this tendency to dispense with "unnecessary science." The chief geologist of the Pure Oil Company in Texas, for example, had told his petroleum geologist colleagues in 1927 to map and curate "a history of the development of numerous inter-fingering and diverging races, an outline of their advances, culminations, and disappearances. We have learned that genera make progress along definite lines." This was economic evolution without the leisure of biological meaning: "Making correlations from field to field and from well to well in the quickest possible time must use all the tools which are available." This kind of thinking permeated the tiny community of conodont workers in the 1930s, encouraging two Oklahoma paleontologists to believe they could see an evolutionary sequence in a series of leaf-like conodonts from the Lower Pennsylvanian (Upper Carboniferous). Here *Cavusgnathus* appeared to evolve into *Idiognathoides,* which in turn became *Idiognathodus,* as a central channel closed and transverse ridges developed to cover the surface of the tiny fossil. They admitted that there was no stratigraphic evidence for detecting this evolutionary direction but "considered it a case of parallelism after the manner of development of mammalian teeth" from a single reptilian cusp to a corrugated form. This was indeed an invocation of evolutionary "progress along definite lines." The argument was soon dismantled by Branson and Mehl, but only through improved collecting. They offered no more sophisticated understanding of evolution.[3]

The conodont workers were, however, aware that identical forms seemed to have evolved repeatedly at different points in time. These were the products of convergent evolution and called homeomorphs. For example, Branson and Mehl noticed that their Lower Carboniferous *Taphrognathus* and Stauffer and Plummer's Upper Carboniferous *Streptognathodus* were indistinguishable though not the same fossil. In 1938, Bill Furnish had said the only way to see beyond such illusions was to unravel the fossils' evolutionary pasts. In other fossil groups such an undertaking might have been seen as a little idealistic, but the conodont workers had come to think this entirely possible after only a few years of intensive study. Branson and Mehl believed they could see evolutionary trends and thereby guess the form of fossils occupying gaps in the record. They were not alone. Sam Ellison's holistic review of the literature convinced him of the conodonts' value to evolutionary studies. The key to such studies, however, was further collecting rather than a more sophisticated understanding of evolutionary theory. It was careful collecting alone that permitted Carl Rexroad, for example, to demonstrate in 1958 that *Taphrognathus* gave rise to *Cavusgnathus,* which in turn spawned the *Taphrognathus*-imitating *Streptognathodus.*[4]

The beginnings of a change in the way paleontologists thought about evolution came when J. B. S. Haldane's 1932 book *The Causes of Evolution* revealed the impact of natural selection on the genetic makeup of living populations. In truth, few geologists or paleontologists initially took much notice of this work, which seemed to signal that geneticists had taken control of evolutionary study. Indeed, Simpson half joked that geneticists thought paleontology too descriptive to be called a science, that by studying the stony remains of fleshy life, it was incapable of understanding the subtle complexities of evolutionary change: "The paleontologist, they say, is like a man who undertakes to study the principles of the internal combustion engine by standing on a street corner and watching the motor cars whiz by."[5]

But Simpson did not stand by and watch his science be pushed aside. Inspired by his colleague Theodosius Dobzhansky's 1937 synthesis of modern genetics with Darwinian theory, he rushed to paleontology's defense, publishing his own book, *Tempo and Mode in Evolution,* in 1944. In it he placed paleontology within this new synthetic view, suggesting that it could contribute "on the rates of evolution, modes of adaptation, and histories of taxa." Together with an earlier essay, this book reintroduced paleontologists to Darwin. The effect on paleontology was revolutionary. Now evolution was to be understood as "genetic mutation and variation, guided toward adaptation of populations by natural selection." This emphasis on populations was a radical departure for paleontologists, who had traditionally thought of species in terms of the most typical individual. American paleontologist Carl Dunbar recalled the revolution in 1959: "This whole scheme collapsed like a house of cards in 1940 when George Simpson published his short but epoch-making paper on Types in Modern Taxonomy."[6]

However, the notion of a species as an interbreeding population proved something of a challenge for paleontologists dealing with fragmentary, sporadic, and variable finds of long-dead animals and plants – so much so that the "New Paleontology" soon became preoccupied with understanding the meaning of variations of form. Uncertainty grew within the paleontological community. At a major symposium in London in 1954, it seemed that species were no longer fixed or definite; some clearly graded from one form into another across time and space.[7] Those attending worried that they could not identify true biological species, only groups of animals that looked morphologically the same. But if paleontologists could not work with biological species, how could their fossils be trusted to show the truth of evolution? Organizer Sylvester-Bradley warned that the "distinction, in fact, between morphological species and biospecies can only be overlooked at the peril of utter confusion." Paleontologists, uncertain what might conform to a true species, preferred to see variation of form as an "evolutionary plexus." The only means to determine the boundaries between species was to collect numerous specimens from a single location and horizon and search for breaks in the distribution of forms. If there had been a crisis of faith in Darwinian evolution before the war, a similar crisis in species gathered strength in the 1950s and would continue to dog the science for the next twenty years.

Some paleontologists were, however, rather less concerned about these problems. They felt simple pragmatism would permit them to sidestep the difficulties of theoretical explanation. Indeed, some saw this population-centered view as liberating. It offered yet another way to make an assault on the proliferation of species names. Ronald Austin recalled how inadequate the literature was at the start of his career: "There was nothing!" A student of Frank Rhodes, he drew inspiration from work done at the Illinois State Geological Survey, where Carl Rexroad, Gil Klapper, Charlie Collinson, and Alan Scott were working on conodonts. Scott and Collinson produced pictorial plots of variability; the variations within them could be subdivided into a number of distinctive morphological types. Austin loved this pictorial simplicity, which surpassed verbose attempts at differentiation in words. Paleontology had always been a pictorial science, and these new diagrams effortlessly showed that species erected on the basis of a few specimens had little validity.[8] They offered a way to circumvent the mire of taxonomic description.

The sheer numbers of rapidly evolving conodonts, their easy extraction using acids, the desires of a new generation on the rebound from the war, the reignition of Darwinian evolution, the fine sequences of Devonian rocks, and the possibilities for a global standard for correlation produced the conditions for a German revolution. Plastic in form and potentially large in number, the conodonts seemed the ideal subject for the population-centered study of evolution. Only in such circumstances might paleontologists believe they were studying true – biological – species. The conodont held the potential to become the palaeontological equivalent of the geneticist's fruit fly, *Drosophila*. But unlike the ephemeral fruit fly, the conodont animal had written an evolutionary diary in stone. Every moment of its existence, it seemed, was there for the reading.

Among those who wished to lead this assault on the German Devonian was, as I have mentioned, Klaus Müller. His East German limestones, left over from his doctoral studies and dissolved in acid out of desperation, produced the first evolutionary study of the dominant conodont genus, *Palmatolepis*. The work was followed up by Walter Gross's assistant at Humboldt University in East Berlin, Jochen Helms, who began his own study in Thuringia in 1958. His first results, published in 1961, teased apart the changes that took place in a distinctively noded group of *Polygnathus*. They also revealed to him the problems of recognizing and then delineating the pattern and flow of change taking place in the animal. The rocks themselves formed the background against which these changes were observed. They in effect represented time, but because they resulted from varying rates of sedimentation, and intervals of non-sedimentation, time was distorted in undetectable ways. The cephalopod fossils traditionally used to track time through these rocks helped little. The timescale they offered was simply too coarse.[9]

Helms understood that these problems would lead to imperfections in his understanding, nevertheless, he knew that these problems would be resolved in time. A greater cause for discomfort was the sheer variety of form, which seemed to conceal the flow and direction of evolution. Eyes alone, he believed, could not make objective sense of this. So he resorted to statistics to produce species and subspecies from the varied forms he possessed. Fortunately the end members of these evolutionary moments – the evolutionary descendents – were much more constant, and he could then trace these forms backward into the morphological soup. By doing so, a picture of evolution, as it had occurred "in Thuringia" hundreds of millions of years before, revealed itself.

Helms could now see a "knotty kind" of *Polygnathus* giving rise to the extraordinarily successful and useful *Palmatolepis*. But, more interestingly, he could see that behind this dominant form, the conservative line of *Polygnathus* continued and, when *Palmatolepis* finally had its day and went into decline, so another branch arose from these roots. This new branch carried the "nodocostata group," which diversified and began to develop forms reminiscent of *Palmatolepis*. It might have occurred to Helms how perfectly the conodont performed as a "fossil fruit-fly," for it seemed to echo performances observed by Thomas Hunt Morgan in his pioneering studies of these insects: "There is another result, clearly established by the genetic work on *Drosophila*, that is favorable to the final establishment of a new type or character if it is beneficial. Most, perhaps all, of the mutations appear more than once. This improves their chances of becoming incorporated in the species, and if the mutation produces a character that favors survival the chance of its becoming established is still further increased."[10]

Other studies were taking place in West Germany, Willi Ziegler having moved to the Geological Survey at Krefeld near Dusseldorf in 1958. Here he began work on his "habilitation" thesis at the University of Bonn, which would permit him to practice as a university academic. It was here that the conodont started to become an essential part of his being and where he would become the ultimate conodont specialist and outcompete his compatriots. Ziegler had already demonstrated evolutionary trends in a number of conodont genera in the late 1950s, including *Palmatolepis*. In his thesis, completed in 1962, he took this

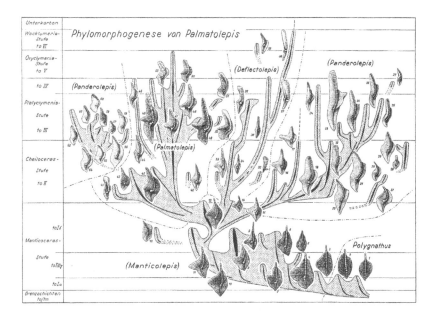

7.1. Helms's iconic representation of the evolutionary development of *Palmatolepis,* "the best of all fossils for the subdivision of the Upper Devonian," published in the 1962 *Treatise.* Ziegler joined Helms in maintaining its accuracy. Time runs vertically, from bottom to top, in this diagram. Reproduced with permission from K. J. Müller, *Treatise on Invertebrate Paleontology,* Part W Miscellanea (1962). Courtesy of and ©1962, The Geological Society of America and The University of Kansas.

work several stages further and produced a high-resolution conodont zonation for the Upper Devonian in its entirety. Based on a study of encyclopedic magnitude in which vast numbers of samples were taken at ten-centimeter intervals, his twenty-four zones and zone divisions challenged the supremacy and current orthodoxy of cephalopod zones. The reward for this herculean effort was the production of a continuous temporal landscape upon which he could plot the trajectories of conodont evolution at its most plastic and eventful. Ziegler was convinced that he now possessed the universal timescale, reassured by tests he carried out in other parts of Europe.[11] It was a realization of what so many had imagined at Dillenburg less than a decade before.

Like Helms, Ziegler found parent forms like *Palmatolepis triangularis* highly variable. Both agreed that this fossil was the rootstock of all successive *Palmatolepis* species. The rapid evolution of *Palmatolepis* was one of the keys to Ziegler's timescale and it gave up its evolutionary secrets rather easily, as did *Spathognathodus*. Ziegler could see change gradually occurring as one species became another as different features were reshaped and resized. He could also detect long-term trends, such as a reduction of surface area and a sigmoidal buckling, which occurred in numerous branches of the evolutionary tree.

However, before Ziegler could publish his results, Helms stole a little of his thunder by giving Müller a diagram showing the results of his own investigations, which Müller added to the *Treatise*, then in the final stages of its publication (figure 7.1). Essentially communicating much of the story Ziegler had also unraveled, Helms's pictorial representation of conodont evolution became an icon for the new science; it would not be surpassed. However, when Helms came to give explanation to this diagram in 1963, he corroborated Ziegler's own interpretations, which had by then been published. Built upon friendly cooperation with Ziegler, and riding the wave of this upsurge in German stratigraphy, Helms celebrated the conodonts' survival in large numbers and their preservation of ontogeny and evolution; they seemed to challenge all other fossils for the stratigraphic crown.[12]

Helms was, however, aware that his diagram showed the evolution of an anatomical part of an animal, not of the animal itself. This was not unusual in paleontology; specialists in fossil mammals, for example, frequently had nothing more than teeth to go on. Of course, this kind of focus was an easy target. As Stephen Jay Gould later remarked, "An old paleontological in joke proclaims that mammalian evolution is a tale told by teeth mating to produce slightly altered descendant teeth."[13] Helms held no such simplistic beliefs. Unconvinced by the universality of his conclusions and aware that a two-dimensional diagram could not capture the complexity of change, he presented that change in prose. Here he discussed the emergence of features, moments of rapid change and periods of character stabilization, in which variable ancestral populations gave rise to distinctive forms.

Ziegler soon overtook Helms in this race for global domination, and it relied not only on the quality of his scientific work, his sheer hard work, and his unparalleled passion for this fossil, but also on his unbridled evangelical zeal. In the autumn of 1961, he visited the United States, where he lectured to Bill Furnish's and Brian Glenister's graduate students at the University of Iowa. Gil Klapper, then a doctoral student working on the Devonian-Mississippian boundary, recalls a field trip to Campbell's Run led by Furnish. With Ziegler's encouragement, Furnish and Klapper took three samples from a measured section of the Sweetland Creek Shale found there. Returning to the university, they processed and sorted this material while Ziegler examined the Geology Department's type fossils. To their amazement, they found three of Ziegler's zones. After Ziegler left to visit other collections and other workers, Klapper and Furnish returned to the section to do a more thorough job and found five of Ziegler's zones. These results, when published in 1963, were a critical breakthrough for Ziegler's global ambitions.[14]

The next major step came a few years later, when Brian Glenister found conodonts in the matrix of cephalopod fossils from the Canning Basin in Australia. Glenister had worked on cephalopods for Western Australian Petroleum (WAPET) back in the 1950s, at a time when these cephalopods were used for international correlation. He passed the conodonts on to Klapper, who then drew upon Ziegler's resources in Krefeld to interpret them. He again found Ziegler's zones. However, when Glenister and Klapper brought this discovery to publication, rather than follow Ziegler's system, thereby assuring it of the status of a global standard, they replaced his names with numbers. This caused Ziegler considerable distress, which became all the more amplified when David Clark adopted the same scheme in a study of the Great Basin.[15] It was, however, a temporary blip, though, as we shall see, it was not the only one to disturb Ziegler's perfect system in the mid- to late 1960s.

The final stage in the conodonts' emergence from obscurity occurred late in the summer of love, September 1967, at an international meeting in Calgary organized to discuss the Devonian. To the surprise and delight of Huddle and others who were there, "paleontologists working on other fossils deferred to conodont age determinations." Klapper recalled this moment as a "historical watershed." The conodont had proven itself on a global scale and overcome any residual fears that faunas might be contaminated or mixed and therefore unreliable.[16] Buoyed up by this recognition, Klaus Müller, Larry Rickard, and John Huddle alighted upon the idea of a society – which was then discussed by twenty-five conodont workers at the end of the conference. On the strength of his seniority, Huddle found himself unanimously elected "president," an office soon to be labeled "Chief Panderer." Charlie Collinson suggested the society's name – the Pander Society – which by the time of the second meeting had raised a few eyebrows (in the United States, "pander" has the "somewhat unsavory meaning" of "pimp"). The society that emerged from this seventy-minute discussion became the research hub for the science. Its membership grew organically and rapidly, its annual newsletter recording little more than a catalogue of publications and meetings, as well as a list of members and their interests: "The 'Pander Society' is informal. It has no constitution, bylaws, or rules and NO DUES."[17] The society reveled in being as enigmatic as its subject matter and claimed that any meeting of three or more conodont specialists could be termed a "Pander Society meeting." Undoubtedly an informal association, the group nevertheless got down to business and in a series of thematic international meetings it tackled the long list of unknowns and concerns that conodont workers shared. There can be little doubt that, with the entry of the fossil into mainstream science, 1967 marked a point of culmination. It also marked a step change in the science's socialization. The new society became a dynamo and catalyst, providing an informed and critical specialist audience, all of which would serve to accelerate the rate of progress. Buoyed up by a new commonality of purpose, which produced shared agenda and a world resource of data, conodont studies entered a new era.

But soon the American-led Pander Society discovered it was not alone; a Soviet group, known as Pander's Grandchildren, had also recently been inaugurated. Indian and other national groups would emerge in the future as momentum grew and the full implications of the globalization of science were felt. By 1972 the Pander Society had a membership of 250 from thirty-one countries.[18]

Ziegler's global ambitions for his Upper Devonian conodont zones were pretty much achieved by the end of the 1960s, and by the time he returned to Marburg in 1969, even the troubling Lower Devonian seemed to be coming under his control. He now became a key participant in those international organizations, that sought to establish global standards for the Devonian.[19] And he also began work compiling the massive and, to his mind, definitive *Catalogue of Conodonts*, a description of every known species.

Although most geologists saw in Ziegler's and Helms's work a more re-fined means to order rocks, these authors' insights relied entirely on earlier prophesies that the key to understanding this animal – at least as an abstract thing – lay in unraveling its evolution. In this, Ziegler was simply the most successful of a generation of workers who attempted to map every root and branch of the conodonts' family tree. Looked at from a distance, this tree suggested that these fossils showed an evolution-ary tendency toward increased surface area. Cones had developed into blades and bars, and these into platforms. This had not, however, taken place as stages on a linear journey, but, as Müller revealed, by different means and along different branches of the family tree.[20]

Müller mapped that tree in 1962. By then he had already began his journey backward along that path – marked by a dashed line in his dia-gram – which took him beyond the Ordovician and into the Cambrian, toward the point of origin of the conodonts. He began that journey on his sabbatical at the University of Iowa in the summer of 1955. Here Furnish had found conodonts low down in the Ordovician. Müller went in search of still older Cambrian forms in the Arbuckle Mountains of south-cen-tral Oklahoma, but these too proved to be Ordovician. So he turned his attention to the Deadwood Formation of the Black Hills of South Da-kota, rocks that had interested the Iowa workers for some time.[21] Using acids on limestones lying below a proven Upper Cambrian shale, Müller at last found his quarry. These conodonts were entirely new, very distinc-tive, and quite unlike any seen in the Ordovician.

Through a comparison of material from Germany, Wyoming, and Utah, he could be sure the rocks in which he had found these fossils were Cambrian in age. Then he began to wonder if these strange fossils really were conodonts, but Furnish reassured him. Regaining his confidence and projecting his knowledge of conodont evolution backward into the void of the unknown, Müller thought his new fossils fitted the bill of

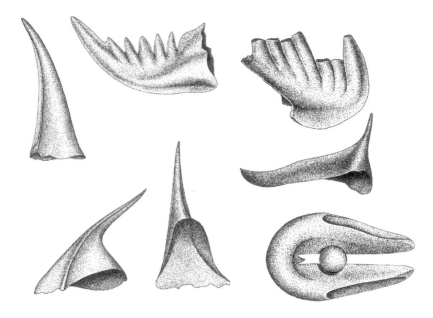

7.2. Müller's oddities. Müller regarded most of his late additions to the *Treatise,* such as these fossils, as fully fledged conodonts. Among them was the extraordinary *Westergaardodina* with its ball. When a revised *Treatise* was published in 1981, nearly all of Müller's oddities were moved into the paraconodonts. This latter group had been redrawn to accommodate them: Rather than being that ragbag of conodont-like fossils Müller had imagined, it was now regarded as a special group of conodonts with large, deep basal cavities, no white matter, and a distinctive pattern of growth. Reproduced with permission from K. J. Müller, *Treatise on Invertebrate Paleontology,* Part W Miscellanea (1962). Courtesy of and ©1962, The Geological Society of America and The University of Kansas, and K. J. Müller, *Zeitschrift der Deutschen Gesellschaft für Geowissenschaften* 111 (1959). Courtesy of www.schweizerbart.de.

conodont ancestors pretty well. They were all simple forms with large basal cavities.

Müller continued to collect Cambrian conodonts and, by 1959, possessed about twenty-eight hundred specimens, mainly from Sweden and northern Germany. These included many species he had seen in the United States. While this extended the possibilities for the conodonts in stratigraphic studies, these discoveries were also accompanied by a sense that conodont science was taking a bizarre turn. Some thought Müller's extraordinary fossils seemed to test the very boundaries of what might reasonably be considered a conodont. Now that there was no animal in mind – the fish and the worm, along with a menagerie of other possibilities, having been consigned to the dustbin by Gross and Wilbert Hass – how could one decide which fossilized things once belonged to this ambiguous and plastic animal? Even so, some may have doubted Müller's sanity when he included *Westergaardodina* among these fossils. A strange horseshoe-shaped fossil, first discovered in 1953, which came with its own detached "ball," it looked nothing like a conodont. Other forms of this fossil were W-shaped. Indeed, they looked like nothing else known. But Müller found some primitive examples of these fossils possessing a distinctive tooth-like cusp, and along with their phosphatic composition this suggested to him that they belonged to an animal that had evolved from conodont stock (figure 7.2).[22]

While paleontologists outside the field celebrated Müller's perceptive association of horseshoe with ball, others doubted his powers of interpretation.[23] Hass, for example, had examined "conodontlike objects" from Müller's beds in Utah and Wyoming back in 1954. Unable to see the laminations typical of conodonts, which he knew might have been destroyed by formic acid preparation, he concluded in the *Treatise*, "It is the writer's opinion that the stratigraphic range of conodonts should not be recorded as definitely extending into the Cambrian until irrefutable, well-documented evidence has been published." Hass's opposition, however, faded with his untimely death. And being one of those tasked with finishing Hass's *Treatise*, Müller chose not to edit Hass's words but to contradict them in a supplementary section in which he described and illustrated his new and controversial Cambrian fossils.

Some of these new fossils were more debatable than others. The paraconodonts, for example, seemed to grow by additions to the lower rim. It remained an open question as to whether they were really conodonts; Müller made no presumption either way. Nevertheless, they sat there in the company of conodonts, carrying a suggestive name and implicitly pushing at the boundary of what a conodont might be.

It was a provocative move that soon met with criticism and disbelief. There were those who wished to return some of these objects and the concept of the conodont to their former meanings. Sweet and Bergström were the first to act, rescuing the useful *Pygodus* from the paraconodonts

and placing it among the "true conodonts." Maurits Lindström was willing to accept the cone-shaped forms Müller had found in 1956, not least
because he had direct information from Cambrian expert "Pete" Palmer
of the U.S. Geological Survey in Washington that these really were
conodonts. By 1964, Palmer possessed a sequence of conodonts from
Wyoming that showed their development from a simple horn-shaped
form to a primitive *Cordylodus* and then on to *Cordylodus* proper in the
lowermost Ordovician. It had been Palmer who had advised Müller on
the Oklahoma rocks and supplied him with material from Wyoming
and Utah. Lindström was, however, rather less impressed by Müller's
more exotic European faunas, dismissing *Westergaardodina* as "highly
interesting and valuable" but "not conodonts, at least not strictly speaking." He doubted that others of Müller's fossils were truly Cambrian.[24]

Müller was rather annoyed at Lindström's dismissal of his work,
based as it was on mere opinion. He remained silent until the Pander
Society's first major symposium in 1969. By then the Cambrian conodont was beyond dispute. Each year had added a new site. Cambrian
conodonts were now known from as far away as Iran, China, and Australia. They remained, however, rare objects in the Upper and Middle
Cambrian, and one doubtful report even suggested they might be found
in the Lower Cambrian.[25]

As Müller almost singlehandedly began the process of giving the family tree its root system, so a rather different argument was taking place
in the upper branches. Here Walter Youngquist, a postwar product of
the Iowa school who for a short time became the most active worker
in conodont stratigraphy, claimed to have discovered conodonts in the
matrix of some Triassic cephalopods from Idaho in 1949. He was pretty
sure these were Triassic fossils, but he recalled Branson and Mehl's response to an earlier German discovery of Triassic conodonts. In 1941,
it had been logical for these two men to consider such fossils as mere
contamination, reworked into these younger rocks following the erosion
of Permian or other strata. But then, in 1946, American D. B. Eicher of
Standard Oil had reported finding Triassic conodonts in Egypt, which

Branson and Mehl had also doubted. The wily Youngquist did not want the same fate to befall his finds, so he sent the same slide of specimens to Chalmer Cooper, Eicher, Ellison, Furnish, Hass, and Mehl, along with a list of pertinent questions: Are these conodonts? Have you seen specimens exactly like these in the Paleozoic? Given evolutionary trends, are these compatible with a Triassic age? Do they look like they have been reworked? With Branson out of the picture, both Ellison and Mehl thought these new fossils similar to the German ones Mehl had earlier dismissed. To a man, Youngquist's respondents thought that the fossils conformed to what one might expect Triassic conodonts to be like, and that none appeared to have been reworked. Cooper and Hass also told Youngquist that the National Museum possessed similar fossils, found just fifty miles away from Youngquist's find location. Any residual doubts these men possessed arose from their ignorance of the Permian fauna, but here Youngquist was a little ahead of the game; he had sufficient information to know his new fossils were different from those found in the Permian. Now he was sure that Triassic conodonts really did exist.[26]

With the conodonts considered rare but proven survivors of the great extinction at the end of the Permian, it was now possible to imagine the animal existing well beyond the Triassic – perhaps up to the next great extinction at the end of the Cretaceous. But Stauffer's reported discovery of conodonts in Cretaceous rocks in 1940 seemed to be a clear case of contamination and Marburg doctoral student and Triassic specialist Reinhold Huckreide's search for conodonts above the Triassic in the Alps in 1955 ended in failure. Indeed, his Triassic fossils showed such senile and variable characteristics that he was convinced he was seeing an animal on the verge of extinction. Then, in 1956, East Berlin's Kurt Diebel reported "authentic conodonts" from the Upper Cretaceous of Cameroon, West Africa. This led to Huckriede's suggestion that the animal must have been restricted to a particular sea, though he had his serious doubts about Diebel's fossils, which looked characteristically Triassic. He simply could not believe that they really were of Cretaceous age.[27]

Diebel's material had actually been collected in 1897–88, and consequently some now doubted its reliability. Hass called for a worldwide search.[28] Others, however, welcomed the Cretaceous conodont. Lindström, who was working away at his conodont book, was convinced by Diebel's finds: "The fauna is too well preserved to have been derived from mechanical reworking through erosion of older, conodont-carrying sediments. Moreover, no such older sediments are available in the neighbourhood. . . . From its appearance the fauna can easily be accepted as Mesozoic and younger than the Triassic." He thought it simply marvelous: "Through more than 300,000,000 years the same structural plan persisted essentially unaltered." Each zero emphasized just how remarkable this unknown creature this was.[29]

Müller attended a micropaleontology conference in Nigeria in 1965 and spoke on Diebel's problematic fossils, but any hopes he might have had to resolve this matter in the field were dashed by "political circumstances." Like Huckriede, Müller thought the conodonts must have had a restricted distribution at this time as they had not been found anywhere else despite a great deal of micropaleontological work on Jurassic and Cretaceous rocks. This encouraged the thought that the animal might still exist, swimming in a sea somewhere.[30] After all, the Coelacanth had risen, Lazarus-like, from the dead as recently as 1938, and it too had been thought to have perished along with the dinosaurs in the Cretaceous. The impenetrable Congolese swamp held many mysteries, including the mythological Mokele-Mbembe. Might another mysterious creature still exist somewhere off the African coast?

Those reading and believing Lindström's textbook saw a "big, blank interval" standing between the Triassic and Cretaceous in which no conodonts had been found. Perhaps they thought that, with a little patience, this gap might also be filled. They did not have to wait long. In 1967, two Japanese workers reported conodonts from the Upper Jurassic.[31] Lindström's optimism seemed to have been rewarded, but in fact this turned out to be the last straw.

Unconvinced by the Japanese and African fossils, Triassic specialist Cameron Mosher led the assault against them. By 1969, he knew sufficient of the evolution of the Triassic conodonts to be able to demonstrate that post-Triassic examples simply did not exist. They were merely contamination. He could show that those from Cameroon were exactly like those found in the Middle, rather than Upper, Triassic. The Japanese fossils looked like those Huckriede had found in the Triassic, but the Japanese workers – convinced by the Cretaceous fossils – had imagined them as a missing link between the Cretaceous and Triassic forms. Now with the Cretaceous fossils in freefall, Mosher and Müller asked the geologists at Kyoto University to reinvestigate the site of the Jurassic finds. The results of that investigation proved inconclusive. This did not matter, however, for Mosher soon unraveled the final stages of conodont evolution in the Triassic. That period showed a rise in the number of highly specialized and rapidly evolving forms that could not have persisted as relics into these later periods.[32] Such forms tended to exploit short-term niches. Ironically, then, the post-Triassic fossils were merely contamination. If anything, it made Branson and Mehl's initial doubting of Triassic forms all the more reasonable. Contamination really was a problem. Perhaps to no one's surprise, no further post-Triassic conodonts appeared, and the family tree was clipped back. The Cretaceous conodont, however, remained where it had been, printed on the pages of Lindström's book, encouraging dabblers and outsiders alike to imagine – well into the 1970s – that Cretaceous conodonts really did exist.

By 1970, the tree might not have been perfected, but who could doubt the strength of the evidence? Who could doubt paleontology's contribution to understanding evolution or, indeed, the importance of that knowledge to the correlation of rocks across time and space? Fluid and organic, continuous and gradual, the conodont evolved as Simpson and others had imagined in the 1950s, and Brinkmann before them. But this work had taken place without high-level evolutionary theorizing.[33] Then, in 1972, paleontology was again shaken out of its lazy suppositions by Stephen Jay Gould and Niles Eldredge. Gould was to become the public face of American paleontology in the late twentieth century. His rise to public notice began now as he took on the role of the "bulldog of evolutionary biology," later to be considered a "living legend" by U.S. Congress.[34] Eldredge had been a fellow student with Gould at graduate school and had become a staff member of the American Museum of Natural History. Together they challenged the belief that species evolved gradually from large populations. They suggested that the gradualism the conodont workers and others had seen was merely a result of their own earlier conditioning.[35] It had long been understood, they claimed, even by Darwin himself, that the fossil record is composed of species suddenly appearing, maintaining some constancy, and then disappearing. The formation of new species required the separation of a species into two or more discrete populations in which evolution could take place independently. In time – and virtually no time at all in geological terms – the populations would become so distinctive that they could not interbreed. Consequently, the fossil record would show the sudden appearance of species; it could not show gradual change. The branches of the tree did not sweep gracefully upward, they said, but appeared suddenly in midair, tentatively connected at right angles to each other. The species so formed continued unchanged. Eldredge and Gould called this their theory of "punctuated equilibria," but their first assault met with considerable opposition, and the two withdrew to regroup, returning in 1977 to answer "all the hubbub."[36]

The conodonts had not been the subject of Eldredge and Gould's attack. Indeed, they admitted to knowing little about these fossils. But Gil Klapper was intrigued. Having worked on species variability, he had, in 1969, supplied Alan Shaw with Devonian conodonts for his presidential address to the Society of Economic Paleontologists and Mineralogists. In his speech, Shaw was at his most provocative, attacking the utility of the species concept and using Klapper's fossils to make the point. What becomes a species, he said, is the result of a "non-objective art": "Each paleontologist puts out his own work of art – that is, his species concept – in the hope, or faith, that his particular form of non-objectivity will find favor among his fellow artists." Success relied upon salesmanship, and lists were only trusted if the author's name was known. Shaw suggested that paleontologists would be better recording morphological change and dispensing with species names altogether. As an oilman, he was interested in the utilitarian possibilities of fossils and was asserting a view that built upon those charts of variability that Austin thought so empowering. It was the antithesis of Gould's view. Gould was absolutely wedded to the notion of species being fixed and knowable, and he and Eldridge ridiculed all the doubting – this crisis of confidence in species – as being "unsurpassed in the annals of paleontology for its pon-derous emptiness."[37]

Klapper, with Glenister, introduced Gould and Eldredge's ideas into their teaching, and with David Johnson, Klapper set about using con-odont data to test punctuated equilibria. They published a description of the Y-shaped branching of populations into two species – the kind of branching epitomized by Helms's diagram – locating intermediate forms that suggested gradualism. Gould and Eldredge responded, pro-posing that these intermediate forms might be hybrids rather than mo-ments in evolution. They suggested that Klapper and Johnson might have engaged in circular reasoning and selected their ancestors having started with this gradualistic trajectory in mind. Gould and Eldredge re-marked, "We cited the evidence of . . . *Polygnathus* in detail not primarily to reveal the fragility of stories built upon it; for most 'phylogenies' based on fossils rely on flimsy data. Rather, we wish to demonstrate that most cases presented as falsifications of punctuated equilibria are circular because they rely, for their gradualistic interpretations, not upon clear evidence, but upon the gradualistic presuppositions they claim to test." They continued, "Klapper and Johnson epitomize their conclusions in an evolutionary tree, unambiguously presented. These are diagrams that work their way into textbooks, there to convince the uninitiated that paleontologists can specify with assurance the (gradualistic) history of life."[38] Though Gould and Eldredge were fundamentally opposed to the views adopted by Shaw, they were no less of the opinion that there had been rather too much interpretive "art." Klapper, who was already convinced that Gould and Eldredge's model operated in some circum-stances but not all, communicated directly with Gould and became even more convinced by these new arguments.

A second paper by Eldredge and Gould in 1977 continued this pro-gram for change. Perhaps they were now less ignorant of the conodont as the fossil equivalent of *Drosophila,* for they wrote, "Populations evolve, not individuals or, still less, anatomical parts of individuals." The universal evolutionary change that underpinned the conodont work-ers' ambitions and had so effectively produced global correlations was theoretically incomprehensible to Gould: "If such problems as these routinely occur when we deal with closely related population samples within a single depositional basin, then the wholesale application of such a research strategy to problems of international correlation (as Shaw, 1969, and Barnett, 1972, have recently done with conodont data) presents many problems and carries little prospect that self-correcting results can emerge. The assumptions underlying such a procedure are simply too vast and too ill-founded. It can't work." Eldredge may, however, have empathized a little with conodont workers, as he too had utilized the "stage of evolution" argument in his studies of trilobites.[39] Gould, however, was firmly of the opinion that such interpretation illustrated "the most pervasive and nefarious influence that phyletic gradualism has had on the development of biostratigraphic methodology."

Gould was at this time pioneering a view of evolution as a succession of particular, unique, and unrepeatable events: "Infraspecific trends in vertical outcrop of one local area may not be repeated in an adjacent region."[40] Replay the history of life, Gould believed, and an entirely different tree of life would take shape. Gould's interest in history would have told him that the past was contingent on particular conditions and events. He saw the history of life in similar terms.

Like many areas of paleontology, conodont studies possessed a resilience in these debates because the rock record provided the ultimate truth. Only here might one locate objective fact. So as these arguments raged, most in the conodont community were untouched by them. That is not to say that they were not interested but simply that the arguments had little direct impact on their work. At that moment, the finishing touches were being put to the second appearance of conodonts in the *Treatise on Invertebrate Paleontology* – now these fossils occupied a whole volume. Here Helms's evolutionary tree, coauthored with Ziegler, was updated but little altered. Gradualism stood its ground here at least, and it did so by turning a blind eye. Conodont workers were already reflecting upon the success that had resulted from their own way of looking: "The first job of conodonts was to demonstrate the value of "nuts and bolts" in stratigraphy. This is being done and on a larger scale than many paleontologists would have guessed. A complete reliable and world-wide zonation of Middle Cambrian to Upper Triassic strata based on conodonts may be possible." But this *Treatise* would suffer the fate of the first one: Delayed in publication, it too would be out of date by the time it finally appeared in 1981.

The revolution that had taken place in the 1960s permitted the conodont to come of age as a scientific object. The animal itself had acquired an evolutionary identity that had been mapped perfectly across time and was second to none. But in this decade, there were tremors that had their origins in earlier times, when Branson and Mehl, and American science in general, decided to deny the truth of the assemblage. Ziegler had risen to the top of conodont stratigraphy by a refined use of the kind of study these Missouri workers had undertaken. But as he rose to the top on the basis of mapping individual conodont fossils, the biological truth of the animal continued to seep out into the everyday of conodont science. It seemed almost inevitable that the abstract basis on which Ziegler's science was based would be challenged.

After a short but violent paroxysm, and about midnight, between the 11th and 12th of August, a luminous cloud enveloped the mountain. The inhabitants of the sides and foot of the volcano betook themselves to flight, "but before they could save themselves, the whole mass began to give way, and the greatest part of it actually *fell in* and disappeared in the earth." This was accompanied by sounds like the discharge of heavy cannon.

HENRY DE LA BECHE,
The Geological Observer (1851)

EIGHT

Fears of Civil War

IN 1967, WILLI ZIEGLER STOOD ON THE SUMMIT OF A utilitarian mountain. Now, as he surveyed the world's Devonian rocks, he fancied that he had within his grasp the means to correlate them all. This mountain had been built through the efforts of generations of stratigraphers who had turned the conodont into an abstract timepiece. Buried somewhere near the mountain's base were Kindle's call to action and Ulrich's erroneous assertions. The greater mass was American and had been shaped by Branson and Mehl and few others. The summit, however – where Ziegler now stood, flag in hand – was largely German. Here, inspired by Beckmann's proof of the conodont's potential in the German Devonian, a whole generation had raced for glory, their heads filled with thoughts of mapping the evolution of animal parts. Only on the upper slopes did Ziegler scramble ahead, driven by ambition, extraordinary resources, and sheer hard work.

The practical science that had built this mountain had found no need for the animal, but after 1950, its biology was hard to ignore; the assemblage had become, for almost everyone, an undeniable truth. It was now impossible to look into drawers of these fossils and not see a deception, an act of denial, a piece of non-science – perhaps even pure nonsense. Matters had not been helped by the willingness of those who had believed in the assemblage to nevertheless toe the line. But then, in the 1950s, everyone also knew that the conodont was important only for the huge stratigraphic potential it possessed. This, however, produced a seemingly insoluble dichotomy: Should they keep the fossil as an abstract tool and in so doing deny the animal its biology or should they adopt the animal as the essential basis for rigorous science but then risk shaking this mountain to its core? The conodont workers had cleverly thought they might "have their cake and eat it." Better to change the law, they thought. But when the parataxa plan failed, Raymond Moore – frustrated by the conodont workers' nonconformism – thrust his own solution upon them, effectively telling them to let the animal sleep and continue their old utilitarian ways. Moore was, however, no conodont worker, and he did not have to deal with the contradictions and unrealized potential that daily faced those who were. Inevitably, among some of those studying these fossils there developed a creeping sense that the science could not go on living a lie. It was the animal itself that told them this, for in tantalizing glimpses it began to reveal sufficient of itself to challenge the charade. Gently, it seemed to push for its own recognition. In time, surely someone would take a stand?

The first cracks appeared in 1954, when Rhodes answered Branson and Mehl's test. It will be recalled that these men had challenged Scott to find isolated conodont fossils in the same proportions as seen in natural assemblages. Branson and Mehl must have known this was a mischievous test as the assemblages were composed of both robust and delicate elements, making it almost impossible to imagine nature preserving their natural ratios so perfectly. Yet against all the odds, this is precisely what these durable and abundant little fossils revealed. It enabled Scott, Du Bois, and Rhodes to emphatically reaffirm the truth. But in 1954, there was no longer any need to answer this test, as Rhodes's arguments had already won the day and, anyway, Branson was dead.

Now Rhodes realized that the test held other potential, for the search for ratios had forced him to think of fossil elements not in terms of species but as components. The animal was, to this way of looking, like an Airfix kit composed of the anatomical equivalents of wings, rudder, and fuselage. One only need find and recognize these components and construction could begin of assemblages that had never been found in any coherent form. So Rhodes divided the fossils into a handful of types, easily understood in the everyday language of bars, blades, platforms, and so on. Seeing these as different kinds of component, he looked again at the assemblages he had found and named, and discovered that each contained the same component parts. This suggested that the basic plan conformed to Scott's *Lochriea,* which Rhodes had rearranged so that it looked broadly similar to Schmidt's reconstruction. The only assemblage that did not adhere to this plan was *Duboisella,* which Rhodes used as the basis for another standard pattern. These now became the equivalents of box-lid illustrations, useful for guiding the construction of previously unseen assemblages from their components. By tracing the history of each component type so as to discover which coexisted, Rhodes could also say, with some certainty, that assemblages similar to *Lochriea* could be traced from the Silurian onward and those similar to *Scotto-gnathus* – which shared the same broad body plan as *Lochriea* – from the Upper Devonian to the Lower Mississippian. The *Duboisella* type, in contrast, had existed from the Silurian to the Permian.[1]

Rhodes knew, however, that this picture was incomplete, that there were other kinds of assemblage for which there was still insufficient data to begin to reconstruct them. No complete assemblages had been found in the Devonian, for example, and Müller was, in 1956, struggling to imagine what they might look like. His collections suggested that the preponderance of elongated blades seen in Carboniferous assemblages had been preceded by a prevalence of platforms in the Devonian animals. But puzzlingly, some rocks from the Middle Devonian produced *Icriodus* platforms and nothing else.[2]

Hermann Schmidt was, in that year, in that same quarry that had first furnished him with assemblages. Here, with the help of three of his students, he spent eight days searching for yet more. In what Müller recalled as a difficult collaboration, he helped Schmidt to interpret what had been found.[3] It was now that Müller realized that his problems imagining Devonian assemblages resulted from the incomplete survival of the different kinds of element. He warned others to beware – building assemblages, in the way Rhodes had began to do, held grave risks.

Schmidt and Müller's paper did little to progress the study of assemblages, not least because these were types already well known. But the work also held other difficulties, for while the men could agree on the basic facts, they could not agree on what they meant. As a result, the paper, like a film with a choice of endings, supported two contrasting interpretations. But Müller did not mind too much; he had already decided that assemblages were relatively unimportant. Perhaps the most interesting outcome was a decision to return Schmidt's *Westfalicus* – a name invented to satisfy conventions being introduced with parataxa – back to its original name, *Gnathodus*. It will be recalled that this name had been chosen by applying the rules of zoological nomenclature. It was once again the only assemblage to be named in this way and a direct challenge to Moore's proposal for interpretive myopia.[4]

Back in the 1950s, Müller had also wondered if assemblages were symmetrical. From the 1920s, it was believed that elements existed in mirror pairs, left and right. Thoughts of the fish made this as an expectation, but Müller believed that Chalmer Cooper had found an unpaired element in an assemblage in the 1940s. It seems probable, however, that Cooper was merely complaining that an assemblage was incomplete. Nevertheless, Müller used this new piece of information to suggest that an unpaired element was missing from Rhodes's *Duboisella*. Soon, and independently, Lindström, Adolf Voges, and Bergström and Sweet were also reporting unpaired elements. It was easy for this idea to take hold now that Walter Gross had demonstrated that the animal was neither fish nor worm, and that the elements were not teeth. Indeed, the thought of unpaired elements encouraged the ever-imaginative Lindström to wonder if the animals were always bilaterally symmetrical. He felt he had evidence to suggest that sometimes they were not. He consulted Carl Rexroad and Sam Ellison, who concurred; they too had platform elements that did not consist of mirror pairs. Lindström was then trying to imagine the animal for his book and speculated, "Some might have floated passively, perhaps even in colonies."[5] He was, however, alone in having any thoughts of the animal.

While Lindström pondered the architecture of the assemblage and what it meant for the biology of the animal, others were trying to see assemblages in collections of discrete fossils. At Marburg, Reinhold Huckriede thought he could see the appearance and disappearance of whole groups of conodont fossils in his relatively sparse Triassic collections. He pulled these associations together, calling them "Satze" (sets) guided by the assemblages Schmidt and Rhodes had described. Similarly, in his 1964 paper on the Silurian of the Carnic Alps, Otto Walliser produced nine "apparatuses," giving each an identifying letter, A to J, but no name. He knew that to name them according to rules of zoological nomenclature, as Eichenberg and Schmidt had done, would be to isolate his work. He also knew that the resulting names would reflect a random and curious history of discovery rather than the zoological logic of the apparatuses. "It would be difficult to convince colleagues," he recalled. "I didn't dare to do this."[6] To advance science, Walliser realized, it was necessary to play ball, even if that meant ditching a few scientific ideals. For the moment his apparatuses remained convenient and practical associations.

Walt Sweet and Stig Bergström also hinted in 1962 that a more natural taxonomy was desirable. Sweet had long been cultivating an interest in conodonts in his students. It was, however, with the arrival of some samples from Arthur Cooper at the U.S. National Museum that conodont science at Ohio State University took a new turn. Cooper had been using acids to extract fossil brachiopods from a thin Middle Ordovician limestone from near Pratt Ferry in Alabama and he sent the residues left behind to Sweet and Bergström for them to pick over in search of conodonts.[7] As the two men examined the fossils, they detected a similar number of "right" and "left" elements and noticed that different kinds of element occurred in approximately the same numbers. With a large number of conodonts fossils at their disposal, these facts suggested that it would be possible to unite the elements in "natural species," but they stopped short of doing so. Instead they held onto the hope that an Ordovician assemblage that would "ultimately indicate which of several possible combinations existed in fact" would be discovered. They had no thoughts of rocking the boat. Now converts of Cooper's "acidizing," Bergström started to digest great volumes of rock.

While these discoveries were being made, Lindström was quietly writing his book, but as he did so he became increasingly of the view that the science could only ignore the animal at its peril. Only a few years before, he had been a strong advocate of the utilitarian approach, but as he looked at his fossils he noticed that some could be arranged into gradational series, each element changing slightly as its lines of symmetry shifted and parts of the element were extended. These were fossils of precisely the same age; he was not looking at evolutionary change over time. Lindström called these "symmetry transition series" and found that they only affected certain element types. They were not found among the platforms, for example. He recognized that these transition series reflected the positioning of each element in the assemblage and suggested that unpaired symmetrical elements might once have occupied the midline of the animal.[8] Lindström's arguments were deep and complex, and a little hard to follow, but they provided a vital key for the reconstruction of apparatuses. They gave the apparatus an anatomical logic but made a mockery of naming individual elements as if they were species. When he told Ziegler this, Ziegler responded, "Yes, so what?" They had been firm friends since 1962, and Ziegler's unwillingness to make any concessions to the biological truths of the animal only encouraged Lindström to take the opposite view.

The problem remained that only from the Carboniferous was there good evidence of the nature of assemblages. Then, in 1964, Carl Rexroad and Robert Nicoll of the Indiana Geological Survey published a paper describing two pairs of elements that had been fused together (figure 8.1). Each of these preserved a different element pairing and seemed to reflect the relative positions of the elements in life. These came from a drill core through the Silurian Kokomo Limestone near Logansport in the north of the state. As no natural assemblages had ever been found in rocks this old, the two men thought these new finds offered important clues to the animal's anatomy. The "fused clusters," as they called them, had survived bulk processing because they were bound together with exactly the same material that made up the elements themselves. Rexroad and Nicoll had never seen anything like them, and they consulted Lindström, Collinson, and Klapper. Only Klapper recalled seeing something similar. Rexroad and Nicoll concluded that the animals must have befallen a "pathologic mishap" and had perhaps suffered a tetanus infection.[9]

By the end of 1964, then, a number of authors were, in their own quiet ways, exerting pressure on the mountain Ziegler was scaling. They

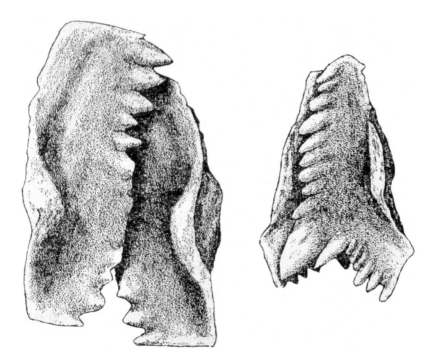

8.1. Rexroad and Nicoll's fused conodont clusters. In one pair the left element overlaps the right; in the other the reverse is true. Reproduced with permission from C. B. Rexroad and R. S. Nicoll, *Journal of Paleontology* 26 (1964). SEPM (Society for Sedimentary Geology).

were producing small cracks and voids. Certainly, we can say that a number of conodont workers were beginning to see their fossils through the lens of the assemblage, and in complementary studies they were acting like catalysts one upon the other. Nevertheless, this remained simply a way of looking; it had not yet become a political movement for change. In that year, no one sensed the coming revolution, even if the germ of that revolution was already in the minds of Bergström and Sweet.

In the summer of 1961, Bergström and Sweet had made a "grand traverse of the Ordovician exposures of the eastern Midcontinent." This gave them a panoramic outlook that would affect their work together over the coming decade. With Sweet's other students, Bergström had been gathering and processing hundreds of samples from the Middle Ordovician Lexington Limestone in Kentucky, Ohio, and Indiana. This rock was extraordinarily rich in conodonts: "There were thousands and thousands and thousands of them." Sweet insisted that Bergström tabulate the numbers and occurrences of those he found, sample-by-sample. This work began but was interrupted, in the fall of 1961, when Bergström had to return to Lund in the hope of securing an academic appointment there. Nevertheless, the two men continued to work on their report by mail. Then Sweet attended a lecture on pollen: "Sometime in 1964, I was struck by the similarity of our work to that of Aureal Cross, a palynologist from Michigan State and I went back to my office after this lecture and began to plot the frequencies of the most common Lexington conodonts on Cross-like relative abundance graphs. The results convinced me beyond the shadow of a doubt that 4 of the element types Stig and I (and my other students) had been treating as representatives of 4 separate species, were, in fact, components of a single 'apparatus' which should be regarded as that of just one conodont species."[10]

The paper was then already close to completion, but based on single element species. Bergström was busy in Lund preparing the illustrations. However, Sweet was now convinced they could and should do the job properly, that they should locate and name the natural species they had found. He wrote to Bergström, telling him that if they continued along their present course the paper would be out of date from the day it was published. When Bergström returned to Columbus from May to September 1964, they undertook a complete revision of the paper, piecing together natural species on the basis of size, color, secondary structures (like denticles), ornament, geographic and stratigraphic range, ratios of components, and so on. As no Ordovician assemblages had been found, they possessed no architectural plan and had to rely instead on the certainties that come with huge collections. In this regard, at least, they were extraordinarily fortunate, for they possessed about a quarter of a million conodont elements! They became connoisseurs: "We found that it was often possible to predict with uncanny accuracy the ultimate composition of a collection after just the first few specimens had been sorted from the residue."[11] The result of this colossal effort was the detection of what they considered to be twenty-three truly biological species. Twenty of these were composed of different kinds of elements and three of only one kind. Far from undermining the simple evolutionary model so effectively exploited in Germany, Bergström and Sweet found these associations gave new evolutionary insights. Now long-ranging elements became inextricably associated with short-lived and rapidly evolving ones. The critical step Bergström and Sweet took, however, was not merely to detect what they thought were natural species but to actually name them. No one had dared do this since Schmidt: "So we messed around with it. We didn't get it all right and there were some people who thought we didn't get any of it right!"[12] The paper was completed in 1965 while Sweet was visiting professor at Lund. It would be published the following year.

Sweet, Sweet's doctoral student, Tom Schopf, and Minneapolis student Gerald Webers gave the first indications of this approaching storm at the annual meeting of the Geological Society of America in 1965. All three had papers in press that in some degree promoted this new way. Between them, they had amassed three hundred thousand conodont fossils – a body of evidence vastly superior to any previous study. Inspired by the statistical predictions of Scott, Du Bois, and Müller, Webers had began his research in 1959 and had, without the slightest hesitation, adopted the rules on zoological nomenclature and used the earliest established element species to name his natural species.[13] He possessed the smallest collection – thirty-five thousand fossils – but this was nevertheless a huge amount of data. He wasted little time reviewing the failed attempts of others to find alternate solutions and felt sure that this new way would find support and generate fewer problems than had been predicted. Webers and Sweet had met on occasion but Webers's project was totally independent. Schopf's project, in contrast, had been suggested to him by Sweet. He possessed some fifty-five thousand specimens in 1962, but while he was happy to discuss recurrent groups, he stopped short of naming natural species.[14]

When Bergström and Sweet's Lexington paper appeared in 1966, it was, for Ziegler and others, as if a 100-megaton bomb had been dropped on the conodont community. The paper itself was about the stratigraphy

8.2. Revolutionaries Walt Sweet (*left*) and, in a more recent picture, Stig Bergström. Photos: Dick Aldridge and Jeff Over, respectively.

of the Lexington Limestone; the new taxonomy was merely a necessary underpinning. To some degree Bergström and Sweet attempted to protect themselves by downplaying this innovative aspect: "The taxonomic philosophy involved is not new, nor is it novel in the interpretation of collections of discrete conodont elements."[15] Nevertheless, "we understood this might be Revolutionary," Sweet later remarked.[16] It shook the science out of its comfort zone and demanded that it think of the conodont as a biological entity. If this in itself was not challenge enough, it also demanded that the science learn a new language and transfer names that once spoke of individual elements to associations that were then to be understood as natural species. As we shall see in the next chapter, this paper was, in all respects, conceptually grand and, in so many ways, startlingly original. It was one of the masterpieces of conodont science.

Bergström, Sweet, and Webers now found themselves "plunged headlong into the jungle so neatly avoided by Huckriede, Walliser, and Lindström."[17] As Bergström and Sweet noted, "The revised taxonomy was not greeted with enthusiasm, to say the least!" But Sweet could not have been surprised. He never seemed to shrink from what he considered the proper scientific course. Indeed, the two men continued their rebellion in the small print, too, pointing out that since the conodont fossils were no longer considered the teeth or jaws of vertebrates, terms like "tooth, jaw, oral, aboral, pulp cavity, fang, ramus" should also be ditched.[18] They now reintroduced the concept of "form species" to describe individual elements; the only proper species names were those attached to element associations.

The mountain was truly shaken. A fault line now ran through it and intersected with the long-ignored animal. Published just four years after the first *Treatise*, this new thinking would render that earlier work completely obsolete if it took hold. Rhodes, who was visiting professor at Ohio State University in 1966, liked the paper and "did a lot to calm the ire of the vast majority."[19] Ziegler, however, was annoyed. He could not help but see this as a challenge to a lifetime's work. It complicated a system that worked so perfectly, easily, and usefully, and it came at a moment when Glenister and Klapper had replaced Ziegler's names with numbers and, as we shall see, when other studies were threatening to topple Ziegler's position. Ziegler must have feared that he would need to start his research from scratch, but he also knew that his collections did not hold sufficient evidence to reconstruct Devonian assemblages. The threat to his position was very real.

Although Bergström and Sweet's stand was greeted with fierce opposition from nearly every quarter, a few looked at what they had done and hardly blinked. Chris Barnes, at the University of Waterloo in Ontario, for example, used Bergström and Sweet's new species to interpret an unexpected cluster of elements from the Middle Ordovician. These convinced him that these four variously sized conodonts of the same type, all oriented in the same way and stacked one upon the other, belonged to the same animal and seemed to reflect their positions in life. The discovery appeared to confirm that Ordovician assemblages were rather different from those that had been found in the Carboniferous. Two years later, in 1969, Austin and Rhodes reported a strange fused cluster from the Carboniferous of the Avon Gorge, near Bristol, England. It was made up of an interlocking arrangement of conodonts of the same kind but of different sizes. They, too, thought these elements had belonged to the same animal and reasoned – like Barnes – that the differing sizes possibly reflected continual replacement of old elements by new or, more likely, the presence of different sized elements in the same animal.[20]

The most significant discovery, however, was published by Friedrich-George Lange in 1968. It reported the finding of seventy clusters bound together with bituminous material. These all came from the Upper Devonian Kellwasserkalk of Steinbruch, Germany, and revealed for the first time the complete natural assemblage containing the element known as *Palmatolepis*.[21] Lange had visited Lund in the mid-1960s, showing Bergström his assemblages. Bergström was convinced that they were apparatuses preserved in their original three-dimensional form. He recalls that Lange was of the same opinion but had to back down in order to get his paper published when Ziegler refused to accept them as anything other than coprolites. This was probably the most important and influential discovery of assemblages since the 1930s, for it showed more precisely the positioning of the elements in Schmidt's reconstruction and revealed that these Devonian assemblages were very much like those seen in the Carboniferous. Ziegler's position was now untenable, and workers asked why his collections did not contain the full range of elements now known to accompany *Palmatolepis*.

Assemblages in the Silurian were only known from Walliser's associations and Rexroad and Nicoll's clusters. However, in 1969, Charles Pollock of Indiana University discussed some fifty-four clusters, mostly from the same Kokomo Limestone. These confirmed and extended earlier discoveries and suggested the presence of two different natural assemblages in the rock. Pollock felt this supported Rhodes's belief that early natural assemblages were composed of fewer element types and arranged fairly simply. (It was not until 1972, when the Soviet Union's Tamara Mashkova found the first complete Silurian assemblage, that Walliser's associations would be demonstrated to be biological species).[22]

Meanwhile, Richard Lane, a young worker at the University of Iowa, was continuing the kind of thinking Lindström had begun. In a paper just five pages long, Lane reviewed the symmetry of conodont pairings. He showed that one could not prove that midline elements existed or, indeed, that any element was unpaired, regardless of symmetry, unless it was found in a well preserved assemblage. Most elements came in mirror-image pairs that were presumed – no doubt based upon the jaw analogy – to have been arranged in life with concave sides facing each other. But by tracing the evolutionary development of these paired elements, Lane could show that some developed from mirror-like pairs into asymmetrical pairings in which each element was morphologically distinct. He pointed out that mirror pairs, asymmetrical pairs, and unpaired asymmetrical elements were known to occur in some modern worms and other marine animals. Lane was not suggesting that the conodonts came from one of these groups of animals but instead pointing out that one could infer relatively little from asymmetrical assemblages. A swimming lifestyle required a bilaterally symmetrical body but not necessarily a symmetrical arrangement of elements. In what became one of the most read papers of the period, Lane fundamentally extended and rationalized the acceptable possibilities.[23]

When the Pander Society met in Iowa in May 1968, it was clear that many had tried but struggled to locate the kinds of natural association Bergström and Sweet had found so easily.[24] But now Sweet made matters even more challenging for his colleagues by bringing along his doctoral student, Joe Kohut. Kohut had, in his studies, come under the influence of Howard Pincus, chairman of Ohio State University's Geology Department. Pincus was a great advocate of numerical methods in geology, and "Joe had Howard's ear and Howard had Joe's firm attention."[25] Kohut soon discovered that he had a natural aptitude for numbers, and he had, by the time of this meeting, already reworked Bergström and Sweet's vast data. Through Pincus he had gained access to the IBM 7094 mainframe computer in the School of Business Administration next door. This was one of the biggest and fastest mainframe computers available in the mid-1960s. Although extraordinarily underpowered compared to a modern desktop machine, a basic system cost more than three million dollars. Kohut converted Bergström and Sweet's form-species names into code numbers and used punched cards and a Fortran IV program to perform a succession of calculations to reveal the joint occurrences for every possible pairing of elements. By this means groups of elements were pieced together, confirming the natural species Bergström and Sweet had found simply by looking: "Application of the quantitative procedures that are outlined in this study have revealed faunal and genetic associations that have been determined previously by careful observation and tedious work over a period of many years. The obvious advantage of the quantitative approach is its rapidity and objectivity, but, like empirical procedures, it must be based on large collections that have been systematically tabulated."[26]

John Huddle heard Kohut's paper at the Iowa meeting and was completely befuddled by it.[27] It pushed what was already a radical argument beyond the everyday concepts and visual evidence that occupied most paleontologists. Few conodont workers were so numerate, and rather than see Kohut's conclusions as affirming the efficacy of numerical methods, they might have felt, instead, that these methods revealed that looking and judging were sufficiently accurate.

Kohut's novel analysis forced many conodont workers to reflect on their own skill set. Some saw an opening door – a new technology capable of advancing the science – and attempted to follow Kohut's lead. By the time of the North American Paleontological Convention in 1969, Sweet and Bergström could claim that Kohut's program had permitted the generation of Rhodes's assemblages from Ellison's 1941 revision of elements.[28] However, this mathematical moment would not last. Kohut found no firm foothold in the tiny field of conodont studies, and with reluctance he entered librarianship. With a sigh of relief, the science slipped back into its more descriptive ways. The discipline had lost its numerate edge.

In May 1969, when the Pander Society met to discuss conodont stratigraphy at Sweet and Bergström's university in Columbus, it was apparent that conodont studies remained locked in a halfway house, neither one thing nor the other.[29] Here converts like Lindström made their commitment to the scheme known; Lindstrom now spoke the new language – as much as he was able. But in what is at the best of times an arcane science, those outside this small group of specialists would have been at a loss to understand precisely what was going on. With hindsight, the language used at this meeting appears to be in transition when really it existed in two opposing camps. Yet very little of this discrepancy appeared on the surface, at least in the published account. Thus Lindström could talk of "faunas" – a word all who attended would understand – as representing groups of conodont form species occurring at one horizon. Yet Lindström would know, and occasionally say, that these were associated with each other because many of them belonged to the same biological species; they had come from the same animal. Others would use the same word but make no reference to this underlying cause of association, believing that such information was not essential to stratigraphic study. Walliser simply updated his earlier work; his apparatuses remained unnamed. Some spoke of "conodont species" and meant species of animal, while others used to same words to mean "species" of element. These differences, however, only become apparent if close attention is paid to the text; no one made a great fuss about them. Those few who, like Bergström, were attached to the new way would describe their new species with a language that had left the everyday. His *Eoplacognathus robustus,* for example, was to be understood as composed of "sinistral and dextral ambalodiform and polyplacognathiform elements." Ziegler, by comparison, could speak of single objects using a single name, such his *Polygnathus varcus,* which required little more than illustration. For the geologist who simply wanted a practical tool, the simplicity of the old way could not be beaten.

The Ohio meeting indicated that conodont studies had entered a kind of limbo. There were tensions, but they never appeared in the written accounts. One night, Sweet, Ziegler, and Klapper stood on the street corner in front of the students' union until 2:00 AM. "We argued and argued and argued that night,"[30] Sweet recalled. Positions were becoming entrenched. Ellison, for example, refused to let his students consider Bergström and Sweet's scheme.

When the conodont workers met again, at East Lansing, Michigan, in August 1970, there was much talk of the Marburg symposium planned for the following year. Here the conflicting approaches were to go head to head. Rhodes was worried: "Neither the dual nomenclature employed by Scott and Rhodes, nor the assemblage Linnean nomenclature of Schmidt,

Bergström and Sweet is entirely satisfactory. . . . The first is illegal, but useful; the second is legal, but ambiguous and sometimes illogical. . . . There is no reason why the change in general taxonomic practice, which recent data now justify, needs to involve a nomenclatorial bloodbath."[31] Ever ready to assume the role of mediator and diplomat, he nevertheless feared "Conodontological fratricide."

In September 1971, between sixty-five and seventy conodont workers arrived in Lindström and Ziegler's home territory of Marburg. They had come from fourteen countries. Huddle arrived expecting the bloodletting Rhodes had predicted, but to almost universal surprise it did not come. Lennart Jeppsson, a relatively new Swedish entrant to the field, thought this owed much to the way the meeting was organized, as presentations were made in stratigraphic order, beginning with those dealing with the oldest rocks. This gave the stage first of all to the leading advocates of change: Lindström, Bergström, and Sweet. There were no keynote speakers, but the seniority of these three men and their positioning in the program assured them of this status. Possessing incomparably huge collections and having pushed hard to progress their interpretations since 1966, they now thought it possible to piece together the evolution of apparatuses in some detail. As a result of this careful planning, by the end of the first day the argument for change had effectively been won. Jeppsson, who spoke toward the end of the conference, felt powerless. His rocks turned up relatively few conodonts and there was little chance he could emulate the Ohio workers. A young Englishman, Dick Aldridge, also found himself at the tail end of proceedings. He stood up, walked to the front without his papers, and said he would no longer give his paper; he had changed his mind.[32]

No blood had been spilt. And despite the fear that the utilitarian mountain would collapse, it did not. Ziegler – who never became fully committed to the new scheme – was satisfied because the proposals seemed to endorse the view, first expressed in the 1950s but implicit in the work of Branson and Mehl, that the most rapidly evolving elements might both define and name the assemblage and also perform in stratigraphic studies. This removed the need for Ziegler to truly know apparatuses.[33] Other apparatuses seemed to be composed of single element types, and these too posed no problem. Ziegler must have been as relieved as anyone: a juggernaut had driven into his university and he had managed to grab the wheel. But those around him were also keen to see the new scheme incorporate rather than alienate him. If he, as the champion and most successful exponent of the old way, was converted, then they knew the opposition would be fatally wounded.

The conodont workers had now committed themselves to pursue a path that twenty years before had seemed impossible. Then they imagined this course of action would result chaos. As it was, they resigned themselves to learning new names and understanding the fossils as components rather than species in their own right. Old words would acquire new and different meanings, and other words would simply disappear. David Clark feared that this "new language" would need to be "mastered in a short time." Lindström and Ziegler, however, preferred to underplay the significance of this change: "We all learned numerous names for fossils when we were students, and a great proportion of these are now obsolete, victims of healthy revision. This is true for all important fossil groups, it must apply in the case of the conodonts, as well."[34] With a little sense, they knew they could rescue a good deal from the ashes of the old regime.

Rhodes was given the role of diplomatic envoy and charged with drawing up and circulating the "Marburg Proposals," which would, of course, reach those who were not there to witness this surprising outcome. Inevitably, there was little agreement among the forty-one responses he received, but Rhodes pressed on, editing these together into what became the third version and final agreement. The aim of the document, as he explained, was not to legislate for others but to provide "useful guidelines for those who choose to follow them." It stressed the need for "great care, constraint and consultation." It also required Rhodes to cook up a little linguistic spaghetti to cover all eventualities.[35]

Having taken this giant leap without the outbreak of civil war, it must have been with a sense of irony that conodont workers read of the intention of the International Commission on Zoological Nomenclature to introduce a scheme of parataxa at the end of that decade. The scheme was proposed by ICZN secretary Richard Melville in 1978 and sought to resolve the same difficulties that had troubled conodont workers in the 1950s.[36] To prepare his case, Melville had reviewed those earlier arguments and alighted upon the conodonts as perhaps the strongest cause with which to garner support. By doing so, he unleashed a hornets' nest of often sharply worded and deeply argued criticism. Jeppsson, for example, felt compelled to send copious letters documenting in detail all that had happened since the 1950s and why the proposed scheme would damage conodont science. Dick Aldridge, who also fundamentally opposed the scheme, took a more dispassionate view. He told Jeppsson how "greatly impressed" he was that Melville, "who has to deal with 'all' systematic zoologists in the world, has involved himself so deeply and thoroughly to bring about a solution of these problems." Aldridge corresponded with Melville, hoping to school him in the niceties of conodont taxonomy, but Melville tenaciously held his ground. While admitting that conodonts were no longer the group in his sights, Melville felt duty bound to serve others who would benefit from the scheme. Aldridge pressed on – and not without effect. The arguments became increasingly refined and arcane. "I am gradually educating myself," Melville told him encouragingly, but then warned, "and must ask you to accept that an old dog has difficulty with new tricks."[37]

When Aldridge attended the Pander Society meeting in Vienna and Prague in August 1980, he feared he was still losing. Among the seventy-one conodont workers from twenty-five countries, parataxa once again became a hot topic, but now it produced absolute opposition. Those present drafted a petition – known as the Wolayer Resolution – which read, "We unequivocally reject these amendments to the Code and urge you in the strongest possible terms to vote against them.... As a nomenclature for parataxa would legalise a dual system of names, paranomenclature is antithetical to the purpose of the Code."[38] Aldridge wrote to Melville to warn him, explaining that he did not know if this was the right way to go about things but that it was now so late in the day that they felt compelled to try anything.

Melville was still willing to learn and asked Aldridge to give him a practical demonstration of the problem that October in London. When the two met, Aldridge found the ICZN secretary amiable and helpful. But as they were discussing the finer points of the proposed scheme, Melville's secretary entered the room with a letter from Walt Sweet. In seven neatly typed pages, Sweet set out a closely argued case against Melville's proposal. In fact, Sweet laid it on rather thick. This was not a letter seeking to forward science so much as win the argument and rebuff an unwelcome intrusion into a field that had so recently emerged from a difficult past. Sweet was manning the barricades again and ready for another civil war. He ended by warning, "Should the Commission approve the amendments involving 'parataxonomy' and 'paranomenclature,' you can expect a somewhat more extensive and more carefully composed response from me in the literature. I hope it will not come to that."[39] Sweet saw in the scheme the potential to undo all that had been achieved.

Melville thought Sweet's letter "magnificent," and he responded at length using a well-honed style that must have seen him through many such arguments.[40] It was smooth and diplomatic, cleverly and no doubt wittily using abstraction and generality to throw a thin veil over his criticisms of Sweet and his other detractors. He told Sweet that conodont workers had made themselves remote from the ICZN, and as a result "the structure of current conodont nomenclature suffers from severe logical weaknesses and is highly vulnerable." Rather pointedly he observed, "If conodont workers in the mid-60's had taken the trouble to keep us informed of the exciting developments in taxonomy that were then taking place, we could have proceeded on a better basis of mutual understanding."

Melville remained resistant to the last, and the commission pressed on and won the vote, but there were enough abstentions to prevent the proposal being integrated into the code. By little more than a technicality, the conodont workers had won the day. Melville reflected, "I was astonished at the strong feeling against the proposal and the – to me – cogent arguments that were advanced. . . . They pooh-poohed the idea of chaos. They believed that approval of dual nomenclature would cause neglect of, and even inhibit, solid zoological studies. Conodonts were used as an example of a group in which advancing knowledge had over-taken earlier and vaguer knowledge and the group is being put on an even firmer footing, without resort to parataxa."[41] Clearly, the conodont workers were turncoats and they had now won their first battle fighting for the other side.

By 1982, the conodont workers considered themselves fortunate to have failed to establish parataxa in the 1950s, for during the following decade the assemblage asserted itself on all fronts. It could not be ignored, and it is very likely that Bergström, Sweet, and Webers would have taken their stand regardless. But with a firm system of parataxa in place there may well have been the fratricide Rhodes so feared. Now the conodont workers had the Marburg Proposals – a kind of peace treaty – which would enable the future to be negotiated with care. In time this shift of language and practice would throw up every expected problem and er-ror, but the mountain did not collapse. It did not do so because despite Sweet and Bergström believing, in 1969, that all conodont workers would be using the language of apparatuses in just five years, thirty-two years later, Sweet and Phil Donoghue had to admit that only a third of genera had been reinterpreted in this way. The mountain remained unmoved because, for the most part, its makeup remained unchanged.[42]

In this country the sun shineth night and day: wherefore this was beyond the Valley of the Shadow of Death, and also out of the reach of Giant Despair; neither could they from this place so much as see Doubting Castle.

JOHN BUNYAN,
The Pilgrim's Progress (1678)

NINE

The Promised Land

PANDER'S ANIMAL WAS AS MYSTERIOUS AS EVER, BUT DURING the 1960s it had begun to take possession of its skeleton. Fossils once considered teeth were no longer to be seen in isolation. For conodont workers this was a move toward biological truth and the only course if their science was to be considered rigorous and legitimate. Nevertheless, many worried about chaos, and some questioned the benefits. It had been the study of isolated fossils – which they were now abandoning – that had made this science so useful and effective. And it was this that had also given the animal a history, or rather, an evolutionary genealogy. Of course, this wasn't really how conodont workers saw it; most were interested only in acquiring a more refined tool. But out of this necessity emerged glimpses of the biological flesh of the animal itself, and it would do so repeatedly as the conodont workers acquired new methods and new ways of seeing.

The adoption of acids had produced a revolution in the study of these fossils, making them infinitely more numerous. Blessed with a wealth of data from different parts of the world, conodont workers began to ask spatial questions such as "Were some of these animals restricted to particular environments or parts of the globe?" and "Did populations of these animals move across the surface of the planet as conditions changed?" And as geology as a whole reached for a grand theory of the earth in the 1970s, so this kind of thinking was swept up into models imagining the global ecology of the planet. It became increasingly possible to imagine millions of these ghost-like animals living quite particular lives. There were, however, very practical reasons for this thinking;

the science remained wedded to its stratigraphic goals. But, once again, the animal itself could not be suppressed. Indeed, the practical science needed to better know this animal if it was to progress.

In the 1960s, many workers believed that the conodont animal was un-affected by water depth or sediment type; its fossils were not indicative of a particular past environment. For Willi Ziegler, with his ambitions for a global stratigraphic standard, this was a matter for rejoicing: "Con-odonts are like God – they are everywhere." This view had been consoli-dated in the 1950s, as the Germans began to see American fossils in their own sections. Indeed, the wide geographical distribution of species, and their frequently reported association with fish and cephalopod remains, suggested that conodonts were swimming or possibly floating animals. Klaus Müller, however, was rather less convinced. Finding them less associated with corals, sea lilies, brachiopods, and reefs, he suggested that the animals *were* susceptible to environmental control and that they did not like strongly oxygenated bottom waters. But in another study in which the distribution of other groups of animals seemed to be con-trolled by environmental conditions, he found the conodonts immune. Frank Rhodes, Walter Youngquist, and A. K. Miller had reported find-ing conodonts in shallow-water deposits. Similarly, Reinhold Huckriede had found that his Triassic conodonts were most common in shallow-water limestones rich in cephalopods, sea lilies, and sponges, and that rocks formed in other environments often contained none at all. Maurits Lindström, on the other hand, had no difficulty in finding contradictory examples: conodonts with shallow-water corals and brachiopods, and in deep-water deposits. Müller, however, remained the doubter, and in the 1962 *Treatise* he suspected that Branson and Mehl's most useful *Icriodus* was ecologically controlled.[1]

These discussions in the 1950s and early 1960s took place as the study of the ecology of the deep past, or paleoecology, finally gained a firm foothold in paleontology. The subject's scant coverage in the 1962 *Trea-tise on Invertebrate Paleontology* conceals this change; Ray Moore was an enthusiast but felt the effort to include it redundant in the light of the publication, in 1957, of the 1,296-page first volume of the *Treatise on Marine Ecology and Paleoecology*. With its integration of both ecology and paleoecology, this new ecological series demonstrated the intellectual necessity of pairing the past with the present but also signaled in its title the relationship between the two, with paleoecology always following, and drawing analogies from, the present.[2]

Among those advocating this new approach was Preston Cloud at the USGS. In the late 1950s, he saw paleoecology as the Promised Land, offering deeper understanding of life in the past: "Some of the most obdurate strongholds of ignorance in geology and paleontology await new or renewed assault by palaeoecological methods." There was no shortage of methodological ideas, but, rather appropriately for a Promised Land, the subject's potential was as much an act of faith as of proven utility. As Cloud admitted, the field was "still groping toward a coherent body of critical observations and specific principles by means of which the evidence can be winnowed and refined, its applicability established, and durable inference reached where data are adequate." British paleontologist Derek Ager saw paleoecology as lifting paleontology out of the realm of "stamp collecting" and extending the scope of the paleontologist to include "the whole world of living nature." Others, however, would look on and worry about the rise of "impractical theorizers."[3]

So as paleontology crossed the 1950s borderland into the swinging 1960s, it was paleoecology that was hip, a new territory that could draw in those who wished to lay down its principles and philosophies. But those who pursued these opportunities did not believe all fossils held equal promise. Of particular interest to them were those that held utilitarian potential or could draw upon modern analogies in order to achieve ecological understanding. In the mid-1950s, for example, paleoecology had already proven its importance to oil and gas prospecting in the Permian reefs of western Texas and New Mexico, and oilfield geologists working with fairly recent strata – of Miocene age or younger – soon found single-celled benthic (seafloor-dwelling) foraminifera to be an almost perfect means to assess paleoenvironmental conditions, because these fossils were similar to living forms. Conodont workers could, by contrast, only mourn the biological ambiguity of their subject: "Determining the ecologic factors that influenced a group of organisms that has been extinct for 180 million years and whose biologic affinities are uncertain is a problem that still challenges students of conodonts. The fact that conodonts were widespread for 400 million years and are superb tools of biostratigraphy during this Cambrian to Triassic interval has compounded the problem."[4]

However, the problem of the conodont's ecology wasn't entirely insoluble; it was simply a matter of finding the appropriate tools, and these were easily located in practices that went back to the early years of the nineteenth century. Paleontologists, like epidemiologists, had adopted the habit of studying patterns in the relationships between two distinct things. The epidemiologist can by these means locate a link between smoking and cancer. Conodont workers such as Bergström and Sweet used this concept to help them construct assemblages from discrete parts, and geologists before them had used it to establish the principles by which fossils can be used to give rocks a relative age. If a mudstone is repeatedly found to contain a particular species of conodont, which is not found in neighboring limestones, then the conodont worker might presume environmental control since the mudstone was produced in an environment very unlike that which generated the limestone. It is a reading little different from that of the ecologist who links eagle to hare and both to an upland habitat.[5] And one extraordinary advantage of the conodont animal in this kind of study is that its elements were resistant to destruction and could be preserved in a range of rock types.

Paleontologists and geologists could now ask questions of rocks and fossils with the aim of understanding environmental conditions and ecological niche. One simply needed to identify and correlate consistent changes in rock and fauna.

In 1957, a number of workers on both sides of the Atlantic began to believe that some conodont animals had restricted distributions. At the Illinois State Geological Survey, Iowa-educated Carl Rexroad decided to test this idea in his local Mississippian rocks, but he found that while the limestones contained more conodont fossils than the shales, they held essentially the same kinds. The animals were not controlled by environmental conditions. But then Rexroad noticed that two genera, *Cavusgnathus* and *Gnathodus,* showed opposing abundances; when one was numerous, the other was not, and vice versa. He hypothesized the presence of two distinct provinces and then collected from a spread of sites to test the idea. The results seemed to confirm his suspicions, encouraging him to imagine an Illinois fauna being replaced by, or mixing with, southern immigrants known from rocks in Texas and Oklahoma.[6] The idea that faunal provinces could be recorded in the rock record was, however, not new. It had been demonstrated many times previously using macrofossils, so it was not surprising that Rexroad interpreted his rocks in this way.

Huckriede became convinced that his Triassic conodonts were also affected by fairly strong provincialism. Lindström too felt he had found a discrete Ordovician faunal province. This had occurred on his honeymoon trip to Scotland. He had left his wife, Ulla, knitting while he went in search of conodonts in John Smith's old collecting haunt of Morroch Bay, but it was she who found them. The fossils were slightly different from those Lindström knew in Sweden, and in a few years similar forms would be found in New England. He wrote these finds up in what he jokingly referred to as his "Rosetta Stone paper," for here he had found conodonts in the hard shales that commonly preserve graptolites but no trilobites.[7] Trilobites and graptolites provided two distinct fossil-based timescales. Now the ubiquitous conodont connected them and threatened to surpass them both. It was this paper that caused Ziegler to joyously exclaim on the godlike omnipresence of his favorite fossil.

There were, then, a number individuals suggesting for the first time that different species of the conodont animal may have been separated geographically. These were, however, little more than footnotes in research that had other goals. A more forceful argument for conodont provinces originated in Walt Sweet's collaborations with a succession of his students and with Stig Bergström. It was in 1957 that Sweet began his long assault on the Ordovician. He had three of his master's degree students – Caroline Turco, Earl Warner, and Lorna Wilkie – undertake a stratigraphic study of the conodonts of the lowermost horizons of the Upper Ordovician where it outcrops along the Ohio River. In all, these newcomers collected and processed some ten thousand conodonts from shale samples. Sweet, who had promised to bring their work together in a joint paper, was not deterred by the large number of long-ranging, and thus stratigraphically useless, conodonts in the samples. Instead, he saw in these long-ranging species a phenomenon that greatly interested him: two distinct and interacting faunas.[8] Drawing on recent work on climate change in the Ordovician, he painted a picture of a mass migration of these unknown animals. The period opened with a north-westward invasion into the area of cold-temperate forms that belonged to an "Anglo-Scandinavian-Appalachian province" known to have existed from studies of other fossils. These conodonts diluted, or coincided with an emigration of, a presumably warm-temperate "mid-continent fauna" from the area. This invasion of coldwater forms was, however, apparently short lived and soon reversed, bringing back that midcontinent fauna that had occupied the area in the Middle Ordovician. To this he could add his own data, which suggested a later "invasion of even warmer tropical or subtropical forms probably from the north-west or north." Sweet's compass directions here refer to the modern-day positions of the rocks. The North American continent has rotated and moved since the Ordovician, when these rocks were laid down. Consequently, Sweet's migration of tropical forms at the time came from what was then the west or southwest. Nevertheless, the varying abundance of different species and their intermixing preserved these dynamic shifts in populations of the animal nearly five hundred million years ago.

Sweet knew that more and better data were required if provincialism was to be firmly established. There were, after all, only three complete conodont sequences known for the Ordovician, all in North America. He was, however, never one to give up on a good idea, and soon help arrived in the form of the meticulous Stig Bergström and Cooper's extraordinary Pratt Ferry residues. Bergström's head was already full of Swedish conodonts and he was intent on studying the conodonts of the Appalachians, suspecting that they might be familiar to him. Back in 1926, U.S. Geological Survey workers had discovered that rocks in the eastern trough of the Appalachian Valley contained fossils with a curiously European aspect. No one had looked at the conodonts. Cooper's Pratt Ferry material came from the southern end of the Appalachians, and in it Bergström had no difficulty finding conodonts he knew from

Sweden that had never previously been found in North America. He thought this astonishing, as it was now possible to locate two sections of identically aged rocks in the United States, just twenty kilometers apart, that could not be correlated on the basis of conodonts. Yet one of these sections could be precisely correlated with rocks ten thousand kilometers away.[9] Could there be more dramatic proof of provincialism?

It was, however, the follow-up paper – Bergström and Sweet's 1966 coup de grâce, discussed in the last chapter, which sought to end old ways and, in some senses, re-establish a biological animal – which really showed the practical benefit of thinking about provincialism.[10] The quarter of a million fossils they isolated in that study told them that the midcontinent fauna was distributed in a certain way. They could now detect provincial encroachments in great detail and utilize the relative abundance of species to powerful stratigraphic effect. Their data suggested three partially open and shifting subprovinces, each perhaps controlled by water temperature and depth. As a result, a species dominant in one area might at the same time be rare in another. Across a network of geological sections, the European fauna could be seen to appear suddenly at different locations at different times – an invading population of animals, whose geographical advance could be read in detail in the sequential appearance of individual species. And here, of course, Bergström and Sweet were writing about *animals* rather than isolated fossils, as they were discussing their new statistical – or biological – species. The picture they created was not so much one of "herds" of conodont animals sweeping across an underwater equivalent of the African savannah as one in which the expansiveness and position of that savannah was shifting over time.

This changing complexity meant one could not identify a species at two different locations and say they represented identical moments in time, as stratigraphers might wish. Nevertheless, Bergström and Sweet saw in their data a way to use these long-ranging species as stratigraphic markers, and here they found another use for pollen expert Aureal Cross's relative abundance charts. The charts, which plot the relative abundance of particular species of plant, represented by pollen and found particularly in postglacial peat, were used to map environmental change since the Ice Age. Once this pattern of change was established through repeated study, it became possible to use it as a relative timescale. Bergström and Sweet took this idea and applied it using the ubiquitous, and often dominant, midcontinent species, *Phragmodus undatus*. The remains of this animal were found in a variety of rock types, suggesting that it was unaffected by environment and thus useful for correlation. So they plotted its relative abundance at every level in every section. The results showed similar, though by no means identical, logs for all sections. This permitted geological sections to be correlated with each other and with sections further afield. Bergström and Sweet also used the local histories of faunal incursions and fluctuations, recorded in these rock sequences, to define local time units. They could not, however, given the long-ranging nature of the species available to them, establish formal stratigraphic zones.

We can now see how fundamentally Bergström and Sweet's remarkable paper challenged everything to which Ziegler aspired just at that moment when he had achieved global success. Not only did it suggest he needed to document the evolution of assemblages, rather than isolated elements, but it also put into question the very notion of the universal conodont.

Bergström and Sweet continued to develop increasingly sophisticated and subtle interpretations of provincialism in their animals. Soon they were "tracing and matching faunal 'tongues' that represent the shifting of provincial and subprovincial boundaries in time." In his paper with Kohut, Sweet could talk of *Phragmodus undatus* retreating northwards while other conodonts were "more tolerant" of change and stayed put. Increasingly he began to think about environments, and as he did so, particular conodont species acquired lifestyles. One, for example, "seems to have flourished in a nearshore, shallow-water environment, perhaps on a tidal mud flat that was periodically exposed to the atmosphere." In contrast *Phragmodus undatus* was "an inhabitant of deeper waters." Gerald Webers, influenced by the Ohio workers, also began to think along these lines. In Ohio, at least, and in the Ordovician in particular, the universal conodont was dead.[11] Necessity had also spawned a rather different approach to stratigraphy. There were no neat time markers here. Rather, time seemed to be marked by the very ebb and flow of life.

In the language of the 1970s, Sweet and Bergström were applying a generalized conceptual "model" – that of the province – and using its perceived attributes to interpret their data. It was a model so enshrined in geological practice that few questioned it. However, at a London conference in 1969, it became apparent that this universal concept was far from concrete. And as the delegates at the meeting deliberated its meaning, so Peter Sylvester-Bradley became increasingly depressed: "I am afraid that we can claim to have answered no problems in this symposium. Quite the reverse. We have dredged up from the stores of knowledge many old problems that had been shelved away and almost forgotten." Those who had organized the conference had to agree: There was no agreement on what a province was, how it could be recognized, or even whether such things existed in the geological record.[12] This did not invalidate Bergström and Sweet's interpretation, but it did mean that the province remained hypothetical.

At that time, the descriptive natural sciences were adjusting to the logic of the computer and the imagined objectivity of numbers. Its world was about to be redrawn in systems diagrams of the type used by programmers and systems analysts in the computer industry. Among those taking a lead was another of Sweet's former students, Tom Schopf, whose *Models of Paleobiology* performed as an evangelical tract for this new way. Schopf argued for a retreat from short-sighted realism and precision, suggesting that generalized theoretical models should be developed to predict, and be tested by, data. He was then at the University of Chicago, and was the latest in a line of academics there who sought to lift paleontology out of those habits of which Ager had been so critical.[13]

The conodont, which in Bergström's and Sweet's minds seemed to exist as a great underwater swarm of expanding, contracting and shifting life, was then being captured by other modelers who made the animal's distribution rather more structured. The origins of this close-up view of conodont distribution lay in the paper Glenister and Klapper wrote on conodonts from the Canning Basin in Australia, which implanted Ziegler's Upper Devonian zonation there. When Glenister and Klapper came to make this correlation, they drew upon a recent study of the basin by Phillip Playford and D. C. Lowry of the Geological Survey of Western Australia. In order to understand these complex rocks, Playford and Lowry developed a model suggesting that the rocks there represented four different reef environments: the back-reef, reef, fore-reef, and inter-reef. To Glenister and Klapper, the distribution of the Australian conodonts suggested they were ecologically controlled, for, like the cephalopods, these fossils were found in the fore- and inter-reef areas but were rare in the reef itself and absent from the back-reef. They were also rare in beds containing brachiopods.[14]

This understanding of ecological control became more specific when English emigrant Ed Druce investigated the conodont fauna of the Bonaparte Gulf Basin in the far north of Australia. Druce was a veteran of the field: "He enjoyed the subtle beauty of the outback and camping appeared to be part of his nature. His mobile conodont lab and field processing of samples meant that field seasons could easily be up to 3 months long. As long as the tea, beer and meat were there, the work would go on." In the Bonaparte, he found conodont faunas restricted to particular parts of the Devonian reef complex. And although Glenister and Klapper had reported no conodonts from the back-reef, Druce found a fauna there dominated by *Pelekysgnathus* and other conodonts not utilized in Ziegler's standard. The fore- and inter-reef were in contrast populated with high numbers of *Palmatolepis*, *Polylophodonta*, and *Scaphignathus*. Here *Pelekysgnathus* was absent.[15] Although these findings were present to some degree in his published data, it was only later that he gave these distinctions clarity. Not long after completing this work, Druce entered the Canning Basin, collecting material from the Devonian reef complex for a doctorate supervised by Frank Rhodes, who had recently moved to the University of Michigan.

As Druce's study of the Bonaparte Gulf went to press, George Seddon was completing a four-year consultancy for WAPET on the Canning Basin.[16] Seddon was unusual in being a member of the Departments of Geology and Philosophy at the University of Western Australia. He would later be celebrated for the range and significance of his work, but in 1970, his mind was on conodonts.

Both Druce and Seddon sought to resolve the outstanding stratigraphic problems of a region that could be linked to Europe but within which many rocks remained uncorrelated because of complexities presented by the presence of reefs. Seddon's aim was not simply to superimpose Ziegler's system on these Australian rocks but to consider the significance of those conodonts that had gone unmentioned in the German scientist's work. Seddon soon came to understand, as Glenister and Klapper had before him, that the distribution of conodonts was controlled by the different environments that made up a reef. Like Druce, but independently, he found two distinct, environmentally controlled, faunas characterized by different form genera. One fauna was typified by Branson and Mehl's favorite, *Icriodus,* together with *Polygnathus* and *Pelekysgnathus.* The other was dominated by Ziegler's most useful *Palmatolepis,* along with *Ancyrodella* and *Ancyrognathus.*[17] The *Palmatolepis* fauna contained examples of the *Icriodus* fauna but in lower abundance. With few exceptions, the *Icriodus* fauna did not include those key genera from the *Palmatolepis* fauna. Seddon imagined the existence of a one-way filter through which *Icriodus* and associated forms could pass but through which *Palmatolepis* could not, though at the time he did not identify what that filter might be. The *Icriodus* fauna was found in the fore-reef but close to the reef itself, while the *Palmatolepis* fauna occurred seawards, from the fore-reef to the inter-reef.[18] *Palmatolepis* was, of course, the key to Ziegler's universal standard, but *Icriodus* was not. However, Seddon's discovery of a filtering mechanism enabled him to relate the two and determine the age of his rocks. To achieve this he created a parallel *Icriodus*-based chronology recognizing that this represented a rather specialist environment. Seddon's paper was published in 1970. It acknowledged Druce's helpful critique of his work, though it seems possible that Seddon knew little of Druce's thinking.

In May that year, both Seddon and Druce presented their findings at the Michigan meeting of the Pander Society. Druce could now talk generally about the depth control of his two faunas and locate parallel examples in the lowest Carboniferous (figure 9.1a).[19] The "exotic and bizarre," he noted, were restricted to deeper waters but had evolved from shallow-water stock. These deeper-water forms were generally more diverse, and it was for this reason they had been so useful to Ziegler and Helms. Druce also noted the same one-way mixing of faunas. That Ziegler's system was again called into question was not lost on the audience, particularly on Ziegler himself. Druce advised his colleagues to record *Icriodus* and other shallow-water forms in order to counter these weaknesses in Ziegler's standard.

Thus far Seddon and Druce had been working in parallel, and broadly with the same geographic, stratigraphic, and intellectual goals. By 1970, conodont ecology was becoming a hot field, not least because it tested the idea of the universal conodont. Increasing numbers of workers began to develop similar and overlapping topics of study. Competition and debate increased. This parallel production of data and knowledge meant that individuals tended to know different things, and this did not depend solely on personal networks and research groups. Some would gather new data or ideas by attending meetings while others would not receive this information until the paper was published – which might be years later or not at all. Some important studies were published merely as short abstracts of orally presented papers. The abstracts were hardly scientific arguments, but they acted as important markers and were often referred to. In 1970, both Druce and Seddon heard each other speak and both published short abstracts of their papers.[20] Druce made a commitment to write up his paper for the book arising from the conference, but this book would have such a catastrophically delayed gestation that the paper would not appear for three years. And when it did, it was not the paper Druce presented at the meeting but one that reflected upon that meeting and Seddon's presentation there. We shall come to that paper shortly.

Seddon, who produced a rather different interpretation of the data for the 1970 meeting, chose not to publish in the conference volume. Instead, he teamed up with Sweet to explain his filtering mechanism in more detail. In order to do so the two men looked for a "likely ecologic analogue" and chose the chaetognaths, or arrow worms. These are typically carnivores, three centimeters long, "that spend their entire existence floating or swimming in the water without relation to the bottom."[21] Although Seddon and Sweet suggested no direct relationship between conodonts and these animals, they could clearly talk of the analogue darting forward to capture prey and visualize their own animal doing the same. There were other similarities too, as one authority considered the chaetognaths something of an enigma: "They may be the most isolated group in the animal kingdom." Doubtless their "paired batteries of anterior and posterior teeth, and grasping spines" were sufficiently unique to support Seddon and Sweet's contention that they had before them a "suggestive analogy." In doing so, they drew upon a number of books to act as their authorities on these animals, many of which came from the 1950s. These sources told them that the animal's wide distribution was controlled by temperature, salinity, and available food, and that species were vertically stratified: Most species occurred in water depths of less than two hundred meters, but more specialized forms could be found between two hundred and one thousand meters, and a few even beyond that depth (figure 9.2b). This suggested a possible mechanism for the operation of the biological filter that affected conodont distribution. It had long been suspected that *Icriodus* was a shallow-water form but now this genus could be visualized as occupying surface waters, overlying the deeper waters where *Palmatolepis* swam or floated. On death, conodonts elements would sink to the seafloor. Those that accumulated in deeper waters would as a result contain both shallow- and deep-water forms.

Drawing upon the contemporary literature, Seddon and Sweet could generalize further and suggest that the shallow zone contained just a few unspecialized conodont species, while at depth, where the environment was more stable, there was greater diversity. Sweet's Ordovician then became the testing ground for this new depth-stratification model, but here Seddon and Sweet could not call upon the relative simplicity of the reef model to infer water depth or relationships between communities. They were, of course, dealing with fundamentally different genera, but they still felt that within the different provincial faunas it was possible to detect this two-way, depth-controlled division of genera. Indeed, they thought the ratio between two particular genera might provide a crude index of water depth, which could be confirmed by lithological data. As Seddon and Sweet were developing this idea, others reported Seddon's distinctive *Icriodus* fauna in various parts of the United States, demonstrating its wide distribution.

When Druce's paper finally did appear, two years after Seddon and Sweet's, it showed the influence of Seddon's work. It further distinguished a third (*Belodella*) fauna, which Seddon had included with *Icrio-*

dus but Druce thought occurred still closer to the reef. Druce sought to test and extrapolate his model across the whole of the Upper Paleozoic and into the Triassic, aware that very similar apparatus architectures and element morphologies had repeatedly evolved across long periods of time. Homeomorphy, the repeated appearance of identical forms as a result of environmentally determined convergent evolution, suggested this close link between the form of an animal and the nature of the environment.[22] This kind of morphological convergence had been seen across the natural world in everything from mollusk shells to teeth. It was reasonable to expect it in conodonts even if the animal remained unknown and unimaginable. How wonderfully informative these tiny fossils would be if this was the case; find a microscopic fossil, and there, in your hand, you have the key to understanding the world in which it lived. Druce was not alone in thinking in this way; Ronald Austin compared Walliser's Silurian species to his Carboniferous forms and, surprised by how similar they were, had similar thoughts.

Ecological models reflected this desire for generalization: They emerged almost subliminally from the data and would then perform as spectacles through which others would look at fossils and the rocks that held them. These workers might accept this new way of seeing, cast these spectacles aside, or make improvements to them. Druce chose to do the latter. He was sure that Seddon's model was better than his own but still not perfect. He knew the *Palmatolepis* fauna was more widespread than that of *Icriodus,* but Seddon's model predicted the opposite. So Druce refined it, concentrating populations toward the shore and effectively superimposing his earlier model on Seddon's (figure 9.1c).

In 1970, the appearance of *Icriodus* and other Devonian conodonts in various parts of the world raised a number of questions about the faunas Ziegler was finding. His collections inexplicably lacked these fossils. Whichever of the models was correct, there was good reason to believe these conodonts should also occur in Germany, and not simply because the conodont was celebrated there as the universal fossil. Druce wondered if some kinds of conodont had not been recorded because they were less exotic, often long-ranging, and therefore less useful to stratig-raphy. Austin and Rhodes had similar suspicions and began to doubt Ziegler's data. They had found that British, Irish, and Belgian conodont faunas were similar to one another but different from those Ziegler had published. (Later these faunas would turn up in Pakistan, Russia and China.) Ziegler found himself increasingly under assault on this point, and attention now focused on the methods he used to process his fos-sils. Doubts had been raised about Beckmann's method, which Ziegler used, as early as 1964. At the time, Beckmann, Ziegler, and all the other British, American, and German heavyweights in the field closed ranks and rejected these claims. But with the identification of statistical as-semblages a few years later it became increasingly obvious that Ziegler's faunas were incomplete. The arrival of this new shallow-water *Icriodus* fauna simply accentuated these doubts.

Ziegler remained resistant to change until Lindström moved to Marburg and used Beckmann's method on his Ordovician conodont samples. To his great surprise, Lindström found the conodonts seriously affected. Indeed, the results were quite spectacular. From two hundred grams of limestone, using Beckmann's method and leaving the sample bubbling away for two weeks, Lindström obtained just six badly cor-roded cone fragments. Repeating the exercise but recovering the residue every two days, he obtained eight hundred identifiable specimens, a few of which were corroded. This dissolution of conodonts was selective, bar-like fossils being particularly vulnerable. This seemed to explain why these elements were missing from Ziegler's collections. In contrast, the platforms, which so interested Ziegler, were much more resilient. But surprisingly, this discovery did not greatly alter Ziegler's use of acids; he simply removed his fossils from the acid solution more regularly. He believed this was fine for stratigraphic work in which only the platforms were required but that where more detailed paleobiological studies were to be carried out, acetic acid should be used.[23] Ziegler was perhaps typi-cal of most conodont workers: If they could find sufficient specimens, they tended to believe the method was fine.

Others were more inclined to frown at Ziegler's disregard for pro-ducing accurate samples. One in particular, Lennart Jeppsson, would take the opposite approach and go to great pains to improve his methods and reduce the risks of loss or distortion. Jeppsson needed to process huge quantities of limestone to get just a few specimens. Sometimes he was literally looking for needles in haystacks.

The ecological models that soon became well known in the conodont research community did not go unchallenged. Glen Merrill, for example, had argued from the early 1960s that conodonts were environmentally controlled. For him, the universal animal was an illusion resulting from inadequate sampling. Having submitted a paper on the subject to the same volume that would eventually publish Druce's, he felt compelled to recall it so he could criticize Seddon and Sweet's model. Merrill's contribution was to point to a phenomenon he had seen in Pennsylvanian faunas and that Chalmer Cooper had briefly mentioned back in 1947. (Indeed, it was not unlike that distribution of fossils Rexroad had spotted in the late 1950s.) Merrill could show that alternative abundances in two platform genera depended upon lithology.[24] One was dominant in shales, the other in limestones. The relationship was both consistent and remarkable.

Merrill knew that the shales were produced near the shore while the limestones represented fully marine conditions farther out to sea. The rocks representing these two different environments were stacked one upon the other, which Merrill suggested represented fluctuating sea levels. As the sea level rose, the shoreline would move inland and these two environments and their conodonts would follow.

Merrill then asked himself how rapidly these changes took place. He could answer that question by using an old idea that required a little lateral thinking. On the presumption that animals were dying at a constant rate and falling to the seafloor to form fossils, high concentrations suggest that the rate of sedimentation has been low and a thin layer of rock represents a long period of time. Finding these fossils heavily diluted by sediment could then be interpreted as evidence of rapid sedimentation. Thus in a rock section, time may be both condensed and stretched. A rough measure it may be, but it permitted Merrill to consider how rapidly the sea had transgressed the land. He found this often occurred so rapidly that it prevented the shale fauna from developing locally. The shallowing (regression) of the sea caused by the growth of a delta was, however, slower, and in these circumstances the expected fauna could develop. Merrill thought he was looking at shallow and very shallow water communities that were not greatly distinguished in terms of water depth. Instead, he believed salinity a controlling factor as the shales contained brackish-water fossils while the limestones were fully marine. He deduced that the animals actively maintained a link to the environment and did not, like chaetognaths, simply waft around in the currents: "The conodont governed its own occurrences in a much more direct manner. . . . It was nektonic; an active organism fully capable of exerting important controls upon its own distribution, depth and destiny." Seddon and Sweet's model simply did not fit the bill.

Merrill believed that the richness of modern data was permitting conodont workers to see subtlety and complexity, displacing beliefs in the universal. He put it metaphorically: For so long it had been possible only to hear the orchestra; now they were beginning to hear the individual instruments themselves, even if "the composition and the composers are unknown."

In the 1970s the science possessed for the first time the makings of a global database of information on conodont distribution in space and time. It was already altering perceptions. For example, simple cone-shaped conodont fossils, which were considered mere contaminants in the Devonian in 1957 and questioned as such two years later, were in 1973 mapped by as a true but progressively declining component in that fauna.[25] In this incremental way the whole data set was changed and extended, knowledge shifted, and interpretations were reformulated according to new and, as we shall see, increasingly spatially aware criteria.

This new trend in conodont studies really took off as geology in general embraced the unifying theory of plate tectonics. It was a theory that gave the earth a fluid geography of mobile continents and spreading oceans. It encouraged the sometimes near-sighted paleontologist to think big as it gave new explanation to the comings and goings of oceans and the appearance and disappearance of animals and plants over time. The idea that the continents drifted over the surface of the globe had been debated for much of the twentieth century, but it was only in the late 1960s that all the components of the theory were locked together into a convincing whole. It became the new paradigm. But not everyone felt the need for an imposing theory. The science had, after all, gained nearly all its intellectual possessions by hard graft in the field or through instrumentation. Conodont worker Anita Harris – a onetime PhD student of Walt Sweet – certainly belonged to this proud tradition. Back when the theory was still fresh, she told popular geological writer John McPhee, in a book in which she was the main actor, "The plate-tectonic model is so generalized and used so widely. . . . People come out of universities with PhDs in plate tectonics and they couldn't identify a sulphide deposit if they fell over it. Plate tectonics is not a practical science. It's a lot of fun and games but it's not how you find oil. It's a cop out. It's what you do when you don't want to think."[26] Conodont science was indisputably practical, and Harris had by then, as we shall see, made a major contribution to that practicality. She had done so by paying her dues in the field, not by theorizing. Her comments indicate how fundamentally geology was changing, and she in time would adjust to it once all the hullabaloo settled down. The new tectonic theory rapidly found its way into the interpretative armory of practical geologists, particularly those concerned with the spatial dynamics of the earth. Leading paleoecologist James Valentine at the University of California in Davis, for example, combined the new theory with evolutionary concepts to revolutionize the way paleontologists thought about the distribution of life over time.[27] Drawing on these new theoretical and interpretive resources, it was again possible to think differently. Geology received a shot in the arm as the developing theory gave explanation to phenomena that had long been puzzling. Perhaps it was coincidence, but just as the conodont workers began to possess global data about their fossils, so the geology as a whole started to think global thoughts.

Chris Barnes of the University of Waterloo in Ontario was certainly among those who thought these new theoretical ideas powerful interpretive tools. Valentine's work was in his mind when he teamed up with Rexroad and Cambrian specialist James Miller of the University of Utah to consider the global pattern of conodont communities and provinces during the period when these communities first appeared and experienced explosive diversification.[28] They reasoned that for this diversification to have occurred there must have been ecological niches available for exploitation by different species. Valentine's refined definitions of province and community suggested a spatial continuum of environments and a way to relate diversity to a changing world. It gave them a means to reinterpret Bergström and Sweet's provinces and subprovinces and attempt to locate communities within them. These communities, they believed, were distributed as a series of lateral bands parallel with the shore and extending into deeper water. Using the brand-new understanding of continental margins, they set these bands in the context of plate-tectonic theory. It gave their ecological niches a theoretical rationale and a new global context. But Sweet and Bergström simply could not agree and recalled a paper in press to add a short addendum critical of this interpretation. It was, however, a sign of the times: "The concepts and ideas of the theory of global tectonics have proven to be a virtual panacea for geologists."[29]

Many of the ideas developed in this first paper were extended and given greater clarity when, three years later, Barnes and Lars Fåhræus of Memorial University of Newfoundland in St. John's reviewed the growing literature to propose a "unifying model of the major habitats" of Ordovician conodonts.[30] Seddon and Sweet had imagined a weakly swimming or floating animal, but Barnes and Fåhræus reassessed the evidence and suggested a bottom-dwelling animal that was in control of its own movements. The lateral banding of fossils was so clear, they felt it was possible to predict neighboring communities. There were, they admitted, some simple species that transgressed these bands that probably represented swimming forms, but these were less numerous (figure 9.1e).

Barnes and Fåhræus then took Bergström and Sweet's two provinces and rotated them in order to understand their relationship in Ordovician times. East-west now became north-south. The Appalachian or North Atlantic Province was understood to represent a normal marine environment with a "virtually cosmopolitan" fauna. The midcontinent fauna was, by contrast, seen as being restricted to a "fairly narrow equatorial belt" and adapted to higher temperatures and salinities.

Using these ideas, they now reimagined Bergström and Sweet's panoramic view of mobile populations of animals with ideas that integrated time and community with the province. The controlling factor was a relative change in sea level. In the lowermost rocks of the midcontinent, the animals held simple cone-shaped elements. They showed no lateral segregation. However, during the Middle Ordovician, major transgressions flooded the midcontinent from the direction of the Appalachian province. Habitats and communities diversified and ecologically specialized faunal belts developed. In this manner, Barnes and Fåhræus gave a sophisticated and detailed reading of the history of a group of animals experiencing environmental change. They, too, were sure that the form and shape of individual conodont elements and assemblages reflected particular lifestyles. Increasingly, they felt it was possible to talk with reasonable certainty about the life requirements of particular genera and species. Drawing upon a sizable literature, they had pieced together a rich, complex, and changing conodont world. One could almost imagine the animal.

Barnes and Fåhræus's paper appeared in April 1975. A month later, Barnes played host to a paleoecology-themed meeting of the Pander Society in Waterloo, Ontario. The meeting showed how the field had diversified.[31] The idea that environments could be read merely from the shape of these tiny fossils was then in the ascendancy.

One of the more extraordinary papers was by Jeppsson, who had been inspired by science fiction writer and acclaimed Ice Age mammal specialist Björn Kurtén. Kurtén had attempted to go beyond the descriptive essentials of paleontology, measuring bones in order to interpret communities. Jeppsson transferred the method to the deep past and took to measuring almost immeasurably small processes on these tiny fossils. He considered the possibilities of the animal having a spawning season and seasonal migrations.[32] Barnes thought the paper both speculative and intriguing. It was typical of Jeppsson's individualistic approach to his science but it also marked a wider desire to give these animals biological clothes of some kind.

While an increasing number of conodont workers were becoming rather theoretically minded, Dick Aldridge of the University of Nottingham in the UK was not alone in treating his piece of geological time with historical specificity. Believing each period to be distinctive and affected by particular influences, Aldridge thought the existing models rather simplistic in their assumptions of cause and effect when so many environmental factors were known to be at play.[33] Many others discussed the various models, though the latest by Barnes and Fåhræus was simply too new to attract much attention. Opinion remained mixed; some testing had taken place, but it was inconclusive.

Karsten Weddige and Willi Ziegler, being dissatisfied with the explanatory power of the models produced to date, came up with yet another (figure 9.1d). They did not believe that depth and distance from shore were responsible for the observed distributions of fossils. Indeed, the models seemed too static; the ocean was a dynamic environment and other factors were at play. *Icriodus,* they said, preferred turbulent, oxygen- and carbonate-rich waters, while *Polygnathus* preferred quieter conditions associated with muddier sediments.[34]

Riding the wave of ecological optimism, Fåhræus and Barnes rushed a second paper into print in the distinguished science journal *Nature.* Their vision was nothing short of oceanographic. The conodont was to be a tool in "studies involving the extent and relative depths of sedimentary basins; for unraveling patterns of transgressions and regressions; as aids in the recognition of depositional environments characterized by raised temperature and salinity; and in understanding palaeogeographic and tectonic changes."[35] They demonstrated this by examining the relative abundance of two genera thought to occupy adjacent and overlapping communities: *Phragmodus,* which occurred offshore, and the nearshore form, *Plectodina.* The changing relative abundance of these in a geological section would indicate transgressions and regressions. It was a view not far removed from Merrill's. But Fåhræus and Barnes now saw the conodont as a precision tool for plotting major global change: "Thus, initial destruction of this ancient continental margin can be dated with considerable precision." In taking this view, Barnes acknowledged the debt they owed Sweet and Bergström whose data they had turned to their own ends.

＊ ＊ ＊

Barnes and Fåhræus were not alone in considering major events in the earth's history. Across the Atlantic, Otto Walliser was having similar thoughts. The science was continuing to change, and as it did so it would move the cutting – or at least fashionable – edge away from ecology. The late 1970s marked a point of reflection for those who had looked at palaeoecology as holding promise. The conodont workers certainly felt they had achieved new understanding, but this was not the view of Smithsonian foraminifera specialist Martin Buzas when he reluctantly took up the task of reviewing the book that came out of the Waterloo conference. In that review, titled "On the edge of the unknown," he admitted to knowing nothing of conodonts or of the rocks in which they are found. He could, however, reflect upon the distribution of foraminifera in rocks, and he remarked how little indication this gave of their distribution in life. He wondered, given the data they possessed, how conodont workers could deduce an open-water swimming lifestyle and distinguish it from an animal that lived on the seafloor: "I conclude conodonts were some strange animals but don't venture a guess as to where they lived."[36]

The initial optimism that had accompanied the rise of paleoecology had begun to wane. One major textbook noted, "Paleoecology, which during the 1960s occupied center stage in the paleontologic theatre has matured as a subdiscipline but has also lost some of its luster; appreciation of the incompleteness of the invertebrate fossil record has led to a general narrowing of goals in the study of ancient marine communities." Joel Hedgpeth, who had pioneered the field in the 1950s, found ecology undergoing significant change and palaeoecology often in possession of old ideas, unable to keep up.[37]

In 1978, Gil Klapper and James Barrick also asked whether it really was possible to infer lifestyle from the distribution of conodont fossils in the rocks.[38] Reviewing a number of marine animals, they became convinced one could not. Seddon and Sweet's favored chaetognath analogue, for example, did not show simple depth stratification after all but reflected the complexities of temperature and salinity, which actually produced *lateral* variation. Indeed, bottom dwellers and swimmers were

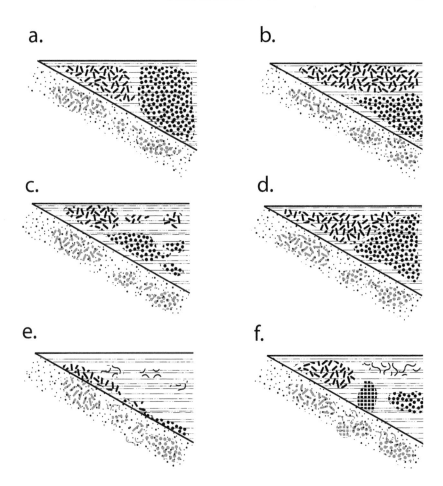

9.1. Modeling the ecology of the animal. Deducting lifestyle from the distribution of fossils in the rocks. The triangles represent the sea in section, showing increasing depth and distance from shore. The patterns represent different genera of conodont animal: (a) Druce saw lateral distribution in his reef limestones; (b) Seddon introduced a one-way depth-controlled filter; (c) Druce adopted Seddon's model but argued that nearshore population densities were higher; (d) Weddige and Ziegler saw water clarity and oxygenation as differentiating *Icriodus* (*left*) and *Polygnathus* (*right*); (e) examining other environments, Barnes and Fåhræus saw clear banding of animal communities, suggesting they were bottom dwellers; and (f) reviewing the evidence in 1978, Klapper and Barrick showed that the distribution of marine animals was governed by complex factors and that the record of the rocks could be explained in multiple ways.

capable of leaving the same record. Only the presence of conodonts in black shales, which were devoid of bottom dwellers, strongly suggested a swimming or floating animal of the open sea. They concluded that since conodonts seemed to be confined to the continental shelf, it would be better to draw on modern analogues to visualize the ways in which they might have been distributed in life. These reflected changes in key environmental variables in relation to the coast (figure 9.1f). Like many living animals, conodonts had a few specialist species able to survive the difficult conditions of the nearshore; their diversity increased away from the shore. This was another model, but it removed the need for a bottom-dwelling lifestyle yet could still produce the observed lateral changes and overlaps. It also captured that reciprocal relationship between genera, which Weddige and Ziegler, Merrill, and even Rexroad had written about.

There were, then, repeated attempts to adjust the theory to the limitations of the data but no wholesale retreat from a line of enquiry that had certainly delivered results. New data continued to reinforce Bergström and Sweet's provinces. Indeed, provinces became established in other periods, too, but turned out not to be universal. In the Devonian, at least, it was understood that conodonts were probably confined to tropical latitudes. Of these, *Icriodus* returned to favor, finding its own parallel zonation to mirror the one Ziegler had developed using *Palmatolepis*. This work revealed that *Icriodus* had become extinct rather earlier than assumed and that an impostor had evolved from *Pelekysgnathus*. The first *Icriodus* was quite well distributed in the upper, highly illuminated waters of the coastal zone, but its impostor, known as "*Icriodus*," was more restricted in its distribution.[39] A reversal of expectation had taken place. Having believed in universalism, workers increasingly expected environmental control.

Near it in the field, I remember, were three faint points of light, three telescopic stars infinitely remote, and all around it was the unfathomable darkness of empty space. You know how that blackness looks on a frosty starlight night. In a telescope it seems far profounder. And invisible to me because it was so remote and small, flying swiftly and steadily towards me across that incredible distance, drawing nearer every minute by so many thousands of miles, came the Thing they were sending us, the Thing that was to bring so much struggle and calamity and death to the earth. I never dreamed of it then as I watched; no one on earth dreamed of that unerring missile.

H.G. WELLS,
The War of the Worlds (1898)

TEN

The Witness

THROUGH THE 1970S, PALEONTOLOGY ACQUIRED AN increasingly global outlook as geology as a whole embraced the unifying ideas of plate tectonics. The conodont workers felt this sense of the global even more profoundly as its field of study spread to every corner of the earth. In this period, the living animal became a mobile entity inhabiting clearly defined niches and repeatedly evolving similar anatomies to deal with the return of particular environmental conditions. Progress for the conodont workers, as for most of paleontology, had been logical and incremental. But then two unexpected events forced them to look and think differently, and even to imagine the unimaginable.

The first, which occurred in 1974, resulted in the discovery of unforeseen utility in the conodont fossils' strikingly varied colors. It indicated that even when dead, buried, and fossilized, this remarkable animal could bear witness to changes going on around it. The second event was considerably more dramatic and occurred at a precise moment in 1980: An asteroid came crashing in, turning the paleontological community upside down. No one saw it coming, but no one could ignore it. Scientists from many different fields came together to understand it and its consequences. In that drama, the conodont played a bit part, valued particularly for the manner of its dying. In this period the geological community as a whole entered its most speculative and imaginative phase, and conodont workers, who were never immune to outside influence, soon developed a wonderful facility for thinking fantastic thoughts.

In this chapter we will explore how this thinking continued to shape the animal in its world. In the next chapter we will begin the final phase of this book and start to follow the scientists as they close in on the animal itself.

In a science so attuned to shades of brown and gray, the conodont's yellows, oranges and blacks, when combined with their beautiful translucence and miniscule but finely detailed form, had an aesthetic effect on all who studied them. These facets contributed to the objects' attractiveness and amazingly – when you think about it – encouraged so many people to place these objects at the center of their lives. Pander had, at the outset, thought the color of conodonts both remarkable and important, but no one made anything of it until the U.S. Geological Survey's Anita Epstein, later Anita Harris, did so in the late 1960s.

Epstein – a product of Sweet's "conodont factory" – owed rather more of her character to her origins in the tenements of the Williamsburg neighborhood of Brooklyn. The daughter of an immigrant Russian Jew, she saw geology as a means of escape from the poverty of New York. Tough and determined, she talked geology as one who lives and breathes it. "She is one sharp cookie, as we say," Sweet remarked. "She has a photographic memory and is as hard a worker as anyone I know." In his road trip with Epstein, described in his book *In Suspect Terrain*, published in 1982, John McPhee introduced the conodont, the subject of Epstein's great innovation, to his general readers using descriptive prose that seemed to be haunted by the ghost of Charles Moore: "At a hundred magnifications, some of them looked like wolf jaws, others like shark teeth, arrow heads, bits of serrated lizard spine – not unpleasing to the eye, with an asymmetrical, objet-trouvé appeal."[1]

In 1967, Epstein had noticed that there was a correlation between the color of the conodont fossils she was finding and the depth to which they had been buried. By buried, here, I mean geological burial beneath perhaps hundreds or thousands of meters of rock. As the conodonts showed a color range much like that seen in butter heated in a pan, she imagined that the color of the fossils might be used as an indicator of the maximum temperature experienced by the rocks in which they were found, as temperatures underground increase with depth. However, finding no encouragement at the Survey, she quickly dropped the idea. It was a chance meeting and conversation with Leonard Harris (later to be her second husband), six years later, that re-awoke this thought with something of a start. An oil geologist, he told her that oil companies had been using color changes in pollen and spores in this way for many years. The color range was exactly like that Epstein had seen in her conodonts. This was a moment of revelation. But when she told him that conodonts performed the same trick, it was he who was surprised.

Epstein's discovery meant that this abundant and extraordinarily ancient time marker might possess a wholly new dimension of meaning. In an era of oil shortages, if Epstein's hunch proved correct, the conodont would offer an easy means to locate rocks that might hold oil. Oil forms from marine algae at depth in rocks exposed to certain temperatures. If the temperature is too low, oil does not form; if it is too high, oil is lost. Epstein now spent much of her spare time experimenting on relatively unaltered conodonts from Kentucky supplied to her by Stig Bergström. These were the same conodonts he and Sweet had used in their groundbreaking 1966 paper. Epstein heated the conodonts to temperatures from three hundred to six hundred degrees Celsius over a period of ten to fifty days, removing samples at regular intervals to record their color. She found that the fossils' color altered in a "progressive, cumulative, and irreversible"[2] way and was the direct product of time and temperature. Reassuringly, the colors produced in those she had cooked were exactly like those found in the field.

Working with her husband, Jack, and Harris, she extrapolated this experimental data so that the relationship between temperature and color could be understood on the scale of geological time. Now the color of the conodonts could be used to indicate the burial history of the rocks that contained them. Announced to the world in 1974, with full results published in 1977, this discovery changed her life and expanded the meaning and importance of the fossil considerably: "This study increases their use from index fossils to metamorphic indexes and demonstrates their application to geothermometry, metamorphism, structural geology, and for assessing oil and gas potential." The scale of color change became known as the "color alteration index," or CAI, and it drew in new kinds of conodont workers. This was yet another case of these tiny things participating in science on the grand scale and, in this case, in science that was really useful to everyday concerns. A fascinating episode in and of itself, which opens up an entirely new story arc, we must, however, leave it here because it tells us nothing of the animal itself. CAI records only aspects of the postmortem existence of the animal. We must turn our attention now to the world of the living animal, though in doing so we will primarily concern ourselves with its death.

The 1980 asteroid, or meteorite, that once struck earth and now, in a metaphorical sense, impacted the scientific community was delivered by Nobel laureate physicist Luis Alvarez and his geologist son, Walter, along with chemists Frank Asao and Helen Michel. It arose from work Walter Alvarez had been conducting at the Cretaceous-Tertiary boundary at the ancient town of Gubbio in Umbria, Italy. Measuring the trace element iridium, which is constantly falling to Earth as meteoric dust, he postulated that its degree of dilution in marine sediments would indicate how rapidly the sediment had been deposited. The sedimentary dilution of a constant – such as Merrill's dying and fossilized animals, discussed in the previous chapter, or meteoric dust falling to Earth – had long been used to deduce the relative rates at which rocks were laid down. The sediment that interested Alvarez marked the end of the Cretaceous, that remarkable moment when the dinosaurs and many other kinds of animal became extinct. However, it was not dinosaurs that first caused Alvarez to stop and think, but the near extinction of those tiny amoeba-like animals with delicate and intricate shells known as foraminifera. At Gubbio, a centimeter-thick layer of clay divides the extraordinarily different foraminifera of the Cretaceous from those of the overlying Tertiary. Alvarez asked, "Has this mass extinction occurred in a human timescale or a geological one?" In order to answer this question, he needed to know how rapidly the clay had been deposited.[3]

As it turned out, the iridium performed better than he had hoped, for rather than simply giving Alvarez the rapidity of change, it also gave him a cause. What he found were extraordinarily high amounts of this element (in relative terms at least). After much deliberation, the team felt this could only be explained by a huge meteorite impact. Astronomy, that esoteric science of other worlds too distant to really know, now became central to understanding the history of life on Earth. Newly globalized, geology now found itself a science of planets.

The idea was a "bombshell" that caused an explosion of papers and conferences – and some bizarre theoretical imaginings, including a death star called Nemesis and an equally deadly Planet x. Supportive speculation and doubting cynicism developed in parallel, but increasing amounts and types of data seemed to confirm this radical and seemingly improbable alien visitor.[4] The impact on science was so great that it is easy to imagine geologists now talking new talk and thinking thoughts that had never previously crossed their minds. But this was not entirely what happened. Their initial response was to work with what they knew, to marry this new idea with existing data. In this respect the asteroid became a new pair of interpretive spectacles through which to look afresh at old things. These glasses would also encourage geologists to seek out obscure and esoteric work that at one time seemed to make little sense. Perhaps it would do so now. In some cases, yesterday's nonsense and self-indulgence suddenly became prophetic. Some on the edge of the community now found themselves treated as oracles and placed at the center of this new debate.

Otto Walliser was among those who welcomed the meteorite, or rather the change of thinking it brought about. He never found a need for the meteorite itself. It might be recalled that Walliser had played a singular role in establishing the conodont in Silurian stratigraphy in the early 1960s. With no good Silurian sections in West Germany in which to continue these studies, he simply left the field. He had moved to Göttingen, and while he retained an interest in conodonts, he did so no more than in the cephalopods that had first attracted him to the science. Trained by Otto Schindewolf, who for a time was the leading light in German paleontology, Walliser had acquired an interest in the global aspects of the science long before the Alvarez meteorite hit. Indeed, back in 1954 and again in 1962, Schindewolf had suggested that cosmic radiation from a supernova could cause catastrophic extinctions of life on Earth. Such extinctions, he said, would be followed by a burst of evolution and diversification among the survivors. Schindewolf claimed that these catastrophic events could be read in the rock record because that record was sometimes complete; there were no gaps in the sequence at those moments of catastrophe. At the time, the idea of mass extinction was too much for many paleontologists. Extraterrestrial causes simply added to a sense that this was mere fantasy. To their eyes, species were lost gradually – as they seemed to be at the present day. Any apparently sudden loss was merely an artifact of missing or eroded strata. Unsurprisingly, then, Schindewolf's views did not find much support, even among those, like Norman Newell, who did much in the 1960s to demonstrate the reality of mass extinctions. Newell had taken a rather different approach and had plotted the diversity of life against time to reveal the truth of these extinctions, though the coarseness of his methods concealed the true prevalence of such events.[5]

Walliser was not convinced by Schindewolf's explanation either, but he knew Schindewolf's data were good and that the phenomenon of global extinction he described was real. That fascinated him and soon began to affect the way he interpreted rocks and fossils in the field, particularly after 1965. That year he visited Iran and discovered a Devonian sequence that was lithologically and paleontologically identical to that back home. He thought this remarkable because it meant that particular changes of environment must have taken place across an extraordinarily wide area, perhaps even globally. What especially fascinated Walliser was the fact that the rocks themselves were capable of showing this change. He now understood that seemingly local phenomena, such as nodules in Devonian strata, were not local at all and that the peculiar characteristics of particular beds in one locality – such as their predisposition to slip and slide (a quality he tested by chewing the rock!) – could be global in their distribution too.[6] Walliser's fascination with these unexplained global phenomena grew.

As he traveled farther afield, particularly to Asia, so his initial findings would be confirmed. He became a connoisseur of rock sequences,

increasingly convinced that nature preserved natural global markers recording moments of transformation in the earth's history. However, his
views were not shared by those more utilitarian stratigraphers engaged
in dividing rocks up into neat, globally recognized parcels. Walliser became an outspoken advocate for locating and using natural boundaries
in rock sequences. His critics preferred to drive their boundary-defining,
and metaphorical, "Golden Spikes" into sequences where nothing much
happened but where the replacement of one species by another could be
recognized globally. He did have some early successes, however, such as
in setting the major Silurian-Devonian boundary at a meeting in Bonn in
1960. But such victories were rarely permanent. Mass extinctions were often associated with difficult lithologies and incomplete rock sequences;
even an imaginary Golden Spike cannot be driven into rocks that are not
there. Consequently, throughout the 1960s, the artificial scheme gained
ground and was adopted by all the various grandly named subcommis-
sions that sought to define and adjudicate on these global boundaries.
Walliser found himself in the minority, frustrated by the victory of utilitarianism over nature.[7] In the story of the conodont, of course, this is not
an unfamiliar theme.

Walliser was not entirely alone. In 1969, his friend Digby McLaren
of the Geological Survey of Canada was at last willing to admit to the
vital importance of these natural boundaries where a range of unrelated animals and plants became extinct, adding, pointedly, "[boundar-
ies] which we are trying to define out of existence." In his presidential
address to the Paleontological Society that year, McLaren said these
boundaries were of two types. The first was quiet and merely man-made
for convenience, but the second recorded some "event" "across which
something happened." It was to explain a boundary of this second kind
– a major extinction in the Devonian known as the Kellwasser Event,
which he had first recognized in the 1950s – that he "landed a meteorite
in the ocean with effects that had been described by [Robert] Dietz in
an article in *Scientific American*" in 1961: "Dietz . . . suggests a giant meteorite falling in the middle of the Atlantic Ocean today would generate a
wave twenty thousand feet high." McLaren pondered the consequences
of a similar event for his now extinct animals, and acknowledged, "This
will do."[8]

Given the reception Schindewolf's alien causes had received, Mc-Laren knew that he risked being labeled a crank. The preferred explanation for mass extinction was the rise and fall of sea level resulting from ice ages and major vertical movements in landmasses. (We need to remember that the vast majority of fossils – and therefore recorded extinctions – are of sea animals.) To this audience, then, McLaren's meteorite was unexpected and unneeded. The response to it is perhaps typified by one English contemporary, Michael House, who believed McLaren had said this "doubtless with tongue in cheek." House viewed the problem through spectacles constructed from geology's most important guiding principle, uniformitarianism, a belief that the world of the past was created by the same gradual processes we continue to see today. Science neither needed nor had a place for catastrophes of this kind. But this was not how McLaren saw it at all: "I do not believe this explanation is far-fetched.... We must look for more than everyday happenings to explain many geological features."[9]

It was, however, simply speculation; there was no evidence of an impact, even if some contemporary astronomers – beyond earshot of the geological community – thought sizable meteorites must have collided with the earth in the past.

These late 1960s discussions of extinction caught the attention of others in the conodont community, forcing them to think new thoughts. Dave Clark, at the University of Wisconsin in Madison, for example, saw in extinction an opportunity for separating true species from their mimics and imitators. This insight arose from work that Clark's doctoral student, James Miller, was undertaking to disentangle the evolution of individual conodont elements in the late Cambrian. With an expectation of finding just four conodont elements in each kilogram of rock, Miller – who now replaced Müller as the most prolific worker on these earliest of conodonts – was nevertheless able to demonstrate that identical elements evolved in quite separate branches of the evolutionary tree. Clark recognized that Miller's great advantage was to study that first burst of evolution, as it permitted him to build upon a clean slate. It suggested to Clark that if one wanted to replicate Miller's trick anywhere else in the family tree, it would be necessary to locate a point of mass extinction and build the family tree from there, in that subsequent explosion of evolution Schindewolf had recognized.[10] Clark's interest in extinction reflected a utilitarian desire to be able to precisely identify his fossils; the attraction was not to indulge in imaginative speculation. He began, then, by asking when extinctions took place in the conodont world. Clark wanted a way to visualize changing diversity beyond the range charts of species and genera widely used by stratigraphers. He wanted to know not the pattern of life but the pattern of extinction.

He began by plotting the number of form genera present in each major period of geological time together with the number that went extinct in that period. He knew the picture he drew was coarse and that his data were imperfect, and he was therefore not surprised when the technique threw up some odd artifacts of method. The Silurian, for example, appeared to be a period of crisis for the conodont animal when, really, the low number of species reflected the period's relatively short duration. As if to record the path he had traveled and perhaps prevent others from falling into the same trap, Clark published the diagram nevertheless. He then improved this picture by increasing its resolution and plotting only those species that first appeared within a given period. Now his plot showed extinction periodically overtaking evolution. In other words, there were moments of decreasing diversity – of impending crisis. Turning these two measures (new species emerging and old species becoming extinct) into ratios, he could then plot what he called an "index of evolution," which showed graphically when the conodont animal was in crisis and when it hit boom times. It was to prove an influential study and one to which Clark would return a decade later, wondering why the conodonts supported two peaks in their life history rather than one.[11]

Inspired by Valentine and the new plate tectonics, Lars Fåhræus saw in Clark's diagrams the opportunity to give these tiny lives a tumultuous Wagnerian interpretation.[12] Everything about the conodont – its form, distribution, diversity, and evolution – Fåhræus believed, could be mapped against an earth composed of violent volcanic island arcs, massive plates of continental crust drifting through climatic zones, and huge continental collisions. It was presumed that the conodont lived,

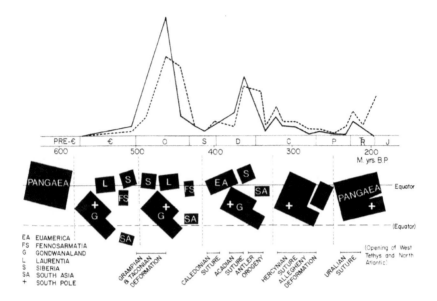

10.1. Fåhræus's choreography. The upper graph is based on Clark's data and similar to the graphs Clark drew. The solid line marks appearances of new species, while the dashed line indicates extinctions. Where the former is below the latter, conodont diversity is in decline. Fåhræus suggested that this tiny animal's success was partly determined by the changing configurations of Earth's landmasses – here indicated by black rectangles – which resulted in changes in environmental diversity. Reproduced with permission from L. Fåhræus, *Conodont Paleoecology* (1976).

like nearly all marine life, at the continental margins and was thus both an opportunist and a victim as continents separated and came together. He constructed a pictorial representation with his rectangular representations of the earth's landmasses performing their ballet to the timing of Clark's extinction overture. The earth's violent revolutions were reflected in smaller revolutions in the conodonts themselves. Or to turn things on their heads, as every conodont worker would, one now might examine a tiny conodont and believe that it reflected in its form the trials and tribulations of a whole planet. Fåhræus's pictorial explanation required little further explanation (figure 10.1); it met Schopf's ideal of the grand paleontological hypothesis better than any other. It linked life to events in the earth's history and did so using plate tectonics, the flavor of the decade.

❋   ❋   ❋

Walliser knew that if the topic of mass extinction was to be tackled empirically, it would need funding. This was big science, requiring collaboration. He managed to convince some of his German colleagues that the project was worthwhile, but when no money was forthcoming their enthusiasm naturally waned. Then, in the late 1970s, a former colleague from Tübingen who had moved to the United States asked Walliser if he would become secretary general of the International Palaeontological Association (IPA). Walliser admitted that he had never heard of the IPA. Indeed, nor had any of his colleagues. On enquiring, he discovered it was a body representing the world's paleontological societies. Given Walliser's international outlook, sociable nature, and desire to pursue his big idea, he realized the organization might be just what he needed.

Walliser's arrival at the IPA resulted in a complete reappraisal of its business and ambitions.[13] The big idea was developing, and with it an international community. But still there was no funding. Then the Alvarez meteorite hit and everything changed. With mankind now imagining something nasty hiding in the night sky, and McLaren remade as a prophet, Walliser at last found his funding. What had once been a rather arcane scientific problem had suddenly become front page news.

He announced what was the IPA's first ever research project in 1982, securing funding for five years from the International Geological Correlation Programme (IGCP) not long afterward.[14] Walliser launched Project 216, Global Biological Events in Earth History, at the International Geological Congress in Moscow in August 1984 and immediately began to build the multidisciplinary community of researchers necessary for this big science. Beginning with his own network of contacts and members of the IPA, who in turn used their own, in time the project would attract participants from forty-nine countries. This really was science on a grand scale.

Walliser's agenda for the group was subtle and complex. It revealed the degree to which so much was unknown or unclear about these natural divisions in the record of life. It was, in this respect, a million miles away from the sound-bite science of asteroid impacts, but inevitably, his group had its asteroid chasers too. Soon the number of extinction events began to swell and geologists began to debate the cause of extinction like never before. Paleontologists remained largely skeptical of extraterrestrial causes at first, though some, like Gould, who saw the meteorite as supporting his punctuated equilibrium, were rather more welcoming.[15]

While meteorites and huge volcanic eruptions grabbed the headlines as the most spectacular theories went head-to-head, many paleontologists remained wedded to a more conservative view of a planet with its constantly reconfiguring continents, its wobbly rotation and changing climate, its ice ages and fluctuating sea levels. This view suggested that there was no need for alien visitors. Nevertheless, the meteorite was felt everywhere. It could not be ignored. Phillip Playford, for example, whose reef model had helped shape George Seddon's interpretation of the Canning Basin, hooked up with McLaren and others and went in pursuit of Kellwasser Event iridium there. They were drawn to the Canning Basin by the secure timescale supplied by its conodonts. Collecting in New York and Belgium had failed to detect a Kellwasser iridium spike, but the Canning produced one at twenty times the background level. This did not, however, land the expected meteorite because they also discovered a concentration of the fossil cyanobacterium, *Frutexites*, which was known to be capable of concentrating this and other elements in its filaments.[16]

Back in the 1920s, Schindewolf and Hermann Schmidt had distinguished Upper and Lower Kellwasser horizons. Now the former was recognized as one of the five greatest mass extinction events in Earth history. It became an important test of theory and of intense interest to Walliser's group. Associated with the Upper Kellwasser horizon are a number of black shale sequences thought to represent anoxic conditions in deep water. As sea level was raised and lowered, so the area covered by this anoxic water expanded and contracted, though perhaps not as simply as this explanation suggests. Walliser imagined that such changes in water depth would have ripple effects throughout the food chain and into other environments.[17] It suggested a silent menace in the depths of the ocean. Perhaps the sky watchers were looking in the wrong direction?

Walliser knew the Kellwasser possessed a complexity that defied easy explanation; there were a multitude of possible causes and many were probably acting simultaneously. Different groups of animal had not been affected in the same way, and both extinction and successive radiation seemed to be stepped. This complexity convinced Walliser that it could not result from McLaren's asteroid impact, pointing out, in 1984, that the iridium anomaly was younger than the main Kellwasser Event.[18] For Walliser, change was slow and accompanied by a few global events, none of which alone had a universal effect on life.

In that same year, Willi Ziegler entered the Kellwasser debate and Walliser's project. Like Walliser, he could now look at Helms's iconic diagram and see a moment of extinction prior to the rapid evolutionary diversification of *Palmatolepis*. Collaborating with Charles Sandberg of the USGS in Denver and Roland Dreesen of the Institut National des Industries Extractives in Liège, they sampled contrasting paleoenvironments in Utah, Nevada, Germany, and Belgium, where they found surprising agreement in the changes taking place. All the sequences signaled a switch from deep-water to shallow-water forms, suggesting a relative fall in sea level. However, there were also some unusual mixtures of shallow- and deep-water species, which encouraged the team to postulate that the Kellwasser Event had caused tsunamis that washed the *Icriodus*-dominated faunas into deeper water.[19] Walliser was not convinced by this sea level–based explanation, even with its rather dramatic finale; if one looked beyond the evidence of conodonts, he said, the picture was far more complex.

Given Ziegler's longstanding commitment to utilitarian stratigraphy, it is unsurprising that he, like Walliser, was interested in events for their evolutionary implications, which could then be fed into stratigraphic study. Ziegler and Richard Lane now re-examined their collections through spectacles made for them by Clark and Walliser. They looked for "conodont evolutionary cycles." Following an extinction event, the conodonts showed a period of low diversity, followed by a short "innovative phase." Here innovative conodonts tended to have larger than normal basal openings. These then formed the rootstock for an evolutionary flowering or a "radiative phase." Finally evolution entered what they called a "gradualistic phase" before yet another extinction event.[20] In the period of time between the Late Silurian and Middle Carboniferous, they believed they could detect seven of these three phase cycles, showing that the conodont was peculiarly adept at providing these high-resolution pictures of extinction and evolution.

Sandberg, Ziegler, and Dreesen continued to work on an explanation for these crises. They came to believe that the extinction that led to *Palmatolepis's* blossoming "occurred in far less than 20,000 years and more likely within a few years or days." Accommodating the complexity recognized by Walliser and others, and using data from six sites across Europe and America, they now pictured a sequence of twelve steps extending over the Kellwasser Event. This event began with a sea-level rise and the drowning of reefs. The sea level then fell, and they theorized that a large "bolide" (an object capable of a huge impact but without presuming to know whether it is a rocky or metallic asteroid or icy comet) passed close by or a small one impacted the earth. Faunas then became re-established on mud mounds but reefs did not make a reappearance. There then followed another couplet of sea-level rise and severe fall. The rise had led to the development of stratified seas and the spread of anoxic conditions in deeper waters. The seas continued to shallow, and now a large bolide impacted, causing storms recorded in the Belgian rocks and the widespread extinction of conodont species. As the seas continued to shallow, inevitably shallow-water conodonts were able to spread. Tsunamis then preceded a new transgression of the sea over coastal lands, but glaciation in the Southern Hemisphere soon caused yet another period of shallowing seas in the north. This was followed by yet another mass extinction later in the Devonian. Of those conodonts that survived the extinction event, there were some, such as *Palmatolepis praetriangularis,* which were "opportunistic" and began a process of repopulation. Two species of *Icriodus* – "strong survivors" – also made it through. These workers located a five-centimeter-thick conodont-free shale situated clearly between their conodont zones in Schmidt Quarry, near Ense in Germany. It suggested that the Upper Kellwasser Event could have lasted no more than 12,500 years, though the preference remained for just a few days.[21]

Those reading this paper may have been rather skeptical of the drama it portrayed, but even Walliser thought it an important contribution to working out the precise chronology of the event. This chronology would be corroborated by later workers.[22]

On the matter of cause, however, the meteorite possessed a sticky aspect. It was hard to shake off and it seemed to attach itself to almost any theory; Sandberg and his colleagues certainly became attached to it. Walliser was both amused and exasperated, recalling that as his project came to an end, many of those who had previously objected to it found themselves increasingly won over. Maurits Lindström admitted to being a late convert. Sandberg, who with Ziegler and Dreesen had merely "theorized" a bolide, became increasingly interested in the Alamo Impact site in Nevada discovered in the early 1990s. Dated by conodonts and then radiometrically, that impact was too early to account for the Kellwasser crisis, but by 2005, Sandberg and his former student Jared Morrow could demonstrate that it was simply one of a number of impacts around this time. Others were known from Germany and Sweden, and the latter seemed to be timed very closely to the Kellwasser Event. This evidence led Sandberg, Morrow, and Ziegler to postulate that in the Devonian's turbulent history comet showers were a probable trigger for major changes in life on the planet. It was while engaged in taking this work further that Ziegler died in 2003. Appropriately enough, for the planet's most single-minded and dedicated conodont specialist, he had conferred with Sandberg from his hospital bed just ten days before.

The full implications of this events thinking are perhaps revealed most strongly in the work of Lennart Jeppsson, a Lund conodont worker who gained his insights and inspiration in circumstances that were in some respects the reverse of those that shaped Walliser's. Both are known for their Silurian work, but Walliser became increasingly convinced of the truth of global events as a result of travel. In contrast, for Jeppsson the world often arrived in the form of scientific papers, and his inspiration came from essentially staying put and working his patch. For most of his career, Jeppsson's focus has been Gotland, the second largest island in the Baltic Sea. With some fifty-seven thousand people living there, this is no Robinson Crusoe retreat but an ancient, sunny, and beautiful place to spend one's life in the field. Here, based at the Allekvia Field Station, Jeppsson spent two or three weeks each summer digging into the Silurian strata and carrying vast quantities of it across the island to the Slite cement works, where it was stored and would eventually gain passage across the Baltic to Limhamn, some twenty-five kilometers from Lund.

The attraction of Gotland for the geologist is the considerable thickness of the Silurian rocks – some five hundred to seven hundred meters. Largely free of the corrupting influences of metamorphism, folding or faulting and gently dipping to the southeast, they preserve almost perfectly a shallow-water platform which in the Silurian lay just south of the equator. Through rapid deposition, the island's thick rock sequences effectively stretch geological time, making it possible for Jeppsson to understand events in the past with the geologist's equivalent of a stopwatch.

Jeppsson came to conodonts as a result of a lecture by Stig Bergström; like Bergström, he had begun his scientific life as a botanist. His curiosity pricked by the mystery of these strange conodont fossils, in 1965 Jeppsson found himself studying Silurian conodonts for the equivalent of a master's degree under Bergström's supervision. He acquired Gotland as a kind of inheritance. The conodonts there had been studied by Anders Martinsson at the University of Uppsala in the mid-1950s, though not as his main interest. Martinsson, who was a regular visitor to Lund and a close friend of Bergström, had set up the field station and remained a leading light when Jeppsson began his first steps in the field. Jeppsson must have shown talent, for Bergström spoke to Martinsson and asked if Jeppsson might take on the Gotland conodonts in a project Martinsson was then running. However, not everything was plain sailing. On arriving on Gotland, he had found himself in competition with the young Lars Fåhræus, who soon took the lead by publishing on the Gotland conodonts in the late 1960s.[23] Fortunately, this turned out to be a temporary problem for both men.

Jeppsson's bigger problem was technical: He discovered that the fossils were far fewer and far poorer than those he had already studied. This meant his task was going to be rather more difficult than he had imagined. The relative rarity of conodont fossils in these time-stretched strata meant that processing vast quantities of rock did not guarantee sufficient material for interpretation. His situation was entirely the opposite of that of Bergström and Sweet, who seemed to be drowning in these fossils. Rather than sifting through hundreds of thousands of specimens,

Jeppsson often found himself looking for a single species recorded by a single specimen. It was a problem that forced Jeppsson and his dedicated technical team to repeatedly study and improve their methods. In 1983, small collections were produced from processing 0.5-kilogram samples. A decade later they were dissolving 20- to 80-kilogram samples of rock in acid. Progressively they reduced the number of conodont elements being destroyed, and soon conodonts were found in rocks Jeppsson once thought barren. His overall productivity increased a hundredfold. Nevertheless, at only ten to one hundred specimens per kilogram, his huge collections of fossils speak of considerable effort and a particular devotion to his subject.[24]

Initially, Jeppsson's task was to describe the conodont faunas. But in such a thinly populated country with few paleontologists, no scientist could consider his subject so narrowly, and from the outset Jeppsson became interested in every aspect of the animal. His research took a particular turn when he discovered that his fossil species were showing episodic disappearance and reappearance, as if rising from the dead – so-called "Lazarus species." He was not unaffected by the ecology and provincialism debates of the 1970s, but by the early 1980s he thought the ecological models of no significance to his work, and like others he doubted the existence of faunal provinces in the Silurian. He gave thought to the repeated and almost instantaneous evolution of identical forms from more conservative stock; perhaps the Lazarus species were really imposters? He admitted this remained a possibility, but he possessed no evidence to support it. Instead he began to consider other factors, such as water chemistry, aware of its extraordinary influence in the modern day Baltic. In 1984, this seemed the most profitable line of inquiry.[25]

Over time his interpretive frame shifted, affected by the massive swell of publications that sought to understand the past globally. Influential in this change of thinking was Princeton's Al Fischer and Michael Arthur. They reviewed data relating to the oceans of the past and considered how radically the science's conceptions of Earth had changed even before the meteorite had hit. Oceans were now transient features, their distributions, depths, submergence of continents, and circulation patterns quite unlike those of the present-day planet. These shifting configurations took place as the chemistry of the atmosphere changed and as Earth gained its succession of animal and plant populations.[26] Fischer saw the possibility of rediscovering Earth history as a singular and specific narrative. The belief that the present alone could be used to interpret the past seemed too constraining; the past was different and exotic.

Yet despite his belief in a singular and particular narrative, Fischer felt he saw an underlying structure to the history of life: "There runs through earth history an orderly thread of changes in the state of the earth as a whole – a thread of complexly interwoven filaments that produce rhythmical patterns." He and Arthur suggested a pattern reflecting alternating periods when oceans were rich or poor in biological diversity. Fischer noted that there was evidence for four ice age periods over the last seven hundred million years, including the one we live in today. In these periods, Earth was in what he called an "icehouse state," with low sea levels (because so much water was locked up in the icecaps), a strong temperature gradient from equator to pole, and an ocean circulation where cold polar waters sank to form the deep ocean. In the intervening periods, Earth was in a "greenhouse state." Ocean waters were warmer and did not sink at the poles. Rather, the deeper waters resulted from the sinking of heavier, more saline waters at rather lower latitudes. Fischer proposed a sophisticated plate tectonic-driven model for the production of these two fundamentally different planetary states, with a key ingredient being changing quantities of atmospheric carbon. He suggested that these changes took place with regularity, that there were supercycles of hundreds of millions of years, with smaller cycles of thirty million years and ten thousand to one hundred thousand years superimposed on them that were reflected in changing faunas, alternating lithologies, and moments when anoxic black shales became widespread.

At Berkeley, Pat Wilde and William Berry were also considering how climatic change could affect ocean conditions. They believed that as glacial or icehouse conditions formed or broke down, so ocean stratification would collapse and result in the upwelling of potentially toxic or anoxic waters and mass extinctions in the life-rich surface waters.[27] They saw these as long-range cycles; most of the time the earth's climate was warm.

These models of the past stimulated some to begin library-based number crunching of the record of life in search of cycles. Others sought to reinterpret their understanding of the record in the field using cyclical eyes. McLaren was certainly one who took Fischer's cycles seriously when interpreting the Kellwasser in the early 1980s. And as we have seen, cycles were also in the minds of Ziegler and his collaborators. Indeed, this cyclical view became embedded in the literature, often picked up by readers who never referred back to the original work.

In 1987, Jeppsson explained a single extinction event in the Silurian, first mentioned by Dick Aldridge a decade earlier, in terms that owed something to this debate: "The extinction . . . seems to have coincided with a change from more oxygenated oceanic conditions to 'greenhouse' conditions, with a final fauna before extinction that thrived during an episode of widespread abnormal sedimentary conditions."[28] By then Jeppsson had global data showing that lithological and faunal changes had worldwide distribution, though not without some local variation. The key seemed to be a global decrease in the deposition of carbonate rocks (limestones) and an increase in the formation of muddy ones. American workers had seen this change at a regional level and interpreted it as a tongue of muddy sediment spreading from an unknown landmass to the east. But now Jeppsson could demonstrate that the phenomenon was much more widespread and had a considerable impact on the conodonts. He believed that the climate in the Silurian was not the consistently equitable greenhouse that had long been imagined. Perhaps conodonts were sensitive indicators of ocean conditions in rather more subtle and sophisticated ways than previously considered. With information on four anomalous episodes in the Silurian, Jeppsson was also of the view that what happened in each was different and could only be resolved using high-resolution stratigraphy. In this, of course, he had the advantages of Gotland's thick strata and the incomparable conodont.

By 1988, Jeppsson had produced a model to explain the effects he saw in the field. The context of his theorizing was the literature speaking of cycles and climatic change. Perhaps inevitably his digestion of field data, and new oceanographic data from workers such as Pamela Hallock, with whom he by chance came into correspondence, led to conclusions much like those of Fischer and others. Jeppsson's P (later Primo) and S (Secundo) episodes were essentially identical to Fischer's icehouse (later O [= oxygenated]) and greenhouse (G) states. In 1990, Jeppsson published his model in full (figure 10.2) and shortly afterward joined

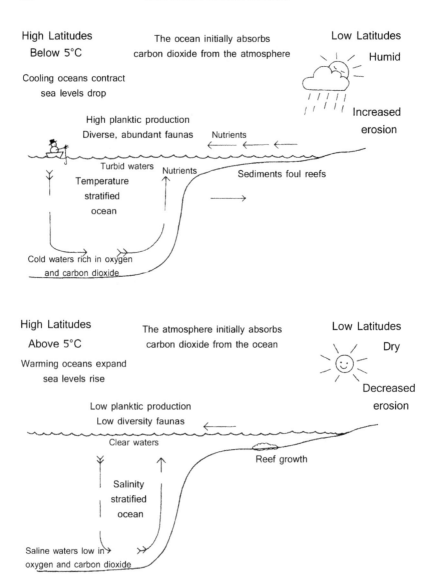

10.2. Jeppsson's oceanic models. Some characteristics of the two states in Lennart Jeppsson's oceanic model: P state (*top*) and S state (*bottom*). Redrawn with permission from L. Jeppsson, *Journal of the Geological Society of London* 147 (1990).

with Dick Aldridge and Ken Dorning to apply that model and name the episodes and events in the Silurian it seemed to describe.[29]

As with Fischer's, the key to Jeppsson's model was the carbon cycle. The cold ocean waters absorbed carbon dioxide and oxygen from the atmosphere and took them down into deep water, effectively lowering atmospheric CO_2 and thus lowering temperatures. In these Primo episodes, higher humidity and rainfall at lower latitudes resulted in runoff, and erosion, with an increase in muddy sediments and the formation of fewer pure limestones. These muddy sediments would foul reefs and lead to the formation of black shales. These were glorious times for conodonts as they diversified to occupy a range of specialized habitats. Not all life was so fortunate: The upwelling of nutrients from depths would kill off any limestone reefs. The low levels of CO_2 in the atmosphere would also mean that rain and surface water would erode limestones on land.

But over time the ocean would be able to absorb decreasing amounts of carbon dioxide leading to a rise in the amount in the atmosphere. Greenhouse conditions would consequently begin to develop and temperature would rise. Now the Secundo episode began, perhaps with a faltering start. The oceans became warmer and cold waters at high latitudes were now less dense than the more saline waters farther south. Consequently, these southern waters began to sink to form the deep water, and because they possessed less oxygen and carbon dioxide, they created low oxygen conditions at depth, resulting in the formation of black shales. Yet, in shallower waters, and in a now drier climate, less muddy sediment was flowing into the sea and reefs and limestones could develop in relatively low nutrient conditions. In these seas, conodonts were rare. But this growth in the production of limestones drew in increasing quantities of carbon dioxide, locking it in the rock and lowering concentrations in the atmosphere. As a consequence temperatures began to decline, leading eventually to a resumption of Primo conditions.

The change from warm to cold oceans would cause a contraction of the water mass and a lowering of sea levels; when warm oceans were reinstated, sea levels would rise again.

The model had an undeniable logic and seemed to unite his new evidence of global changes in sediments with the widely discussed importance of the carbon cycle. To this Jeppsson added the Milankovitch cycles – variations in the orbit, tilt, and wobble of Earth – which had been key explanations of cyclical climate change since the mid-1970s. It seemed that the regular changes in the tilt of the Earth matched the changes Jeppsson saw in the field, and he imagined the Milankovitch effect nudging one unstable system into another but in complex ways. By the time Jeppsson published his first paper with Aldridge and Dorning, he was very busy drawing out aspects of his model in order to show its applicability outside the Silurian.

Of the events discussed by Jeppsson, none was more important than the Ireviken Event, Jeppsson's paper on which, although submitted for publication in 1990, languished in press for seven long years. The Ireviken was the Silurian's most impressive event, killing off 80 percent of conodont species worldwide and 50 percent of trilobite species on Gotland. In Jeppsson's mind it marked a period when Earth switched from Primo to Secundo conditions. The event lasted two hundred thousand years, and in most geological sections it occupies just one meter of rock. On Gotland this stretched to nineteen meters. Using conodonts he could at times attain a resolution of one two-hundredth of the overall event. It meant he could talk in terms of periods of ten thousand years, the geological equivalent of microseconds. He now understood that different conodont lineages reflected preferences for different ocean conditions. The stepwise extinction of conodonts produced eight variably spaced datum points he believed he could see in papers describing other sections in Alaska, Canada, California, New York state, Virginia, Greenland, Britain, and Australia. This high-resolution work was soon taken up by two Estonian colleagues, Peep Männik and Viive Viira, and before long Jeppsson was pressing ahead with establishing a new conodont zonation based on extinctions rather than appearances. Soon Jeppsson's model was appearing in textbooks as an exemplar of what could be achieved with high-precision paleontology, and by the mid-1990s it became a standard explanation for change in the Silurian.[30]

In life, in death, and in dying, conodonts bore witness to extraordinary events. That at least seemed certain. The nature of those events remained rather more a matter of interpretation and personal conviction, but they undoubtedly made possible the seeing of new things and the thinking of new thoughts. In this respect, they acted as catalysts or interpretive spectacles. In themselves, however, their most important aspect was to make possible connections between data: data of different kinds and data produced at different locations. This made a small contribution that permitted the science to generate new hard evidence. This science did indeed discover large numbers of meteor craters and make geologists think of their subject in planetary terms. Major events in Earth history became a reality in the science's thinking.

All this work, of course, added only small, and rather implicitly received, knowledge concerning the animal itself. That discussion was going on elsewhere. Following Walter Gross's destruction of the fish in the 1950s, the animal had not been much discussed. Instead, the conodont workers had put their heads down and worked hard to develop a very practical and connected science. However, at the end of the 1960s, the animal again began to emerge and this aspect of conodont studies once again became a hot field. It is now time to unveil the animal.

That's one small step for man . . .

NEIL ARMSTRONG,
on landing on the moon, July 21, 1969

ELEVEN

The Beast of Bear Gulch

MANY WHO HAD HEARD THE CONODONT'S STORY DOUBTLESS imagined that impossible day when the animal would be found. In a corner of so many minds, there was intense curiosity about these tiny things. They were, after all, as Maurits Lindström had put it in 1964, "the biggest and most important group of fossils about which the zoological relationships are entirely unknown."[1] The fossils seemed to defy comprehension. That special day did, however, come. It was September 5, 1969. That, at least, was the day of realization. Less than two months after Neil Armstrong stepped onto an alien world, paleontology produced its own alien and it was utterly bizarre. Before long, the news spread across the same networks that had covered the moon landing, though rarely warranting more than a column inch.

The animal's reception – at least in the scientific world – was, appropriately enough, that reserved for aliens in those classic American sci-fi films of the 1950s: From the moment of its innocuous arrival, mankind, so it seemed, sought the alien's destruction. And in keeping with that tradition, it seems appropriate that our hero is also our antihero, his integrity doubted, his actions condemned. His name was Harold Scott. It will be recalled that he, as a young man in the 1930s, had made that giant leap to reveal the animal's complexity – a discovery that ultimately turned the science on its head. In 1969, he was at the other end of his long career and nearing retirement, yet his role in the drama was no different. As in the 1930s, Scott was again making assertions few could, or wanted to, believe. His critics thought this latest conodont animal was the product of a fertile imagination. Scott however, had good reason to believe that he would again prevail.

The great moment had been billed to take place at a meeting of the North Central Section of the Geological Society of America in Iowa in May 1968, but this proved to be a false dawn. Rumors circulated suggesting that Brian Glenister possessed a communication from Scott that would reveal all. It was to be read out at the final session of the meeting. The room was packed to standing. Anticipation was intense. As Glenister began to speak, the room became silent, but the atmosphere soon changed as the audience began to realize that this was not the great moment at all. The disappointment was palpable. What Scott had found, in those same Montana rocks that had furnished him with his assemblages, were "blebs," carbonaceous or asphalt patches two square millimeters in size. He had found eighty of them and they contained conodonts. Seemingly impossible to photograph or draw well, the Iowa audience looked and saw the same rather patchy material their intellectual parents had seen when Croneis had presented Scott's results thirty-four years earlier. As with the previous disappointment, Scott was not there to witness the cynicism, and he pressed on regardless with a paper that claimed "Discoveries bearing on the nature of the conodont animal."[2]

When Scott looked at those blebs or blobs, he saw the animal's head: "These conodonts have been held in the cartilaginous material of the head of an animal, not in gills; the cartilaginous head material has been thicker and stronger than the remainder of the body and upon death the conodont teeth remained 'stuck' in the cartilaginous substance. As the cartilaginous material altered to a bituminous or asphaltic base the conodonts became twisted, intertwined, and occasionally broken." He continued, "We cannot fully judge the position in the mouth, but the evidence points more and more to a circum-oral arrangement as rights and lefts rather than uppers and lowers and functioning as strainers in a mouth-esophagus rather than as gill-rakers."[3] But the teeth in the blebs were puzzlingly small. The "black, glossy, asphaltic patches" looked, to Scott, like skin: "The blebs do tell us that the head of the conodont-bearing animal was at least partially covered with reticulated skin. Also, they tell us that at least a portion of the head consisted of cartilaginous material, and that this cartilage was thick and strong enough to be pre-served, thereby raising hopes of future discoveries."[4] Of all this, Scott was quite certain – as certain as he had been, decades earlier, that the animal was a worm. Now he was willing to speculate that it was a cen-timeter-long ancestor to the primitive jawless fishes, the lamprey and hagfish.

Confirmation of Scott's recent discoveries soon arrived in a paper from German paleontologist Friedrich-George Lange. It said that Lange had found fossilized excreta or coprolites containing conodont assem-blages, but as we learned in chapter 8, Lange had only used this expla-nation in order to get his paper published, as Ziegler had objected to the interpretation of these fossils as accurately preserved apparatuses. Of course, Scott did not know this, and on reading Lange's paper, Scott was convinced that Lange was mistaken, for Lange's finds, like his own, each contained a single unmixed assemblage. This made Lange's cop-rolite theory bizarre. Was Lange suggesting that the animal "waited to excrete the remains of the one victim prior to eating another"? Scott had seen coprolites and knew they were stained "lumps" containing broken conodonts; they were not like Lange's finds.[5] Scott now used Lange's fossils to support his own ideas, believing that evidence was mounting and that he was on the brink of a major discovery. This was, however, a new direction for him, for although he had published a few important papers three or four decades earlier, conodonts had not become his life. Now he was back but rather out of touch with all that had gone on since. At the start of 1969, he applied for a National Science Foundation (NSF) grant to continue the search for the animal. But he was unaware that others were now on his patch, and it would be they who would take the next big step.

In March 1968, an undergraduate student, Douglas Wolfe, brought two fine fossil fishes to Bill Melton, curator at the Geology Department of the University of Montana in Missoula. They had been found by quarry

owner Charles Allen and Ralph Hartin in the local Bear Gulch Limestone, a rock that lay, so they thought, within Scott's Heath Formation.[6]
Melton was impressed. Complete fish of this age were rarities, so he
gathered up some students and returned to the quarry in search of more
specimens. Another fish popped out, along with other fossils. What
Melton had come across was a "Konservat-Lagerstätte," a deposit containing exceptionally well preserved fossils. Melton thought the discoveries unusual and significant, and when Eugene Richardson of the Field
Museum of Natural History in Chicago read about them, he thought so
too. Richardson was then organizing sessions for the first North American Paleontological Convention due to take place at the museum in
early September 1969, and he invited Melton to give a paper on these
new finds.

Melton spent the summer before the convention in the field with his
assistant, Jack Horner, looking for fishes in what was known, or became
known, as Surprise Quarry. They found sixty-five. But these were not
all they discovered. In the first week, Melton later recalled, "we found a
curious, carbonized impression of an animal that I could not identify in
the field. Several others were found in the next five weeks."[7] When they
at last got these enigmatic fossils back to the laboratory, they discovered
they contained conodont elements.

Melton turned up at the convention with his fishes and strange fossils. Soon fish specialists were swarming over them, and one of these,
noticing the conodont-bearing specimens, called for conodont specialists from the audience. Huddle, Collinson, Lane, Scott, and others
crowded in for a brief look. "Plans were quickly changed," Melton recalled, "and a photograph of a specimen was made." Melton then presented his paper, titled, appropriately enough – but with more under-
statement than he knew – "Unusual fossils." It was a prime spot: The
first paper of this first convention, and in it he made mention of the
"soft-bodied animals." However, the full impact of the discovery was
not realized until Scott took to the stage immediately afterward. Apparently nominated by his fellow conodont workers because of his seniority
in years and his groundbreaking work in Montana in the 1930s, Scott's
appearance was already a break from the published program. He did
what it was necessary to do on this great occasion, something conodont

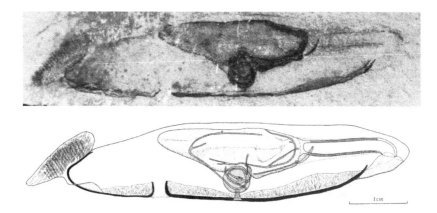

11.1. The beast of Bear Gulch. Melton's headless animal was extraordinary – it was like nothing previously seen. The tail is on the left. Reproduced with permission from S. Conway Morris, *Philosophical Transactions of the Royal Society, Series B* 327 (1990).

workers had rehearsed hundreds of times before. He told the story. With the enormity of the enigma in the audience's mind, Scott then said, "I have just seen the conodont animal!" To some in this audience the news must have seemed more remarkable than the moon landing. "It was a most exciting discovery," Scott later told a friend. Stephen Jay Gould was there, "a wet-eared, first year professor" impressed by the paleontological talent in the room: "If the Russians – or the Chinese, or whoever wanted to destroy this entire profession, one bomb . . ."[8] One young Japanese scholar, Kenji Konishi, told Scott five years later of "the still unforgettable pleasure" of hearing Scott's opinions on "this epoch-making find." This was September 5, 1969. The "conodont animal" had landed (figure 11.1).

That evening, delegates gathered around the piano for a singsong:

Ah, sweet mystery of conodonts, I've solved thee!
So you really had a body after all.
It was firm and roly-poly, flat and flabby;
'Twas like a worm, echinoderm, or jelly ball.
Now the guessing game is over and we're certain
You were sexless, winged, six-sided, more or less.
Did a notochord support your velum curtain?
Yes, we're certain you were just a mess!![9]

Now the problems began. Scott's prominence in the announcement had little to do with his own work and nothing to do with his blebs. And he certainly could not stand there as a modern conodont worker. What happened next, then, was extraordinary. Indeed, it is commemorated in a rumor that has probably altered with every retelling: Scott bribed a student (Melton) with a degree in order to gain possession of the fossils, which he then locked away for years, permitting no one to see them. I should say here, however, that Scott's friends, those who knew him best, like Frank Rhodes, never believed the rumor. They knew Scott to be an honorable man who was, like so many others, "dazzled" by the puzzle of the conodont animal.[10]

The story of the animal, then, became clouded in rumor and suspicion, in untruths and secrets. To those outside the frame, it was easy to imagine something underhanded going on, but Scott left behind a transparent record of his actions that show this was not the case. Thanks to that record, we can unravel what went on and see something of the social underbelly that is found wherever science is performed. We shall see it again in the closing moments of this book, but now we must return to Scott, Melton, and those strange specimens.

Even then Scott may have been considering his own mortality. While we cannot know precisely what was in his mind, he did, indeed, see in Melton's specimens a discovery as enormous as any in the history of paleontology, and he did act to take Melton and his specimens under his wing. Given that many then working on conodonts would know Scott only as a historical figure in their field, these actions immediately turned many American workers against him. They felt a sense of indignation: Scott had taken possession of and then hoarded treasure to which he had no moral or professional right. To the conodont research community, which had been forming cohesive ties since the mid-1950s, Scott the outsider had stolen the equivalent of a religious idol.

From Scott's perspective, however, things looked entirely different. He had, briefly and by chance, been a pioneer in conodont research. His was one of the landmark discoveries that had stood the test of time. He also had recently demonstrated his continuing activity in that field, and he had an NSF grant in his hand to prove it. He was the only worker searching for the animal at that time, and his research project had, with the discovery of the animal, now made a giant leap forward. The animal was in every way in his territory and on his research trajectory. It was easy for him to justify his actions. Indeed, he felt he had predicted the discovery.

Beyond the recollections of those then active in the field who now took a clear stance on Scott, a later generation of scientists also had to confront what Scott saw in the new animal. One even wondered if Scott was "mad." He was not. Scott's correspondence tells us something of what he was thinking and what he wanted others to think. He must be read like any other actor in this story: complex and changing, not entirely selfless or selfish. One can, of course, look through cynical eyes imagining manipulative actions, but his relationship with Melton shows that he was compassionate and supportive, self-assured but also vulnerable. The story of the beast, then, is not simply a tale of the discovery of the conodont animal – as Scott saw it – but is also revealing about the roots of rumor and implausible explanation.

Scott was sixty-three that September in 1969, and after spending most of his career in Urbana, he had just escaped a culture change there that was abhorrent to him. Now he was the new departmental chairman at Michigan State University in East Lansing. His good friend Frank Rhodes was then just up the road at the University of Michigan in Ann Arbor. Unknown to Scott, however, this move meant he had to reapply for the NSF grant he already possessed. But this was a fortuitous turn of events because now he could include Melton in the project. Indeed, it seems very likely that Melton joined the project because Scott had the money in hand. Melton's own university was often "on its uppers" and he desperately needed funds for the fieldwork he evidently loved. With Melton on board, and imagining one hundred conodont animals hidden in the quarry, Scott drew up an enlarged bid. He invited Melton to prepare an appendix asking for whatever he wanted: "May I suggest that you include the cost of a rock saw, money for two or three assistants, money for yourself, extensive travel money, and an enlargement figure for publication costs." Scott was leaving nothing to chance and planned to fly to Washington to present his request in person.

Scott's control of the grant application might suggest that he was also now in control of the project, but he was happy to modify his plans in any way Melton saw fit. Despite Scott's seniority in years and elevated position, it was, from the outset, a collaboration on equal terms, with Scott, the experienced manager, taking on the administrative chores. It was not the kidnapping some imagined. Melton, for example – the "student" of rumor – had been born in the early 1920s. He was a veteran of the Korean War and had as much life experience as Scott. He had been vertebrate fossil preparator at the University of Michigan from 1957 to 1966 and was clearly a man of far greater field expertise than his collaborator.[11] Melton was there by choice. Given Scott's interests, historical place, current funding, and present work, it was the logical place to be. For his part, Scott was generous and caring. He also liked to act as wise council. As a result they soon developed a good, friendly working relationship.

But not long after the convention, Melton began to receive mail highly critical of Scott. These were not correspondents Melton knew – and they knew nothing of him, his plentiful scientific connections, or his background in science. Bewildered, Melton sent copies of the letters to Scott, who was at a loss to explain them. In a reversal of their ages, Scott now used Rhodes, the rising university administrator, as a wise uncle, asking him for advice: "I can tell you that they do not seem to recognize that you and I have ever studied conodont assemblages." Scott considered contacting John Huddle at the National Museum in Washington, as he at least would remember Scott's earlier work and might support his actions. But then he learned that Huddle was one of Melton's correspondents. Scott felt isolated. In his defense, Scott told Rhodes that the impoverished Melton, with dependents in tow, was an active participant in their project and worthy of Scott's patronage: "Mr Melton had an unfortunate experience when he was preparing for a Masters degree. He left school with most of his work completed but was denied the M.S. degree. Sometime after that, he was in the Korean War and suffered a skull wound which affected his speech but not his thinking. Therefore, he speaks slowly but thinks quickly." It is possible that the promise of the degree was already a subject of rumor, and indeed, Scott was already investigating a transfer of credits from the University of Michigan, where Melton had studied, to Scott's university in the hope of giving Melton his degree. Scott laid it on a little thick in order to convince Rhodes of the legitimacy of his actions. Perhaps he felt a sense of unease. Perhaps he knew he was punching below the belt and had been caught out. Perhaps he believed the magnitude of Melton's discoveries warranted these actions. Maybe it was Melton who had actively sought the degree. It did take Scott's full professorial weight to force the matter, but we cannot really know his motives in full.

Around this time, the *Pander Society Letter* – a roughly typed and Xeroxed annual newsletter and who's who of conodont research – arrived in Scott's hands. It included a hurried report on the convention, which admitted to the sensation of Melton's discovery but made no mention of Scott or his speech. Its editor, John Huddle, concluded, "The specimens were examined by most of the conodont specialists at the convention. Many were not convinced that the conodont-bearing animal had been found and thought that the animal ate the conodont-bearing animal."[12] In October 1969, this was not a political statement, but a year later, when the animal had been studied and presented to the world, it would be. At this moment, then, it did not disturb Scott.

Scott suggested to Melton that the site of the discovery remain secret, at least for the moment, to prevent "people interfering." That interference was as likely to come from trophy-hunting collectors as from scientists. As a field man, Melton entirely understood, and it was natural for Scott to fear that the project might be wrecked or the opportunity stolen. They would come to realize, however, that the threat lay closer to home: in the quarry itself.

Although the site was to remain hidden, the discovery itself needed publicity. In mid-October Scott sent a short note to the editor of *Science*: "This is an exciting moment in the history of paleontology, and a preliminary examination of the animal indicates that it holds unusual interest to all biologists because of its unique characteristics. We are not in a position to say at this time what this animal really represents; but I can tell you that, if my suspicions turn out to be true, it may well hold a very unique position in the evolution of life." The mythology of the conodont, which always fed anticipation and never seemed to result in resolution, was still being written. This, it seemed, was the final chapter, and for a biological sensation of this enormity it was a fitting conclusion. In Scott's mind the discovered animal was everything for which he had hoped. It could not be easily understood. It looked entirely new.

Perhaps strangely, and certainly intentionally, the note was not authored by both men but by Scott alone. Its authorship permitted Scott to observe Melton's triumphal discovery, for which Melton was given full credit. But it also allowed Scott to exercise his authority and legitimize his participation by relating the discovery to his existing research. To this Scott added a small note purportedly authored by Melton on their plans for the future. As these were plans concocted by Scott, it seems that Melton's authorship was a political move to demonstrate that Melton was acting under his own free will and had not been kidnapped. That Scott felt the need to publish two notes rather than one is in itself interesting and reinforces the sense that a political message was being communicated to his critics. It is certain that Melton was involved.

In late October 1969, Melton arrived in East Lansing, all expenses paid by Scott's department, to spend two weeks getting to grips with the conodont animal. As they did so, an artist prepared sketches of the specimens. At some point over the coming months, Scott produced a formal proposal that divided up the work of producing the paper: Each would deal with parts of the zoology of the animal, Melton would also cover the stratigraphy and wider fauna and help Scott with the ecology, Scott would deal with the taxonomy and chemistry, and Melton would handle the comparative anatomy. Initial progress was rapid. They discerned the animal's distinctive three-part form, its mouth and anus, a fin, and a dorsal nerve cord, and thus its orientation. They found the "conodonts" making up Scott's *Lochriea* assemblage in the gut of the animal. Scott imagined that it was an unspecialized ancestor of the fishes, a chordate but not quite a vertebrate.

With his blebs paper having just been published, and perhaps confusing readers of *Science* who thought it announced the animal discovery, he told *Geotimes* to tell the world the animal had been found. Again he related the discovery to his earlier work, making it a natural culmination. Now the only five specimens of the animal in the world sat in his

office, the ultimate proof of all he had previously said. He told his growing number of correspondents this but added, "The animal is headless,
and the teeth are part of the digestive tract rather than part of a 'head.'"
These correspondents might well have thought this a significant change
on Scott's earlier views: He had seen his blebs as the remains of the head.
They might also have wondered what a strange form of life Scott possessed – headless with teeth in the gut! Rhodes asked Scott to tell all at
a GSA meeting he was organizing for East Lansing for May 1970. Scott
agreed to do so, believing the paper would be in press by then.

Scott willingly sanctioned limited publicity, always keeping enough
secret to prevent someone usurping the discovery by naming the animal
or identifying its affinity. That would pull the rug from beneath their feet.
He worried increasingly about this and eventually wrote to Richardson,
who possessed the only illustration of the fossil, made during the North
American Paleontological Convention, asking him not to publish it.
Richardson was then preparing an announcement in *Earth Science,* but
Scott told him that "prepublication" of the image "might give someone,
especially foreigners, an opportunity to use some name to designate this
animal." Scott dressed this up a little, telling him the form classification
used by "most workers (other than Scott) . . . has often reached stages of
absurdity." The "other than" here disguised the fact that Scott had not
needed to use it as he had not been publishing on conodonts! Richardson
respected Scott's wishes.[13]

That was a close call. With alarm bells ringing, Scott wrote a single
sentence to Melton to seal any future leak: "I am impressed with the
need for us to protect in any way possible the distribution of any pictures,
sketches, or figures which will give anybody the opportunity to mess this
matter up." Over the coming months there would be frequent requests
for reports and illustrations from encyclopedias and magazines. They
refused them all.

In mid-November, Scott met with the NSF, which asked him to re-
draft the proposal, integrating Melton's appendix into the main body
of the bid. Scott told Melton that "they encouraged me to believe that
we could have anything that we requested within reason. The only item
which I think might be questioned would be a new truck." Scott now
planned to join Melton in the field in the summer of 1970. Melton was al-

ready there and had turned up tiny objects he thought might be younger stages of the animal. Meanwhile, Scott continued to work on the technical details of their report. He told Rhodes that its assemblage and therefore its name was *Lochriea wellsensis*. This, of course, was the illegal terminology Scott had invented that, apparently outside Scott's field of vision, was already being disassembled by the likes of Bergström, Sweet, Lindström, and others. However, the final battle in that distant war had yet to be won. Scott wondered if the separate system he had invented might be carried throughout the classification. Perhaps Eichenberg's Conodontophorida could be kept for the isolated fossils, while his animals might be known and grouped by a separate system. The conodont animals might be known as Lochriates. Scott was perhaps aware that he was being led by his heart on this matter but asked Rhodes for his opinion all the same. Never once did he consider using Pander's "Conodonta.". It was probably at Melton's suggestion that they considered calling the animal *Panderi Rhodesi,* but Scott thought that impossible.

As the animal was being shaped in Michigan, across the Atlantic Maurits Lindström was in the rather impossible position of writing a paper on the affinities of the conodont animal for the East Lansing meeting. Rhodes had just told him that Scott possessed "what must be very seriously considered the probable conodont animal."[14] Lindström asked Scott for "inside information," aware that he might, for good reasons, not wish to give it. Scott sent him the same few lines he had sent others. Although Lindström was not disappointed, it was hardly enough to help him. All he could do was work from the long-known fossils, telling Scott, "What you are telling me in your letter sounds quite fascinating and will no doubt create great sensation." Scott, doubtless feeling the weight of expectation, now planned an "elaborate display including three-dimensional models" around which participants would engage in open debate about the animal. He never doubted that the animal would be accepted and that all would share in the triumph. He asked Rhodes if he might speak for longer so as to fully describe the animal. This was indeed to be a moment of sensation.

Scott and Melton's world was, however, enclosed and invisible. Beyond it, speculation about the animal continued, and it was not based on the specimens shown at the convention. Nevertheless, it was with some surprise that Scott read of an ongoing debate about the animal in the December issue of *Science*. This centered on some extraordinary photographs that had appeared in the magazine the previous January. Published by Rupert Riedl, a zoologist at the University of North Carolina, they showed conodont-like structures in poorly known animals called Gnathostomulida. These animals had previously been off the paleontologists' radar, but when they saw the pictures many immediately thought they saw a solution to the great mystery. Some wrote to Riedl, others to *Science*. Chris Durden at Texas Memorial Museum in Austin and John Rogers at Yale, for example, were quite convinced by the similarities. Perhaps conodonts belonged to the same group of animals? Durden imagined them as bottom-dwelling "worms," the conodont elements forming the cores of papillae used to tear up algal mats and fungal hyphae. Both Durden and Rogers recognized that there were significant differences, particularly in size and chemistry, but were nevertheless convinced by the morphological similarities. Riedl was elated by the response and admitted that he too had pondered the conodont connection but felt unqualified to comment. Now he thought these imaginings justified, aided by Durden's reference to an obscure paper by Wetzel, published in 1933, that described Cretaceous "microconodonts" from the Baltic.[15] Durden knew these were not truly conodonts, but for Riedl they were a missing link and he began working with Sam Ellison to explore the relationship further.

Scott felt compelled to respond. He did so as one who had "studied conodont assemblages since 1934" and as the "designer" of the name "scolecodont." His authority established, he closed the debate: "I can assure you that there is no relationship between Gnathostomulida and conodonts, whether direct or indirect. They are not related in any manner." He now turned Melton's puzzling animal into definitive evidence. It became an exclusive rod of office: "The reason I am so sure of this is that I have in my possession five whole conodont animals."

Scott was firm but courteous. He copied his letter to Durden and Rogers. Durden, who had done some work on the matter, confirmed that these were different groups of animal and said that he and Rogers had simply wanted to stimulate debate. However, he thought the microconodonts might belong to Riedl's animals. Riedl responded to Scott's "most challenging letter," equally courteously, enclosing a picture of the 1933 microconodonts. Scott told him they were most likely the remains of worms. That was not, however, the end of the matter, as some months later, Serge Ochietti, of the Université du Québec, sent Scott a French-language article he had written with André Cailleux. They had also been inspired by Riedl's illustrations and proposed a link to conodonts. That the elements in question were composed of chitin (the material that makes up the skeleton of insects), were 25–250 times smaller than conodonts, grew differently, and came from animals that live in the interstices between sand grains in sediments did not seem to matter. They were convinced that nature could support such diversity without destroying the link. Scott wrote to Ochietti and in a sentence dismissed their claim. He had the animal and that was the end of it.

Meanwhile, Melton and his assistants were braving the Montana winter in the quarry, where a bulldozer had been of little help in making the conodont animals appear. He had just two animals and was now investigating the older sections of the quarry where the weathered shale split more easily. This move into a disused area might also have been a retreat from quarry politics, for the quarry owner, Charles Allen, had asked him for one thousand dollars to cover the disruption to his quarrying operations. Melton was told that someone from the University of Montana had offered him such a sum. Melton told Scott and this turned February 9, 1970, on its head. Was the project to be "messed up" here, on site? They had been blind to the possibility and, thinking the University of Montana story untrue, imagined that a private collector might be muscling in. Scott doubted Allen and felt there was dishonesty at play. Anyway, one thousand dollars was a large sum and could not be raised so easily. Scott also worried about the ethics of it all and considered withdrawing. It was a classic clash of cultures, of differing risks and rewards, of differing ways of doing things, and Melton and Scott simply could not see the world from Allen's perspective. Scott felt he was being asked to buy the specimens at an exorbitant rate. Besides, Scott did not yet have the NSF grant so little could be done but delay the payment, if a payment had to be made. Nevertheless, it raised the question of property and ownership, which caused Scott to draw up an agreement: "All fish fossils obtained from the quarry shall be given to and shall remain the property of ~~William Melton~~ the University of Montana. All conodonts shall be given to and shall remain the property of Harold W. Scott, Michigan State University." Melton deleted his own name. As a curator he could not compete with his institution. In contrast, Scott was willing to give up all other finds in order to be the personal beneficiary of the conodont animals. This action was not particularly unusual. Most conodont workers remained in possession of those materials they collected and studied. They would want to know if someone else wanted to examine these objects and perhaps draw other conclusions. Excluding others from studying the objects, *after* these workers had made their own views known, might be fairly objected to, but this was not the case here. Scott's intentions become clearer later on.

Scott's fretting over the payment made work in the quarry difficult. To break the deadlock, Melton gave Allen a down payment of $575 out of his own pocket. His department bailed him out for $200 of that. Melton explained to Scott, "It is a sticky situation and I don't particularly like it but feel obligated to finish it now that I have started on it." Scott felt bad, and considered paying Melton, claiming it as the cost of a "bulldozer," but he did not do so immediately. He had, however, "forced the issue" on Melton's master's degree and it had now been awarded. So at least the exuberant Melton might get a salary increase. But then Scott sent him $275. Melton was puzzled. Was this for the existing conodont animals? If so, he corrected Scott by pointing out that these belonged to the University of Montana. Melton asked Scott if he was planning to keep that material, but Scott explained that he simply needed a reason for paying a bill – for sending Melton money – as they could not pay themselves. It seems this was a bungled attempt to cover Melton's losses without admitting to it in the accounts. Scott said that he hoped in time the material would go to the National Museum.

Melton was not alone in the field. He was aided by Montana-born Jack Horner, a student not long returned from Vietnam who would later gain international celebrity as a dinosaur expert.[16] Jean Lower of Michigan State was also there and would soon be offered an assistantship to work on the fossils. There were volunteers from Iowa and a party of two from the University of Pittsburgh who were looking for fossil fish but had turned up a conodont animal. Scott remained attentive to Melton's needs while desperately trying to catch up with the conodont literature on classification. On this subject he had puzzled a great deal. Then, in March, he realized that he had two different assemblages and thus two species of the animal.

Scott was also starting to think about where their now completed paper would be published. He wrote to the Paleontological Research Institute, which published two major U.S. journals, selling the paper's significance and the need for copious illustration. "I do not believe that I am egotistical," he said, "when I say to you that this is probably the most important paper ever prepared on conodonts. You may know that we have searched for this animal since 1856; and now that our search is realized, the paper deserves unusual attention in reference to publication." He considered it "almost certainly [to] be the most referred to and possibly debatable paper of all conodont literature in the future." The editor told Scott the paper would have to wait in line and that would mean that it would not be published until sometime in 1971. This level of delay was normal, but it was not unreasonable for Scott to think that a publisher might prioritize this paper simply to obtain it. As no such offer was forthcoming, Scott decided take his paper elsewhere. It was a decision that would cost him dearly.

Near the end of the month, Scott received a telephone call from the NSF. It was not the news he expected: The foundation was willing to pay him fourteen thousand dollars, half what he had requested. Scott told Melton to cut back; they would need to concentrate on excavation. Of that money, twelve thousand dollars was for fieldwork: two to three assistants for two months, one from Montana, one or two from Michigan; two rock saws; money for the "bulldozer" and "bulldozer" money for paying off the quarryman; truck rental; food and other necessities; and volunteers. Some money was also set aside for publication. The work would begin in June and run through July. Preparations now began, and Melton sent Scott photographs of the locality. When he saw them he was surprised and told Melton, "We may have to hire a man to use explosives."

※　　※　　※

In the spring of 1970, at the East Lansing meeting, Melton and Scott prepared to reveal all: "The sensational paper of the morning however was the one by W. Melton and Harold Scott on the progress they had made in studying the 'conodont bearing animal' found in the Early Pennsylvanian of central Montana. Harold Scott presented the paper, reconstructed the animal, named parts, and proposed physiological functions for the parts. He suggested the conodonts are 'stomach teeth' supporting a digestive organ. 3 specimens contain conodonts, but only one has the conodonts arranged like assemblages."[17] By the repetition of structures, chemistry, knowledge of assemblages, and an awareness of "similar" creatures (like the famed near-vertebrate amphioxus), they had made sense of the flattened, seventy-millimeter-long, cigar-shaped animal. Their beast was "not comparable to . . . any living or fossil animal." Its most standout feature was a distinctive black spot at its center. Above this was a triangular darkened patch within which the conodonts were found. This was the animal's gut, and in it the conodonts supported a filtering structure. From the animal's symmetry they imagined it "in almost perfect balance; very little energy would be required to move the animal in any given direction." Melton and Scott supported their claims with a chart that gave percentage values to the accuracy of their interpretations. This revealed that they were certain of the orientation of the animal and that an organ holding the conodonts existed in the animal's gut. What they were uncertain about was the form of this structure. As they put it in the published account, "These facts, in addition to the generalized surviving shape of the animal, make it a distinct possibility that some ancestral conodont animal could have given rise to the vertebrates and possibly to the protochordates, though perhaps all protochordates had a common Cambrian ancestor." The term protochordate was used to describe the non-vertebrate chordates such as amphioxus. In drawing these conclusions, Scott found British vertebrate paleontologist Bev Halstead's view that conodonts might belong to a filter-feeding, planktonic protovertebrate most useful.[18]

Some delegates expressed skepticism, believing this was an animal that ate the conodont animal. Others felt the fossils were inconclusive. One of the doubters, John Huddle, imagined Melton and Scott were disappointed by these interpretations, but the two men showed no sign of this after the event. Melton was thinking about the animal's biology on the flight back, clearly having enjoyed himself. He wrote to Scott to suggest that the tail of the animal must have contained stiffening rods to enable it to swim. Its midregion would flex a little perhaps and the head area not at all. Melton's arguments were precise and informed by what they could see in the specimens. Clearly the animal was still developing, and this required reasoned speculation. Having rented his house to the incoming departmental chairman, Melton now returned to the field with sufficient monies to continue the work.

Scott could do little else but think about the meeting. Rhodes, who had written to him to congratulate him on his "excellent paper," was now planning to produce a book to be published by the Geological Society of America and assumed their paper would be included. The date for submission was December 1, 1970, giving Scott and Melton an opportunity to benefit from the excavations that were shortly to begin. Scott, however, was rather less concerned about adding to what they knew, preferring instead to see the earliest possible publication. To that end he considered publishing the paper through his university. He told Rhodes he could have it, if it did not take two years to appear – in which case he would go elsewhere. He pressed for publication before mid-1971 and requested copious illustration.

If the conodont workers imagined Scott in a state of disappointment, they could hardly have been more wrong. He made no mention of any criticism, having evidently dismissed it as ill-informed. After the meeting, he received so much public attention that he could not have failed to feel a sense of triumph. A flurry of letters and cards requesting a copy of the paper arrived almost immediately from amateurs, professionals, and an interested public. Melton and Scott's paper had hit the media network, appearing in the national and local press across the country and attracting attention internationally.

One of these letters came from Springfield, Ohio, and concerned the illustration of the fossil possessed by Richardson. It was apparently drawn by the author of the letter. The headline in the Springfield paper had read "Missing Link," which stimulated the writer to give the terms by which the illustration might be used: $30 per single use and $150 for full rights. Scott, no doubt with a little glee, responded that the illustration was "technically incorrect and cannot be used." A cardiovascular professor heard of this "ancestor of primitive fish" in the *Denver Post*. As a leisure-time paleontologist excavating and publishing on ancient fish, he wanted to know more. He asked Scott for "technical information" and Scott replied, "It has about twelve fish characters and about an equal number of non-fish characters. Perhaps its single most interesting feature was its ability to produce calcium phosphate in the form of filter-feeding structures." Scott said the paper describing the animal would be published in a year. The editor of *Perimeter* – "A Journal of Human Frontiers" that concerned itself with the sometimes wacky edge of science – had read about "the 'minnow-like fossilized creature' which reportedly links invertebrate and vertebrate animals" in the *Washington Post*. He wished to publish a report. An author preparing a popular book on evolution, a schoolgirl in New Jersey, and a researcher in oral tissue had all read a similar announcement in the *New York Post* and sought further information. A reader of the *Oakland Tribune* in California – which reported "Another Link – 400 million years ago, simple forms of sea life grew tails and rudimentary backbones, and a Michigan professor thinks he has found a fossil of one that fills the gap between vertebrates and invertebrates" – believed he possessed similar material and offered to send it to Scott. Someone in Des Moines, Iowa, wrote to Scott thinking that Scott was probably the best person to identify the fossils in her table top. She sent photographs. A community college teacher in New York wanted a slide for teaching his general zoology class. A geology graduate drafted into the armed forces and serving in Vietnam was attempting to keep abreast of his science and wanted to know more.

The public had their curiosity pricked. As one put it, "The subject fossil find was reported in a local Washington newspaper about a month ago and ever since, like thousands of others I'm sure, I've found it difficult to contain my curiosity." An avid consumer of *Scientific American* and

Natural History Magazine, she had telephoned staff at the Smithsonian to get more information, but they knew nothing. The article itself, like all these articles, failed to mention the conodont, a term too arcane for mass consumption. But this reader was sufficiently perceptive to guess: "The newspaper article description sounded like a primitive amphioxus and I'm wondering if there is any relationship to that animal. Also, I'd be interested in knowing if Mr. Melton's find may turn out to be the long-suspected bearer of the conodont. I know that the conodont is thought by many to be a filter-feeding mechanism, and the newspaper account did say that the fossil under investigation has a plankton-straining digestive system." Some had clearly been hanging on for further details in the popular scientific press. This was certainly the hope of a reader of the *San Francisco Chronicle* who understood that Melton was the finder and Scott the interpreter of the fossil. Scott answered all the enquiries and put most correspondents on a mailing list for the paper when it was published in 1971. Others asked for his *Micropaleontology* paper on blebs, which Scott had set up as successfully predicting Melton's find.

One interesting aspect to this sensation is that there were actually two sensations masquerading under the same façade. One concerned the long-term enigma of the conodont. The other concerned the discovery of a "missing link," the "missing" suggesting scientific prediction and final resolution in the story of life. The idea of "missing links" had long existed in the popular imagination, most importantly in the search for human origins. It was an easy notion that could translate the arcane into the popular. Of course, the idea that the conodont animal might exist in that ambiguous borderland between vertebrate and invertebrate was also not new. In Scott's youth, if the animal was a worm, it was one with some vertebrate attributes. It was logical, then, to look for vertebrate and invertebrate characters. The truly bizarre nature of the animal did nothing to prevent this, though this way of seeing was entirely due to Scott's belief that this really was the conodont animal. It was those same reasons he had fired at Lange – most notably the presence of single assemblages in each of the animals – which meant these fossils could be nothing other than the animal itself.

Scott continued to work on the paper, correcting and rewriting sections in the light of the meeting, planning to enter the field on June

24. He had also got a handle on the budget and gave Melton instructions: $400 for a graduate assistant, $400 for expendable equipment and supplies, $1,033 for Melton's salary, $1,000 for food and living, $550 for truck rental, $1,000 "to pay for services rendered to the quarry man," and $1,000 maximum for the rental of the bulldozer. Scott was now convinced that they could make better progress with a little nitroglycerine: "Believe me, Bill, a man who knows his business could blast that off almost layer by layer; and a great quantity of material could be quickly and readily examined." He wrote to Jean Lower, who had evidently had had an accident, to wish her well. She responded: "My bruises are healing and the scar tissue is forming – at first my knuckles looked as if they have been completely skinned." She continued: "The fish and other not yet identified critters (worms and plant material) are coming out rapidly and several specimens which closely resemble our animal but as yet, we have only the one smudge containing three different elements."

Their relationship with Allen had not improved. They found themselves told off daily for covering up stacks of his good building stone, stone that Lower and her colleagues could not remember seeing. Horner believed "some guy from Lewistown made off with it." The complaint meant that they might have to spend days uncovering that stone. Scott, who had formed his opinions on the basis of rumors and personal communications, remained doubtful of Allen, but Lower reassured him: "By the way, he does sell this stuff because I have talked with two people who have bought from him, so he is legitimate, at least that far." Scott also tried to pay the $575 to Melton, wondering if through a little manipulation of receipts this might be possible, but it would require the quarry owner's involvement. And that relationship was still cold. Allen had, however, reluctantly done some blasting, but Scott felt the task required greater expertise and he made arrangements for an explosives man to visit the site.

At the end of July, the money spent, Scott planned the return of the truck. Melton told him that would be the end of the field season as he had already broken the axle housing on his own car. It had been a disappointing year. The fossils were interesting but not numerous, and now Horner had cut his hand badly and, after the stitches were removed, was told not to work for a week. Melton sent Scott what he had: "I found something that looks like a large beetle and some water mites about ½ to ¾" across. Finding ones that are recognizable makes it easier to explain some of the little blebs but not others."

With Melton's family now moving back into their own home, Scott relented and found another month's truck rental. As he did so, problems with the quarry owner came to a head. These called for a telephone conference between Scott, Melton, Horner, and Lower.

None of this activity was being communicated outside their circle. Scott was upbeat, telling correspondents that they had found two more conodont animals during that summer's excavations. A little optimism was politically astute if he was to make another grant bid, but the reality was rather different. Melton wrote to Scott on the last day of August, "I am sending you the conodont that I found last week. It was a very frustrating summer. I had expected to find at least eight more conodonts and found only one real good specimen." This recent find had come from exactly the same place Melton had found his previous fossils. It seemed to add to the detail that could be observed in the other specimens, and Melton postulated a respiratory function for some of the observed structures.

Melton had by then dealt with the quarry owner, who would "be no trouble any more." He had "got the Bureau of Land Management to check his claim and it will be declared invalid." He later said, "I had the land withdrawn from any mining use" at the recommendation of the Bureau, even though that organization would have liked quarrying to continue. To collect from the Bear Gulch strata on government land one would now need a collecting permit, but Melton had spoken to all the landowners in the area and reported "it will be all right to collect more if we want to." Melton now began to count the fossil finds for reporting purposes. Among them were still unknown bits of potential: "I will send all of the bleb material to you. Some of it is larval forms of something, I am sure." When the boxes arrived, Scott admitted that they were indeed very interesting fossils.

Throughout the summer, Scott continued to work out the evolutionary relationships of the new animal. He contacted Chuck Pollock, who had recently discovered fused conodont clusters, asking for his opinion on the makeup of Lower Silurian assemblages and admitting, "I am head over heels in studying assemblages." Pollock's response, when it eventually came, was rather less illuminating than Scott had hoped, but Pollock admitted his hands were tied: "I hope this helps you to some extent. I'm sorry I couldn't be more specific on some points. Being with an oil company my time is limited in the time I can spend on these interesting aspects of the conodonts." Scott then tried Lindström, who responded with the data he wanted. This made the Lower Ordovician assemblages like those he knew from the Carboniferous. The next question was "Did *Lochriea* possibly extend into the Triassic?" He asked Cameron Mosher at Florida State University in Tallahassee.

On September 14, Melton left the field, his expenses fully reimbursed and unrepentant about the fate of the quarry owner. Scott sent Melton the final copy of the conodont paper, admitting that the classification aspect was highly complex and that he might wish to protect himself from having to defend it by noting this was the work of the junior author (Scott). Melton felt no need for protection and was satisfied with the paper, though he still did not know Scott's publishing plans. Another complication was the age of the deposit. Back in June, Melton had told Scott that they had it wrong. In September, the USGS returned cephalopods they had sent and they were all labeled Mississippian. Melton thought the fish were like those from the "uppermost lower Carboniferous" of Scotland. In October, Scott realized he needed to do something about it, not least to "safeguard" his "personal historical position in this matter." He sent replacement pages to Rhodes that were rather noncommittal on the point.

In September 1970, before the paper was even in press, that year's *Pander Society Letter* arrived. It pulled the rug from beneath their feet. It may have been a rather hurriedly produced newsletter, but it reached nearly every conodont worker and in that respect was opinion shaping. Here, Huddle captured the sensational aspect of the East Lansing meeting but repeated his earlier claim that until proven otherwise this was the animal that ate the conodont animal rather than the elusive animal itself.[19] The report ended with a cartoon that had one of Melton and Scott's animals say to another, "Did you have a good breakfast?" To which the second responded, "Fine, but it was full of little bones." Beneath it, Huddle wrote, "Remember Branson and Mehl's reaction to the assemblages reported by Scott and Schmidt?" Was Huddle reporting on the irony of Scott's situation, here, or engineering it? Of course, Huddle had been on the side of Branson and Mehl in that argument, until Rhodes had proven Scott's point. Scott could not help but feel that his science was being ridiculed, seeing it as the work of those who had sought to disrupt his plans after the Chicago convention. Elsewhere in the newsletter, a strange, fanged, alien caterpillar-like cartoon animal peered menacingly at readers. Titled "Another candidate for THE conodont-bearing animal," it depicted the "aboriginal leech" said by distinguished geologist Joseph Peter Lesley in 1892 to have possessed the conodont fossils as teeth. Were readers to imagine Scott as a modern day Lesley?

Scott wrote a formal letter to the editor. "This implication is false and needs to be referred to," he said. He pointed out that five tons of rock had been excavated in 1968 and one hundred tons the following year; thousands of square feet of bedding plane had been examined without finding a single isolated conodont. The only place they were found is in the animals themselves. To Scott, this was a striking observation, for how could a conodont animal eater exist where there were apparently no conodont animals to be eaten? He pointed out that the opening detected in the animal was too small to permit a conodont bearing animal to be swallowed and that the conodonts themselves were unbroken and undamaged by digestive action. But most convincing was the fact that the assemblages were not of mixed forms – they conformed precisely to those he and Rhodes had described years earlier. Scott's interpretation had been arrived at by careful examination of the evidence: "Any suggestion that the conodonts have been ingested is wholly untenable and there is no evidence whatsoever to support such a conclusion. New evidence based on the 1969 finds wholly supports the original position."

Melton, not being a member of the conodont community, remained unaware of this development. Scott, however, prepared a note to be appended to their paper. This Scott later modified, probably at Rhodes's advice, to a rather bland statement regarding doing their best with the interpretation and being aware of other interpretations.

Scott's correspondence with his public now became rather less emphatic: "Briefly, they are protochordates; and you may rest assured that they were free-swimming and in my judgment mostly planktonic forms. They are about 2½ inches long, and the conodont elements were part of the digestive tract as part of a filter-feeding system (in my judgment)." The enquirer, realizing he had missed a momentous event at East Lansing, then asked Scott if he now understood why the platforms evolved so rapidly. Such enquiries might force Scott to ponder previously unasked questions. He postulated: "In my judgment, the bladed forms were probably vertically arranged in a basket and were relatively stabilized whereas the platform elements were horizontally arranged and were active 'food movers and sorters.' Therefore, they were subject to active evolution ."

Given the failures of the previous year, Melton wondered if the case for the next field season might better succeed if made on the basis of the fish, which were jumping out of the rocks in large numbers. The conodont animal had once again become a rarity. They could get, he believed, "at least two or three. . . . Fifty would be nice but in the realm of improbability." The past season had produced 90 fishes, 79 shrimps, 17 brachiopod slabs, 25 cephalopods (straight and coiled), 18 slabs containing "grass," 27+ coprolites, 6+ worms, 2 clams, 2 mites (which Melton thought fish parasites), 11 slabs of bryozoans, and 37 slabs of organic material. Much was new, and some material extraordinary. The season had delivered three conodont animals. By December, Scott had news that a bid for further funding had been successful – enough to give Melton and Horner one thousand dollars each plus expenses. When the season came, Scott was attentive to Melton's needs, wishing him to eat and sleep well. He retained a cushion of five hundred dollars just in case of medical expenses and other contingencies and told Melton to call if it was needed at any time. Scott was then preparing a second paper on the animals for a little geological series published by his own department. Melton, hearing that Scott was preparing for a European tour that fall, which would include attendance at the Marburg conference on conodont taxonomy, told him, a little prophetically, as we shall see, "Incidentally from the literature I would say that the Glencartholm section around Edinburgh would be the next most likely place to look for conodont animals."

In June 1971, the East Lansing book seemed to be progressing well. Melton was in the quarry sheltering from the summer rain, catching up with his field notes and dropping Scott a line on progress. Fish were being found, but the year was proving colder and even less productive. The most interesting fossil was a "worm like thing shaped like Amphioxus" that Horner had found. Melton calculated that it took forty hours to find a fish and an impossible number of hours to find a conodont. The quarry owner was, from Melton's perspective, "still causing trouble. . . . I thought I was rid of him but apparently not." He thought some kind of feud between the local and state offices of the Bureau of Land Management had permitted the owner back.

Melton was also now planning for funded excavations in future years, and at the same time he worried that "the legislature is in a hassle about how to finance the state for the next two years." No one at the university had contracts. In Michigan, too, there was much financial uncertainty. The summer malaise was broken, however, when, in mid-August, Melton sent Scott four conodont assemblages (three individuals) Horner had found.

In late September, Scott considered the future of the animal fossils. Other than those that belonged to the University of Montana, they were, so he believed, his personal possessions. Scott now drew up a statement of his wishes regarding them in case of his "death or incapacity to act." This gave full control of the specimens to Rhodes to do with them as he wished. Rhodes thanked him and reassured him: "I am sure that there is no question of any need for this, but please be assured that I am honored that you should put such trust in me." What had prompted Scott's actions is unclear. Perhaps he had returned from the Marburg meeting realizing that the world had now changed from the one he knew. Rhodes wrote to lift him, "I could not help feeling in retrospect how much the whole convention owed to you, for the initial discovery in 1934 and still more for your ongoing leadership in the field of conodont research. It must be a matter of great satisfaction that the pendulum has swung so far away from the initial skepticism of some conodont workers about the validity of natural assemblages, so that now a group of 70 experts meets together with the sole purpose of facing up to the taxonomic reflections of these assemblages." Eight months later, Scott contacted the National

Museum in Washington to carry through what he said he would always do: transfer the fossils to the central repository where they could be available to all and properly protected. It is clear that Scott felt the weight of responsibility of possessing objects so unique and important. Whether he was prompted by rumors is unknown, but it would be natural for a scientist to retain material until after the paper was published and then find a permanent repository.

December 1971 arrived without the paper appearing. Scott sent his Christmas letter to his old mentor, Carey Croneis, enclosing a description and sketch of the "missing link": "I thought you would be interested in these comments and the enclosed picture of what may be one of the most important finds in paleontological history. I say this because I am of the opinion that this animal may have been an important link in the early history of the vertebrate group." Soon 1972 arrived. The sensation of the discovery and its publicity in 1970 was now a fading memory, and the paper describing it was still locked in the GSA publishing house. Melton was back in the field with Horner, unfunded but pulling out fish faster than ever before. He promised to send Scott any conodonts they found.

In May, Scott heard from Charlie Collinson that Melton had given permission for him to publish a figure of the conodont in a field guide he was producing. Scott expressed his regret that this would preempt publication of the paper, which was now expected to appear sometime between August and November. Scott asked Collinson to include a statement that "reproduction or reprinting is specifically prohibited" but doubted it would have any validity. He asked Melton not to circulate any further pictures, but Melton corrected Scott on the details. He had not given Collinson the drawing – it was the one made at the Chicago meeting. "Since it appeared that he had it, I could see no reason to add to the confusion by being credited with finding the thing and having it published upside down." Melton continued, "I do not particularly like the idea since I was not sure that he was asking me or telling me and there was an implication that we were withholding information which is not true." The animal had been found more than three years earlier, and the supposed secrecy was entirely due to the slowness of the GSA's presses. Scott told Melton, "I think you had no other choice and handled it well."

Later in the month, Scott heard from the National Museum and began the process of allocating numbers to the conodont animals, asking Melton if he approved of this plan. Melton told him not to use the National Museum numbers for the type specimens – the most important specimens – as they were to be retained by the University of Montana, "at least for the present." This, too, was a natural curatorial reaction. He knew he could care for them and also that they would give his department celebrity and importance. Melton also told Scott that they had found another four conodont fossils, including two complete animals. By the end of the season Scott had nine, all confirming their earlier conclusions. Scott felt Melton should publish them, but Melton was happy to see Scott do it, if only "to convince those who don't believe in them that they have to at least consider them." Scott produced a report but it was not published.[20]

The East Lansing paper remained in the doldrums, the GSA editor now replaced. Scott's second paper, to be published by his own university, was waiting in the wings; it could not be published first. Unknown to Scott, Melton had published a local account, certain their paper would by then be out. He now sat on a heap of reprints awaiting that day. Then, in December, a gale blew the roof off the GSA editor's office. As he told Rhodes, "The next day the snow melted and my entire office was inundated, causing considerable damage to my furniture and the papers spread out across my desk. Among these was the material for Special Paper 141 which I was assembling for you." Promises of speedy resolution never materialized, and in the first months of 1973, Scott found himself answering correspondents who blamed him for forgetting them. The paper was becoming as elusive as the animal. Then news came that it would appear on the last day of May 1973, three years after Melton and Scott had presented it. They finally saw it in July. After all that delay, two figures – one of the rock face and the other a reconstruction of the animals swimming in the sea – were published upside down, making both incomprehensible. The publisher also had resized all the images without stating the new magnifications, giving the animal a variable and inexplicable scale. This was the only paper botched up in this way, but no conspiracy theorists took up the cause. This was not the luxuriant publication Scott had imagined, but as it was edited by his longtime friend, he said nothing. A list of errata was pasted into every book. All in all, the publication of this, billed as one of the most important papers in the history of paleontology, had been an unmitigated disaster. He had imagined this as a glorious personal moment, but the whole exercise had done him considerable personal damage. It was as if the world had conspired against him, through little fault of his own.

In November 1973, Sam Ellison, a near-contemporary of Scott, wrote a very brief and formal note thanking him for the paper. Its main purpose was, however, to state categorically, "I remain one of those that believe the animal you describe is an animal that devoured conodonts because from the balled up nature of the complex in the middle of the fish-like remains." Scott responded, "Also, I take this opportunity to send my personal assurance to you, for whatever that may be worth, that the Conodont animal is for real. I am fully aware of the 'stomach-ball' canard which was started and I am equally fully aware of the reasons for which this misrepresentation took place. I have always been surprised to what extent scientists will sometimes go to broadcast concepts even though not based upon fact. You may remember that some people did not like my discovery of 'conodont assemblages' back in 1934, but they have been subsequently found by many people and are accepted, I think, by all." Scott then gave another reason why he believed the conodont elements belonged to these animals: "Also, every summer's search has uncovered a few additional specimens including some half-size and three quarter-size animals. . . . I call to your attention that the Conodonts in the half-sized animals are one half size of those found in the adult." The assemblages within the animals were not mixed but distinct, even though two different kinds of assemblage had been found. "Sam, I write in some detail to you because of our long friendship," Scott added. "Otherwise, I would probably not write at all." The defense of his interpretation was full and revealed the considerable logic behind his conclusions. Scott had had other conodont workers visit and leave convinced.

Fortunately, Scott was now in receipt of so much appreciative correspondence that any heartache he felt must have subsided. One young

Swedish researcher told him the paper made "very 'thrilling' reading." Bill Furnish wrote to thank him, remarking, "I am now one of the few who can recall the excitement of your discovery of assemblages, 40 years ago. It has been even more exciting, I know, to work with the complete remains. These are fine publications to document their exact structure and how they occur. You are to be congratulated on having completed this painstaking research." One writer complained with tongue in cheek, that he would no longer be able to use Scott's 1930s paper as an example of paleontological reasoning in his classes because the animal had now been found.

In January the following year, Scott was still researching these strange fossils and had written to the Department of Fisheries at the University of Quebec to obtain longitudinal sections of larval lampreys, believing they might have something in common with conodonts. Melton was expecting a visit in March from Preston Cloud, who wished to see the animals and he asked if Scott could return them, adding, "Pres can be a good advocate for them if he is as convinced as we are that the conodonts are part of the animal. I don't think that anyone can look at them carefully and not be reasonabl[y] convinced." Melton had heard from Ellison, too, who had told him he had "seen the tails of these animals in Kansas City and other localities and didn't know what they were." He refused to believe in the animals, and Melton wondered if he believed in assemblages. Scott did not contradict Melton; he had long known of Ellison's opinions and told Melton, "It is hard for me to call a man a scientist when he refuses to look at the facts." As Scott told another enquirer at this time, "There are those in America who would have been happier if the animal had never been found. . . . The canard, which I mentioned earlier, was started for reasons of pique and in my judgment, was very unbecoming anyone who calls himself a scientist. I beg of you to believe in their existence and that things are reported accurately. Of course conclusions derived from the observations are subject to interpretation. . . . I assure you that in due course of time, the truth will be accepted." Scott was rather less convinced that Cloud was sufficiently knowledgeable about conodonts to form an opinion that would be influential but nevertheless believed that he should have access to the specimens. Scott sent them off, telling Melton not to return them but to think about sending them to the National Museum – indeed, with the papers now out, the National Museum requested the material. Scott made assurances to the museum that he would send his material but said that despite repeated attempts to convince Melton to do likewise, the best specimens would remain in Montana. Scott was now preparing to retire, but before he did so he began to adapt to the new taxonomy and accepted that *Lochriea* was to be a synonym of Schmidt's *Gnathodus*. Meanwhile, in Japan, there were preparations for the translation of the paper.

In June, Scott retired, intending to continue to work on the animal and believing that history would eventually prove him right. Bear Gulch had, by this time and as a result of work Scott had directed, become internationally important. The outcome for the animal was far less clear, and many – perhaps most – joined with Ellison and others in being disbelievers. Scott imagined this the work of conspirators upset by his actions following the 1969 convention, but Lindström was not among these. He had listened to Scott at the East Lansing meeting, looked at his material, and immediately become a disbeliever. Scott could only see this doubt as politically motivated rather than reasoned. As he told Ellison at the time, "I suppose that once a false idea catches on, it will take years, if not a generation, for that idea to be lost. . . . In closing, may I say that I too am often a doubter and at times even a heretic. Insofar as the Conodont animal is concerned, I do not fall in the group of doubters and neither will anyone else in due course of time." Scott considered taking a lecture tour with specimens in hand to all the centers of dissent in order to quash the "meat-ball" hypothesis. As he told another correspondent, "It becomes a mathematical impossibility for these to represent anything except the true animal. For these reasons, I think the search for the animal is over."

How can I describe my emotions at this catastrophe, or how delineate the wretch whom with such infinite pains and care I had endeavoured to form? His limbs were in proportion, and I had selected his features as beautiful. Beautiful! Great God! His yellow skin scarcely covered the work of muscles and arteries beneath; his hair was of a lustrous black, and flowing; his teeth of a pearly whiteness; but these luxuriances only formed a more horrid contrast with his watery eyes, that seemed almost of the same colour as the dun-white sockets in which they were set, his shrivelled complexion and straight black lips.

MARY SHELLEY,
Frankenstein, or the Modern Prometheus (1818)

TWELVE

The Invention of Life

TO MOST OF MELTON AND SCOTT'S CONTEMPORARIES, THE
conodont animal from Bear Gulch seemed impossible, even ridiculous – the latest and most spectacular addition to a heap of such impos-
sibilities. To Melton and Scott, and a few others, of course, it looked
entirely plausible; the reasoning that had brought it into existence was
sound enough. Most conodont workers, however, liked to imagine the
animal existed elsewhere. Some even thought it might already exist
but remained lost in a paleontological blind spot – known but not recognized. A long list of outsiders had taken this kind of thinking to the
extreme and imagined the key to the mystery already existed out there
in the zoological world. With this thought in mind, we might ask if it
was really so ridiculous for the young Klaus Fahlbusch to propose, in
1963, that conodonts were secreted by algae? Lindström, who had just
sent his conodont book to the publisher, simply could not believe it,
and Ziegler, who examined Fahlbusch's material, told him not to. In his
naïvety, Fahlbusch had hit a hornets' nest, and almost immediately the
swarm (Beckmann, Collinson, Helms, Huckriede, Klapper, Lindström,
Rhodes, Walliser, and Ziegler) was upon him, stinging him with accu-
sations of poor science. Later, Lindström would feel nothing but regret
for this incident, but when he did, he had perhaps forgotten that "conodontology" was not, in 1963, the respected science it was to become. It
was still scrambling for recognition. In time, however, Fahlbusch would
find some relief, for it was in this paper that he also told his seniors that
their methods of acid preparation were damaging their fossils. On this
point, too, they were outraged, but here Fahlbusch was to be proven

right. And, as it turned out, he was not the last to look at this group of fossils and see plants. In 1969, Felton Nease published a paper suggesting that bar-like conodonts formed the midrib of aquatic plants found in the Chattanooga Shale, plants he called *Conodontophyta chattanoogae*. It was a suggestion treated with laudable seriousness by Huddle in the *Pander Society Letter,* although the idea must have tickled the conodont research community, which was then sufficiently mature to be unruffled by such outlandish ideas.[1] Many years later, conodonts would again be mistakenly identified for plant remains, but on this occasion, as we shall see, it led rather unexpectedly to discoveries of huge significance in the hunt for the animal itself.

The 1960s are remarkable for the degree to which the nature of the animal was *not* discussed. This was a period of intense conodont research, but Wilbert Hass's and Walter Gross's work had made the subject of the animal a taboo. The animal had become unimaginable and the fossils themselves simply provided insufficient data to resolve the matter. Of the new generation, only Lindström broke this silence. The first to write a monograph on these fossils, in 1964 he had no choice but to ask what form of life had possessed them. It called for some deductive reasoning, but this is an art in which Lindström was to excel.

He started with what had become a fundamental question: Are the conodont elements internal or external structures? Hass had argued that the manner of their growth meant they must have been surrounded by tissue. If they, like teeth, then emerged, growth would stop and wear would begin. But Hass found no wear, only breaks that had been repaired; as the broken part had not been lost, it seemed logical to believe it had been retained in a fleshy covering. Frank Rhodes had not been entirely convinced by these arguments and thought the hardness of the conodonts prevented significant wear. He also believed broken parts had been lost. He, too, thought the assemblages must have formed "ingestive aids" but felt the conodont elements were probably exposed and attached only at their bases. It was Gross's later study of the base that finally put paid to this idea.

Lindström believed the fragility of the conodont fossils made them unsuitable for active food gathering and thought it more plausible that they were covered in flesh. What stood in the way of this interpretation was evidence of lost parts and wear, but Lindström thought reported losses could be explained by parts being "resorbed" or expelled. Wear was more of a sticking point, but there was no agreement on whether it could be observed. Lindström searched for an alternative explanation, suggesting that what looked like wear might instead represent "retarded growth."[2]

The animal Lindström was to build relied as much on these explanations or interpretations as on the facts as they were then understood. But, then, many of these facts were also interpretations. What Lindström was searching for was firm ground – a group of judgments that might legitimately be used as building blocks for the imagined animal. He began by establishing that the conodont elements may have been embedded in tissue throughout the life of the animal. That was the new orthodoxy. If so, then how did they function? That, he reasoned, might depend on the animal's mobility, and this in turn might be discerned from its symmetry. If bilaterally symmetrical, it was reasonable to believe the animal had been a swimmer. But some workers had reported asymmetrical assemblages, which encouraged him to think that some conodont animals may have been colonial, floating passively in the sea. This was, however, just a possibility.

The animal that began to form in his mind possessed numerous protective spiny elements: "The broken and regenerated denticles show that protection was wanted but also that the animals could escape alive after being attacked." But, he asked, were the preserved conodont fossils really fit for a protective role? He could imagine more formidable architectures if this were their primary function. He decided that he needed to think beyond these active roles and so returned to Hass's suggestion that they were supporting structures. What could one imagine if one looked at the range of elements said to be possessed by a single animal? Depending on the shape of the underlying element, one could visualize finger-like projections of flesh perhaps with tiny nipple-like coverings or fringes. If the morphology of the elements closely reflected the form of these fleshy extensions, some could be brush-like. Lindström also imagined other tentacle-like projections unsupported by conodont elements, or perhaps supported at one point in their evolution, then lost, then reappearing later on, as the fossil record seemed to suggest. Why, then, he asked, are there no instances of elements fusing together as the animal evolved? Lindström asked Rexroad if his fused conodonts indicated an evolutionary development but Rexroad thought they did not. Perhaps this meant that the organ of which they were a part had to remain flexible? Perhaps the animal could thrust its denticles outwards in defense? Perhaps the whole organ could expand and contract to draw in and expel water? The possibilities were considerable but were any really plausible?

The exact form of an organ composed of all the parts of an assemblage would, he knew, depend on its functions: food-gathering, respiration, defense, etc. The only test was to attempt the construction of what appeared to be a plankton feeding organ using the best information then available concerning the relative position of the parts. Cilia were then introduced to direct food along grooves that converged on the mouth. What he produced and illustrated in elegant detail was a hypothetical lophophore, a usually horseshoe-shaped organ of ciliated tentacles found around the mouth of certain aquatic animals that strained particles out of the water (figure 12.2a). By imagining a lophophore, he shifted his hypothetical structure into the realm of known biology, giving it greater plausibility. Lindström knew that brachiopods had lophophores with mineralized skeletons to support them and that lophophore-possessing animals were particularly common in those seas that held the conodont animal.

Even if the animal that appeared in Lindström's book had no effect on his fellow conodont specialists, others looked at it and thought they saw the state-of-the-art conodont animal rather than one man's thought experiment. There would not be another conodont monograph for a quarter of a century, and the next *Treatise* was still seventeen years away. Lindström's book became a key point of reference, especially for those who were not conodont workers. British paleontologist Bev Halstead, for example, took Lindström's animal and, reversing the flow, believed he saw a convincing filter-feeding animal. It was this that impressed Scott.

Lindström's book did nothing to disrupt that long shadow cast by the microscopy of Hass and Gross. Indeed, Lindström had constructed his own interpretive spectacles from their studies. They had demonstrated that fishes, worms, and teeth were illegitimate thoughts, so Lindström had imagined something else. Indeed, Hass and Gross's publications had made an area of investigation unattractive, as the effort required to take these studies further would be colossal, require extraordinarily well-preserved fossils, and carry a high risk of turning up nothing new. Consequently, their views stood without further testing. This set of circumstances gave their ideas a natural resilience. Unable to modify or overturn them, those who wished to progress the science of the animal had to accommodate them.

In time, however, technologies improve and open up new ways of seeing. Each generation that had used the microscope to study the conodont knew there was a resolution of observation beyond which they could not go. As if in a fog, and just beyond view, there were always tantalizing glimpses of structures awaiting technological advance. Here too one could imagine one saw El Dorado faintly in the distance. In the first *Treatise,* written in the 1950s, the conodont fossils were magnified up to 420 times and illustrated as two-dimensional slices or thin sections. By the time of the second *Treatise,* written in the 1970s, three-dimensional images were possible with magnifications of up to 8,000x, though this extraordinary degree of enlargement was rarely needed. This leap in the ability to see had profound effects on the nature and form of the object; a new object – a giant object – emerged through the fog. Now conodonts were, effectively, twenty times bigger, and it was possible to see into their deepest recesses. The science had advanced into the world of the tiny and so acquired a language of "micromorphology" and "ultrastructure."

This breakthrough, which occurred in the late 1960s, came with the arrival of affordable scanning electron microscopes (SEMs). Seeing a new territory open up, which was certain to deliver new knowledge, the conodont workers rushed in. At Marburg University, Lindström, Ziegler, and a track record in conodont research proved sufficient to get one of the first SEMs installed; the university had been willing to give Lindström

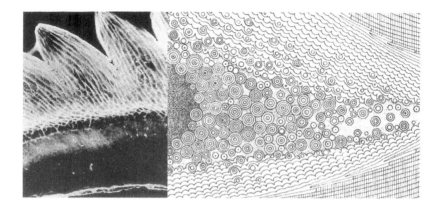

12.1. Müller and Nogami's art. These workers produced extraordinary SEM images of conodont fossils (*left*) and some of the science's most artistic representations, such as this drawing of spherical concretions in the basal plate (*right*). From K. J. Müller and Y. Nogami, *Memoirs of the Faculty of Science, Kyoto University* 38 (1971). By courtesy of Kyoto University.

anything he desired in order to secure his services. The Marburg instrument found its main use in the study of conodonts, with Ziegler being among those who, in 1968, first published an SEM study of conodont structure and chemistry.[3] He and Lindström became fascinated by the new patterns they observed on the surfaces of the fossils: polygons, furrows, tiny denticles, pits, and so on. One structure looked like a plaited loaf. Was this the remnant of an organ of some kind? They never quite knew what the SEM might throw up.

Klaus Müller, then in the West German capital of Bonn, who had taken good pictures of conodonts using a lensless pinhole camera in Iowa in the 1950s, was also among those who saw the potential. He acquired his first SEM via a military contact, its DM 150,000 price tag sufficiently small as not to raise any military eyebrows. Müller's first sortie with his new weapon was published in 1969. Working with Yasuo Nogami of Kyoto University, Müller also used the latest optical microscopes to push at the boundaries of understanding conodont fine structure. Illustrated with copious photographs and extraordinarily fine and beautiful drawings (figure 12.1), their paper showed that conodonts had probably evolved from organic structures that had become increasingly mineralized over time. Progressively, the crown and base of the conodont had become distinct, the distinguishing feature of the base being the retention of a far higher proportion of organic material. They also described the mysterious and opaque "white matter" (because it looks pale in reflected light, though paradoxically it is black in thin section) first noticed by Pander and then by every conodont microscopist since. White matter was concentrated in particular locations, and they regarded it as an advanced development. They believed lamellae had been dissolved and recrystallized to produce this substance. They also found numerous examples of the tips of conodonts being dissolved and smoothed before a new and distinct tip had grown on top. Their work was the first to begin to produce doubts concerning the Gross orthodoxy.[4]

Recognizing the power of technology, Müller had his Bonn colleagues parcel up 120 papers produced from the SEM's use and sent them to the university's administrators. Not long afterward, a new and improved model was installed.

In Ohio, Walt Sweet built an alliance with a colleague in biology in order to secure that university's first SEM. Installed beneath the university's clock tower, it operated twenty-four hours a day and had its own dedicated staff member. It was by this means independent of any departmental control or politics. Sweet thought this ideal. Chris Barnes in Canada also became active in SEM studies, these helping him understand the relationship between structure and ecology. In 1970, for example, he and his collaborators came to believe that robust Ordovician fossils with a shallow basal cavity and little or no white matter were characteristic of near-shore, hypersaline, carbonate environments.[5]

This explosion of activity peaked in the early 1970s. It brought to light a host of new structures and supported studies of evolution, classification, and the mechanical properties of conodont fossils, but it left many of the most fundamental gaps in understanding unresolved. As Rhodes reflected at the time, "The use of scanning electron microscopy has revolutionized our knowledge (though not yet our understanding) of the internal structure and surface architecture of conodonts."[6] As a physical object, by the early 1970s, it certainly seemed that the conodont element was known in every detail, yet the science was still no nearer to understanding how these elements, or indeed the assemblages, functioned, and consequently it had no chance of comprehending the nature of the animal itself. The work of Hass and Gross had been superseded, but these new studies did not immediately displace the constraints they had put on the fossil. Of rather more importance to the future of the animal was a general unshackling of the imagination arising from the cultural shift taking place in the science in the early 1970s that encouraged paleontologists to become freethinkers, modelers, and theorists. It was this that would allow conodont workers to again think creatively about the animal much as Lindström had done in 1964.

In this respect, Melton and Scott's animal, with its bizarre stomach teeth, had certainly done the science a service because it put forward the most bizarre of explanations for how the conodont elements functioned. Lindström, of course, had been asked to reflect on the nature of the real conodont animal for the meeting at which the Bear Gulch beast was presented to the world. He had to do this without actually knowing much about that animal, but as soon as he saw it, he was perhaps relieved to know that it was not really the conodont animal at all. Indeed, in the book edited by Rhodes in which Melton and Scott's animal was described, he mourned the fact that so little progress had been made. The new statistical work of Bergström and Sweet, and others, seemed far more illuminating, for it had revealed that the animal had only a limited number of assemblage architectures, each linked to the other by an evolutionary thread. The existence of a clear evolutionary relationship between these different assemblage types encouraged Lindström to believe that there must also be a functional relationship. It was logical to consider that they all did roughly the same job, just as the teeth in our mouths do pretty much what they did in the mouths of our ape ancestors and their ancestors before them. This idea – that these elements had particular functions – was also supported by the knowledge that similar architectures and characters evolved repeatedly. It was already being established that shape correlated with lifestyle, and this further enforced the view that function was directly linked to form. This thinking led to the asking of new questions. For example, platforms, which look most like crushing teeth, sit alongside elements lacking a single plane of growth. If these latter elements could not have closed as one tooth does upon another, how could the platforms next to them do so? More to the point, platforms seemed to have evolved from conodonts of this other kind.[7] This sort of reasoning suggested even the most tooth-like elements could not have functioned as teeth. Even apparatuses composed entirely of simple conical elements could not, Lindström thought, have functioned as jaws, as the elements were too long, too thin, or simply wrongly shaped.

With teeth out of the picture, Lindström reflected on those SEM images, which showed a honeycomb pattern of pitting on platform surfaces at right angles to the main denticle row. These looked similar to muscle attachment pits, encouraging him to imagine the conodont elements supporting tentacles. Fused clusters were now of more use to him; perhaps these conodont elements had been brought together through the contraction of muscles on death. Using this kind of reasoning, Lindström refined his view of the oral cavity of the conodont animal, picturing "a wormlike, soft-bodied animal, moving or floating in most cases freely in the sea." At five to fifty millimeters long, he reasoned that "it probably fed on microplankton that it strained from the sea-water by means of a lophophore, which was supported by phosphatic elements and may have formed a ring round the mouth" (figure 12.2b).

Lindström's biological thinking had, since his arrival in Marburg, been greatly affected by weekly lunches with distinguished algal biologist Hans-Adolf von Stosch. He also began to spend two or three weeks a year with his students at a marine station in Roscoff, Brittany. There, surrounded by an outstanding library, he looked in detail at the full range and morphology of marine life through the eyes of a conodontologist. With these increasingly biologically informed thoughts in his mind, the final stage in the development of his personal interpretation of the animal resulted from an invitation from the International Palaeontological Association to give its prestigious annual address at Burlington House in London in March 1974. After several decades of serious but muted geological research, the newsworthy sensation of Melton and Scott's beast had returned the conodont animal to its rightful place as paleontology's greatest and best-loved enigma. Behind this sensation, however, stood a period of rapid progress that had stabilized the language, revealed an advanced understanding of the animal's evolution and microstructure, and proven the fossil's unrivaled stratigraphic utility. As Lindström spoke it would have been clear to this diverse and informed paleontological audience that he possessed arcane knowledge, that conodont studies possessed their own complexity, sophistication, and expertise. Lindström would also have come across as a man who delighted in the thought experiment and lateral thinking. As a speaker he was doubtless aware of how compelling such skills become when attempting to unravel a seemingly insoluble puzzle.

Lindström challenged his audience to imagine the functional interrelationships between the different fossilized parts of the animal, each of which was in a process of evolutionary remodeling. Surely they could see, as Lindström could, that the morphology of the elements, with denticles arranged along different planes, meant they could not be teeth. At the time, the arrow worm or chaetognath was a popular model for the conodont. Seddon and Sweet had used it in their ecological model as a behavioral indicator, and Siegfried Rietschel, at the Senckenberg Museum in Frankfurt, having compared conodont assemblages with the radulae of gastropods, jaws of polychaete worms, and grasping apparatus of chaetognaths, thought the latter provided the best anatomical model for those conodont animals possessing simple cone-like elements. Rietschel admitted that this model could not be extended to more complex forms, but Lindström thought the argument fundamentally flawed: Conodonts fossils were often too thin, curved, or complex to have been used for seizing or holding food. He was supported in his doubts by the new evidence of a fossil chaetognath found in the Pennsylvanian Mazon Creek Lagerstätte in Illinois.[8] Here, iron-rich nodules, exposed during strip mining and left on heaps of spoil, were found to contain a multitude of fossil worms, including an arrow worm looking very much like its modern-day cousins. Although its head was poorly preserved, the find suggested that chaetognaths had coexisted with, rather than evolved from, the conodont animal.

Lindström now adjusted and developed his original model, attempting to accommodate all that was securely known and believed. There was no better person to do this. Lindström was considered a consummate conodont connoisseur – even Ziegler tipped his hat to him. Nevertheless, there had to be some art to clothing bones in flesh; this could not be achieved through objective science alone. Lindström reasoned that the structure of the conodont element must have had a bearing on the fleshy structures attached to it. Conodont evolution showed that the surface area of elements increased over time, but what function, he asked himself, would warrant this? Breathing, excretion, and the uptake of nutrients – these all seemed possibilities, but only the latter had specific requirements of shape. If tentacles were attached to muscles on the surface of the elements, they could be extended and retracted. The sharp and pointed faces of the elements were, he reasoned, pointed away from each other because, covered in flesh, they could not have articulated one against the other without causing damage. This, he thought, was reflected in the assemblages that had been found.

He considered other possibilities but settled on a barrel-shaped organ. It was still a lophophore. For Lindström, Melton and Scott's conodont-eating animal – as he saw it – limited the overall size of the conodont animal to seven to nine millimeters and constrained its shape: "The animal is more likely to have been oblong or even barrel-shaped rather than long and worm shaped." Now the animal emerged: "The conodont animal thus sketched was not necessarily very mobile. It might even have been a passive floater, relying on its battery of unpalatable denticles for its protection. Some conodont animals may have formed colonies. . . . The food gathered by conodonts might have been both microscopic particulate matter and dissolved material." This animal reflected Lindström's awareness of modern forms of life possessing lophophores and an earlier view that conodonts and brachiopods might have a common ancestor.[9] With the animal and its lifestyle out in the open, he placed it in a real-world environment: "If this is true we might expect the conodonts to occur most plentifully in areas where such nutrients are abundant, as for instance in environments characterized by upwelling deeper ocean water. Such environments may occur on the margins of oceanic troughs or along submarine rises. The occurrence of conodonts in certain fossil sediments (black muds and trough-rise limestone facies) appears to agree with this prediction." Lindström's "barrel" or "gooseberry" was now established, a sophisti-

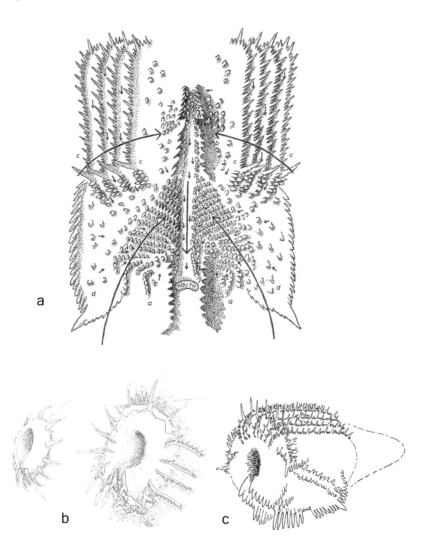

12.2. Lindström's imagined animals: (a) the conodont lophophore is born in 1964; (b) it developed to fringe the oral cavity of the animal in 1973; and (c) in 1974, this became Lindström's barrel. Reproduced with permission from M. Lindström, *Conodonts* (1964), Elsevier; M. Lindström in *Conodont Paleozoology* (1973), and M. Lindström, *Palaeontology* 17 (1974).

cated reading of the evidence, clothed in flesh and dropped in the ocean (figure 12.2c).

The paper demonstrated Lindström's mastery of his field, but he left certain things unsaid. It had occurred to him that the animal forming in his mind looked something like a spiny ostracod. Perhaps the ostracod was mimicking the conodont animal with its armory of tough spines. With this thought his mind, the conodont animal acquired its barrel – or ostracod-like – shape. It was an idea to which he intended to give further thought, but when he did so, it occurred to him there were no predators in that sea capable of seeing with any precision an animal so small. So what use could there be for this kind of mimicry? He laughed to himself and forgot about it. He remained attached to the idea that the conodont elements might support tentacles on a lophophore, but for him the barrel was a rather more ephemeral imagining. It continued to exist, however, in the published paper, there to be the subject of serious debate and occasional ridicule (as Lindström's "toilet roll") long after that thought had, for Lindström, passed.

Simon Conway Morris was among those who took Lindström's barrel seriously as he pondered a curious fossil from another – the most famous – Lagerstätte: the Middle Cambrian Burgess Shale of British Columbia, Canada.[10] He had chanced upon a sawn slab and put it to one side while searching through the huge Burgess Shale collections at the National Museum of Natural History in Washington. A short while later its counterpart turned up. Conway Morris, who was to build his career on bringing to light the weird and wonderful Burgess Shale fauna, knew this fossil was something special: "The specimen had evidently never been noted by any other worker. No other specimens have been found." Poorly preserved as a thin film darker than the surrounding rock, this gelatinous, flattened, worm-like animal was a contemporary of the earliest known conodonts. It lacked a distinct head but possessed a curious, almost figure-eight-shaped apparatus within which could be detected the impressions of twenty-five "teeth" very similar in shape to conodont

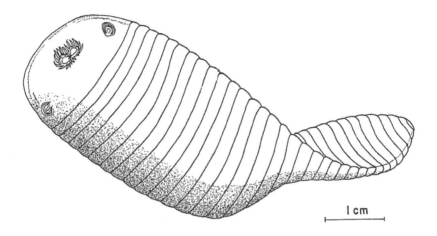

I cm

12.3. Conway Morris's animal. The conodont elements fringing the mouth supported a tentacle-covered lophophore. This lophophore had been reasoned into existence by Maurits Lindström. Reproduced with permission from S. Conway Morris, *Palaeontology* 19 (1976).

fossils of that period. Unfortunately, most of what had caused these impressions had leached away – a common problem with calcareous fossils in the shale. Nevertheless, Conway Morris thought them conodont-like in size and shape and noted that they too possessed a basal depression and showed some variation in form around the apparatus. He thought this reflected a possible "symmetry transition," an idea from Lindström's book, which Conway Morris used as his main guide to the conodonts. The counterpart specimen lacked these teeth but instead seemed to show the decayed remnants of tentacles of a kind seen in some other rare fossils. But unlike paleontologists studying common fossils, Conway Morris's science lacked the firm footing of familiarity. He had to decide what constituted informative data on the two specimens and what might have resulted from post-death compaction and alteration. His interpretations depended on his own connoisseurship of fossils and on the work of others – such as Lindström's barrel and transition series, and a 1932 interpretation of tentacles – work he could not know well and relied upon the interpretations of others. These were his building blocks for imagining the animal; in effect, Lindström offered him both spectacles and blinkers. He could not have imagined the animal he now drew without them.

Conway Morris produced an animal that swam by contracting muscles along its sides much like a modern nemertine worm (figure 12.3). The "teeth" were not really teeth. They were too fragile to produce a rasping lophophore, but they may have supported tentacles on a lophophore covered in fine cilia, which directed food toward the mouth. He looked for a modern-day animal group with this same feeding structure and found it in brachiopods, the colonial bryozoans that resemble tiny corals, and a group of sessile worms that live in chitinous tubes. These animals confirmed to Conway Morris that his fossil did possess a lophophore, but they also told him that his animal, which he called *Odontogriphus omalus* (from "toothed riddle"), was not related to any known type of lophophore-possessing creature.

Agreeing with Lindström, Conway Morris dismissed Melton and Scott's animal as a predator of the conodont animal. Now Conway Morris believed *he* possessed the conodont animal. It was rather larger than Lindström had imagined, but Conway Morris felt Melton and Scott's predator could accommodate it. And while Lindström's animal floated, Conway Morris argued that modern animals of similar size possess locomotion. He also did not like the way Lindström had placed his lophophore on the exterior of the animal; he thought this improbable, but he accommodated this by imagining Lindstrom's more complex lophophore as an evolutionary development on that seen in his own animal. By these means, Conway Morris gently shoehorned the nose of his worm onto Lindström's barrel, claiming without fanfare to have found a conodont animal – an animal that could not be allied to any existing group. He saw his animal as ancestral to the conodont animals that blossomed in the Paleozoic and, so he thought from reading Lindström's book, continued up to the Cretaceous. And now a recent discovery in Kazakhstan seemed to take the conodont even further back in time, to the boundary with the Pre-Cambrian.[11]

Gould later remarked on Conway Morris's discovery: "What a potential coup for a beginner – to discover the secret of secrets, and resolve a century of debate!" Whether Gould was ever really convinced by this animal is uncertain; most conodont workers were not. This was, however, a science of improbabilities, so ably demonstrated a few years later by a little debate that emerged on the pages of *Science* concerning "conodont pearls." Tiny phosphatic spheres up to 0.7 millimeters in diameter, each possessing a dimple, had been found in their thousands in the United States and Australia, associated with conodonts and composed of the same layered structure. Brian Glenister, Gil Klapper, and Karl Chauff thought they resulted from irritation caused by detritus or parasites.[12] They were, it seemed, the equivalent of the oyster's pearl and perhaps akin to the balls associated with Müller's *Westergaardodina*. Duncan McConnell and David Ward at Ohio State University were unconvinced, reasoning that these spheres were the nautilus-equivalent of bladder stones, well known in the modern animal. Glenister and company stood their ground but the topic faded.

It was now possible to think freely again about the animal, to reason from what was known. The field was heating up and the animal was once again a hunted species. But with no fresh discoveries to hand, the hunt took place in acts of quiet contemplation. It was time to imagine.

Julian Priddle, a young zoologist at the University of Reading in the UK, reflected on the similarity of some conodonts to the horny eversible "tongue" of the hagfish. He thought conodont elements acted as skeletal supports beneath a layer of secreting flesh which produced a horny surface layer (figure 12.4b). While hagfish lacked this underlying structure, lampreys possessed it. And while the lamprey showed a radial arrangement of these elements, in the hagfish they were arranged symmetrically. Priddle also noted that the structure of pits and striae revealed by recent SEM work was strikingly similar to that found in unworn mammalian teeth. In mammals, this pattern was caused by the cells responsible for the secretion of enamel.

A model of concise reasoning, Priddle seemed to suggest that one could have one's cake and eat it: "This reconciles the internal nature of conodonts with a tooth function (whereas it was previously generally assumed that to function as teeth the elements had to be external . . .)." Priddle sent Scott a manuscript copy of his paper while the paper itself was in press, but Scott was now in defensive mode. Implicitly, Priddle's model doubted Melton and Scott's discovery, and Scott responded, believing Priddle's mind had been contaminated by the "canard": "I want to assure you that the animals are for real and that the assemblage of conodont elements occurs precisely where it is reported." Scott had a series of animal fossils now that showed that the conodont elements grew in size as the animal grew. "For these reasons," he argued, "I think the search for the animal is over. The questions now are, how did they work and what is their evolutionary significance?"[13]

Conway Morris's fossil had emerged in some senses from a blind spot – it had been collected and placed in a museum but had not been studied or analyzed. Günther Bischoff, now at Macquarie University in Australia, also thought he had uncovered a blind spot. He puzzled over some long-known spiked bars and ribbed plates. Walliser had noticed that these spikes mirrored the interior shape of the conodonts' basal cavity. That encouraged Bischoff to develop the idea that these bars were "conodont supporting elements," which joined together into ribbed plates and together made up the mysterious Conulariids. These animals existed over the same time period as the conodonts and remained difficult to interpret. They are considered part of a group of animals that includes corals and jellyfish. The conodont elements were to Bischoff, then, a covering of the exterior shell of this animal.[14]

Jan Hofker at The Hague in the Netherlands compared the structure of conodont apparatuses to those found in a diverse group of microscopic animals known as aschelminthes, which include the tiny soil-dwelling rotifers. By careful rearrangement he tried to show that they were homologous,[15] using that same visual trick executed by Macfarlane half a century earlier. Few conodont workers would be fooled by such lookalike reasoning.

These views indicated that the enigma was again reaching beyond the closed community of conodont workers and again generating solutions from those who had little sense of how sophisticated the field had become. Within the conodont community the search continued for new ways to look at what was already possessed. For example, the base of the conodont had remained, since the time of Gross, largely ignored. In part this was understandable, for no basal body was generally associated with

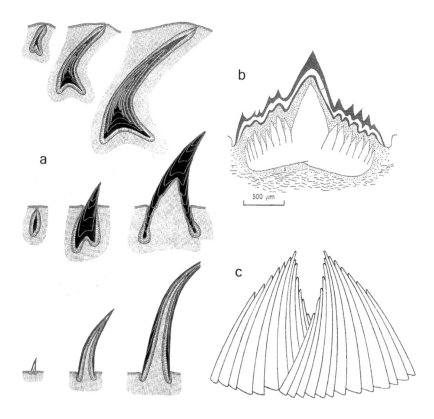

12.4. Conodont teeth: (a) In the mid-1970s, Stefan Bengtson gave conodont workers reason to believe again that conodont elements might be teeth. The series of images on the left, bottom to top, show an evolutionary relationship between protoconodonts, paraconodonts, and true conodonts based on changing patterns of growth. The true conodonts were enclosed in pockets from which they protruded when in use. The argument had a beautiful logic that convinced many. (b) Inspired by the anatomy of the hagfish, Priddle had earlier imagined conodont elements covered in flesh that carried a horny surface layer. (c) Landing suggested that the simple cone-shaped paraconodonts and protoconodonts might have operated together as superteeth. Reproduced with permission from S. Bengtson, *Lethaia* 9 (1976); J. Priddle, *Geological Magazine* 111 (1974); E. Landing, *Journal of Paleontology* 51 (1977). SEPM (Society for Sedimentary Geology).

those later conodonts that had most interested conodont workers. But Cambrian conodont specialist Stefan Bengtson, in Uppsala, thought the base ripe for renewed investigation and thought it might throw new light on how the conodont elements grew.[16] This had, after all, been one of Gross's main objections to the tooth hypothesis. Pointing his SEM at the problem, Bengtson developed a model that saw the "crown" of the conodont (the conodont fossil as commonly understood) attached to the base and sitting within a pocket in secreting tissue. Bengtson's key innovation was to suggest that when in use, the conodont protruded through the tissue; when at rest, it did not – it was protected and could continue to grow or be repaired. It was a neat solution that had parallels in the anatomy of arrow worms. Bengtson arrived at this model, however, not by analogy to any living animal but by studying those Cambrian fossils that looked very similar to conodonts. Within these Cambrian forms he distinguished two types on the basis of their growth patterns: a newly named group he called "protoconodonts" and Müller's paraconodonts. Both had a rather contentious relationship to true conodonts, but Bengtson suggested these other fossils were ancestral to them. With this in mind he considered how their growth differed. His protoconodonts, for example, appeared to have grown on the surface of the secreting tissue and were progressively extended by additions to the base. In the paraconodonts, these structures began to form within the tissue, emerging as they grew. The true conodonts, however, could not grow in this fashion as additions were made to the whole of the structure and not just to the base. To resolve this problem he had placed them in fleshy pockets. It was an elegant model, not least because it drew upon real fossils and made no appeal to modern analogy (figure 12.4a). While the more speculative animals were met with silence, at least in press, Bengtson's model stimulated considerable debate, forcing Bengtson himself to make accommodations.

Perhaps the most important outcome of this model was to give the conodont workers back their teeth, which for so long had be relegated to a supporting role.[17] The implications were considerable, for teeth connect the animal to its environment, permitting the interpretation of feeding habit, ecology, and lifestyle. Teeth are fundamentally more exciting than hidden supporting structures of unknown function. They are visible and comprehensible – more so than any other skeletal component. When Hass first pushed the conodont beneath the "skin," giving it only a vague function, it was as though something had been lost from the science.

Sweet thought Bengtson's solution ingenious. Almost immediately conodont workers began to imagine teeth again. As had been seen repeatedly in this animal's history, it was not the evidence of fossils alone that shaped the animal in the heads of scientists, for this evidence was never unambiguous; the animal that had thus far lived in those minds was the result of reason, familiarity, ignorance, imagining, and perhaps even desire.

Bengtson's solution was little affected by Müller and Dietmar Andres's discovery of a fused "protoconodont" cluster from the Upper Cambrian shortly afterward. The cluster's most interesting feature was the size range seen in its six pairs of simple curved and cone-shaped elements. It suggested that elements might have been added during life – a radical and potentially disrupting notion, particularly for those who built apparatuses statistically. A year later, Peter Carls interpreted these same fossils rather differently, suggesting elements might also be lost.[18]

Ed Landing at the University of Michigan also possessed some thirty-five fused clusters of essentially the same type. These came from a conodont animal that was unique for being known from individual elements, natural assemblages, and fused clusters.[19] Landing's finds, however, forced him to see these fossils differently. The elements were placed so close together they seemed to form two opposing "superteeth" capable perhaps of grasping and capturing prey (figure 12.4c). He imagined these superteeth grew by the addition of conodont elements, as Müller and Andres had suggested, and thought they might be a model for the paraconodonts and protoconodonts but not for the conodonts proper. Landing also poured cold water on Conway Morris's creation, doubting that any animal could possess so small a lophophore in comparison to its size or that true conodonts had actually evolved at that time. If these were conodont-like objects they would have to be protoconodonts.

Just as lines of support had once formed behind the fish and worm, so now a division appeared between those who welcomed the return of the tooth and those attached to the filter. The latter group was headed by Lindström and supported by Conway Morris. They were joined by Robert Nicoll, who found a complete conodont apparatus in the gut of a fossil fish from the Canning Basin. Nicoll painstakingly extracted the conodont elements, each complete with a basal plate. He then argued against Bengtson's pockets, believing the conodont element was covered in tissue and that the basal plate had been attached to muscle fibers. Reversing the direction of the apparatus – something Jeppsson had done earlier – Nicoll proposed a ciliated filter feeder.[20] Situated in a groove beneath the animal's head, the different element types picked up food, directed water currents, and sent food into the mouth (figure 12.5a).

Shortly afterward, Victor Hitchings and A. T. S. Ramsay from the University College of Swansea in Wales joined this group by reinterpreting Schmidt's apparatus arrangement. They imagined Schmidt's jaws opening and closing, while his gill elements became a food-filtering basket (figure 12.5b).[21] Scott was also thinking along similar lines (figure 12.5c) and gave Ronald Austin an unpublished illustration in 1978, before these British workers had published, that also imagined a filtering mechanism. Presumably, this filter was to exist within the Melton and Scott animal. Within a few years, then, science had furnished the animal with every possible filtering arrangement. The kind of reasoning that had led Lindström to produce his animal now infected others.

On the opposing side, the tooth theory had been strengthened considerably by the evidence of objects that were not true conodonts but that Bengtson said evolved into them, the paraconodonts and protoconodonts. As yet, no one had confronted Lindström's important sticking point that conodonts were simply too complex and fragile to have functioned as teeth. Then fellow Swede Lennart Jeppsson entered the debate. He had long pictured teeth. Indeed, he would have preempted Bengtson's model by several years had not an editor required him to cut an inordinately long and complex paper down to size. The pages he removed were insufficient to form another paper, so they were put on a "to do"

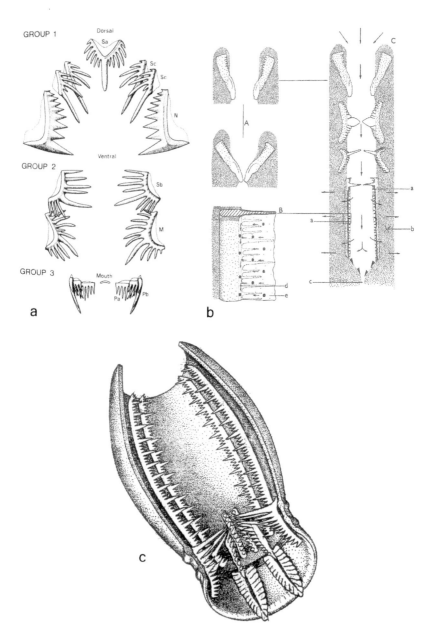

pile for later. With Bergström now back in the United States, the young Jeppsson was geographically and intellectually isolated. This made it difficult for him to judge the significance of his unpublished work. He eventually published his thoughts on the functioning of the conodont assemblage in 1979, and here he followed a familiar logic: Conodonts look like shark teeth, and like teeth in general they have a base and a crown.[22] Why should they be so divided if merely internal structures? Jeppsson carried through his arguments using photographs as indisputable visual proofs. Even the most complex of conodont elements – those with problematic multiple planes of growth – seemed to have equivalents in the dental elements of fish. In essence, Jeppsson was saying if element shape is a product of function, how can we deny that conodonts are teeth if their shape so perfectly matches that of known teeth?

His concept of "teeth" was necessarily broad, including not just those of vertebrates but also the mechanical food-processing units of a range of invertebrates. He felt these functional analogies could be taken even further. The different shapes of shark teeth, for example, reflect their different functions – slicing, crushing, grasping. Perhaps it was possible to use this information to better understand the functioning of individual conodont elements and the apparatus as a whole. Grasping teeth in sharks, for example, had to resist forces operating in a number of directions. This seemed to be replicated in conodont assemblages, and it was the presence of the mineral apatite that permitted these elaborate constructions in both animals. Jeppsson's arguments

12.5. Conodont filters: (a) Nicoll's filter is here seen head on and slightly tilted so as to show the three groups of conodont elements one behind the other. These would have been covered in flesh and perhaps tentacles. Those nearest to us picked up the food, those in the middle directed water currents, and those furthest back directed the food into the mouth. (b) Hitchings and Ramsay's filter turned Eichenberg's gill-supporting elements into a food-absorbing basket. (c) Scott's filter is similarly arranged, though drawn the other way up: food entered through the lower opening. Reproduced with permission from R. S. Nicoll, BMR *Journal of Australian Geology and Geophysics* 2 (1977). © Commonwealth of Australia (Geoscience Australia) 2012. This material is released under the *Creative Commons Attribution 3.0 Australia License*; V. H. Hitchings, and A. T. S. Ramsay, *Paleogeogeography, Palaeoclimatology, Palaeoecology* 24 (1978). Scott's filter courtesy of Ronald Austin.

were about understanding function, not, as had so frequently been the case, an argument for affinity on the basis of similar shape. His paper called for the conodont tooth to again be taken seriously; its destruction had been premature.

This paper now started a debate in the pages of the palaeontological journal *Lethaia*. Conway Morris was the first to enter the fray. His filter-feeding animal could not coexist with Jeppsson's theorizing or with Landing's supertooth. For Conway Morris, the "peeling back" of tissue to reveal even a simple cone – as Bengtson suggested – posed space issues. He found Jeppsson's case for teeth suggestive only as a result of selective reasoning and argued that the significant differences of scale in the objects he was comparing would affect their function. With Scott-like certainty he claimed that his was "the only fossil in which soft parts have associated conodont-like elements arranged in a pattern that is consistent with a viable hypothesis of conodont function and as such merits attention." Jeppsson thought this latter assertion something of an overstatement – there was no certainty that Conway Morris possessed a conodont animal or had interpreted the fossil correctly. There was a resemblance but nothing more. Jeppsson simply could not see any underpinning logic in many of Conway Morris's assertions and felt they did nothing to undermine the tooth. Bengtson thought Conway Morris's arguments "traditional" (in other words, addressed in the past) and his animal improbable.[23] He argued that evolutionary evidence was growing to support his model and that common examples from zoology could be used to counter the argument about space.

Hubert Szaniawski from the Polish Academy of Sciences now teased apart the curved, spine-like protoconodonts from which Ed Landing had made his supertooth and built from them an ancestral arrow worm or chaetognath. Müller and Andres had noted how chaetognath-like they looked but thought this to do with convergent evolution. To Szaniawski, even the internal structure of the protoconodonts was arrow worm-like. There were other actual and hypothetical similarities too, as arrow worms also have their spines in two groups and enclosed in a sheath when at rest. Only the phosphatic composition seemed to argue against the elements belonging to chaetognaths, but Szaniawski could show that this was a common condition in animals at the time. As Conway Morris had noted, in the early Cambrian there was an excess of phosphate in the environment. With the seemingly constant problem of chemistry dealt with, the case for the arrow worm seemed remarkably strong. The supertooth may have been demolished, but it had flagged up the difference between these more primitive forms and conodonts proper. The question remained whether Bengtson's model of evolution was still plausible and thus whether chaetognaths and conodonts shared a common ancestor. The research landscape had again shifted. Old objects had been renewed; once again there were new things to prove and disprove. Walt Sweet found himself "particularly attracted" to Szaniawski's arguments: "And, with the appearance of Szaniawski's elegant study, many concluded that we had finally discovered the conodonts' roots."[24]

Doubtless many conodont workers dug into their bookshelves to discover a little more about these relatively obscure animals. Ralph Buchsbaum's classic *Animals without Backbones* included this vignette: "In the open ocean we find transparent, slender animals, usually 1–3 inches long, that look like cellophane arrows as they dart after their prey. Though at certain seasons they occur in incredible numbers, and at such times form a large part of the food of fish, the arrow worms are members of a phylum, the CHAETOGNATHA, which has relatively few species. The name means 'bristle-jawed' and refers to the curved bristles, on either side of the mouth that aid in catching prey. The body is divided into head, trunk, and tail and has finlike projections, which probably serve as balancers. The brain is well developed, and there is a set of eyes. The anus is situated at the junction of trunk and tail, about a third of the way from the posterior end. The three body regions are separated internally by transverse partitions, and there is also a longitudinal partition which separates the coelom into right and left halves. The animals are hermaphroditic: both male and female sex cells arise from the lining of the coelom. The body plan is so different from that of other groups that it is difficult to say what relationships they have to other invertebrates. In certain details of development the chaetognaths resemble some of the members of the phylum to which man belongs."[25] One could imagine the conodont animal again.

By the end of the 1970s, the enigma of the conodont had reached the peak of anticipation. The desire for, and elusiveness of, this real animal, had caused the production of numerous imaginary ones. In paleontology courses around the world, lecturers would regale their students with the tale of the science's great enigma. It demonstrated better than anything else that nature still possessed great mysteries despite the best efforts of science to defrock them. In the middle of this decade of imagining, the second *Treatise* was written, devoted this time wholly to conodonts. In it, Müller provided data essential to the conodont myth: In the course of 120 years of study the conodont had generated fifty-three possibilities for what it might be. All were in some respects products of the imagination, invented forms of life. As Müller concluded, the true nature of conodonts remained "one of the most fundamental unanswered questions."[26] He could not know then that the answer was just around the corner.

. . . and the discovery of the Golden City, or El Dorado – believed
by him to be situated in Guyana – and the conquest of that country,
occupied his mind; but which appear to have been some time
before in his contemplation, and required only the circumstances
in which he was now placed, to give them life and activity to exert a
controlling influence over his thoughts.

Account of Sir Walter Raleigh's search for El Dorado,
JACOB ADRIEN VAN HEUVEL, *El Dorado* (1844)

THIRTEEN

El Dorado

FOR THOSE WHO WENT IN SEARCH OF PANDER'S EL DORADO, that distant city of gold was where the extinct mythological beast lay at rest, its flesh sufficiently preserved to at last reveal the truth.¹ In 1923, Macfarlane had dreamed of such a place, "that some layer of subaquatic volcanic ash may yet be discovered." Many had dreamed, but the animal had not revealed itself. Few, if anyone, had imagined that this sacred place might be a shelf, box, package or drawer. But there it was, this Holy Grail of science. And there it had been *for some sixty years.* Hidden from view and beyond the reach of all earlier attempts to find it, it might just as well have been lost in the mountainous jungles of Guyana. Then, in 1982, Euan Clarkson found it – though at first he did not know precisely what he had found. A paleontologist at the University of Edinburgh, he had been searching for fossil shrimps in the Granton Shrimp Bed. This rock outcrops where that city meets the sea, but Clarkson was not braving the Scottish weather. He was working his way through old collections held by the British Geological Survey in Edinburgh, much of which had been collected by the Survey's fossil collector, David Tait, early in the twentieth century.

A shrimp bed had been first discovered at Cheese Bay in East Lothian, some twenty-five miles from Edinburgh, in 1903, but was soon exhausted as a result of the attention paid to it by collectors and the Survey's paleontologists. It contained a single kind of shrimp but no other fossils. Collectors then made their way along the coast in search of further outcrops. In 1917, Tait at last found one: a finely laminated limestone, forty-five centimeters thick, in a coastal section dominated by Carboniferous black shales. The limestone was unique and quite different from that at Cheese Bay, for despite its considerable age, it preserved the soft and delicate structures of shrimps, worms, and a host of other animals.[2] Tait gathered up small slabs of this precious rock and placed them into the protective care of the Survey, there to serve science as and when they were needed.

At that moment, of course, on the other side of the Atlantic, Edward Kindle, E. O. Ulrich, and the others were just beginning to believe that the conodont might resolve the dispute over the black shales – rocks roughly contemporary with those at Granton. But in 1919, those thoughts were still immature and micropaleontology itself a thing of the future. Ulrich and Ray Bassler had not then split the conodonts into a myriad different fishes, nor had they or Ted Branson and Maurice Mehl proven their utility. For Tait, working in the practical world of Survey geology, the conodont lay beyond his field of vision. Among all those smudgy suggestions of life, the tiny animal was in so many ways invisible.

Clarkson lived in a different world. He belonged to that postwar generation that had aspired to a new paleontology and in recent decades had gained a new optimism following the discovery that Lagerstätten were not as rare as previously thought. A science built largely on the evidence of bones and shells, these remarkable deposits provided "windows" through which lost worlds could be seen in all their ecological and anatomical glory. They made it possible to both imagine a deep past clothed in flesh and aspire to the invention of that long desired science of palaeo*biology*.

Among those who had pioneered this new vision was Clarkson's collaborator in the shrimp project, Derek Briggs, a paleontologist at Goldsmith's College in London. In the 1970s, Briggs and Simon Conway Morris had been part of a three-man research team, led by Harry Whittington, that re-examined the oldest and most extraordinary Lagerstätte then known, the Burgess Shale of British Columbia in Canada. Using the evidence of rare and fragmentary fossils, they conjured up a previously unimaginable world, for the rocks seemed to record a great biological experiment that had taken place at that very moment when life had acquired its anatomical complexity. The old orthodoxy, which they were now displacing, suggested that life had been born into a number of biological groups (mollusks, echinoderms, annelids, etc.) that persisted through the millennia. Animals had continued to evolve, but they had done so within the constraints of this natural order. It was this knowledge that encouraged paleontologists to pigeonhole their finds and believe that all life must fit somewhere in this ordered world. And it was this expectation that had made the conodont so remarkable, for it had repeatedly resisted all attempts to pin its biology down. Many of the new Burgess Shale animals were similarly resistant, prompting the idea that their early extinction recorded "failed experiments" in the diversification of life. This brilliantly evocative notion would find support over the course of the following decade and become celebrated in 1989 in Gould's bestseller, *Wonderful Life*. Immortalizing Briggs, Conway Morris, and Whittington as the heroic architects of this new vision, Gould promoted the idea that life in the deep past was considerably more exotic and varied than it is now, and that it was only a matter chance that those animal groups we know today survived this moment of experimentation. If one could replay time, he said, a rather different biological world might evolve – one that was truly alien. There was no *natural* order and evolution would not repeat itself. This was a radical and exciting new way to think about the past, and it gave the conodont a new lease of life, for it could now be considered a strange survivor from that moment of experimentation.

Clarkson had tried to find the Granton bed from about 1980 onward: "Went searching on Sunday mornings when my elder sons were playing rugby. I found many loose blocks and then the 'mother lode.' Briggs and I had collected quite a lot of material before I searched the Survey collections in 1982." When he did search those collections, Peter Brand, the curator, handed him this curious specimen. As Clarkson looked closely, he could see a wormlike animal – just four centimeters long and two millimeters wide, preserved on opposing limestone slabs. It had what looked like fins, a bi-lobed head structure, and tiny teeth. "Was this an ancestor to the modern hagfish or lamprey?" he wondered. If so, then it would be an important find as these boneless animals had left almost no fossil record. Briggs sent Clarkson a paper describing a fossil lamprey from the Mazon Creek Lagerstätte in Illinois. As it too had come from the Carboniferous, it was reasonable to expect that the two fossils would show similar features, but Clarkson's had none of the head cartilage seen in the American specimen yet nevertheless looked rather more lamprey-like in overall morphology. Clarkson was also perplexed by his animal's teeth, which were quite different from the peg-like structures found in modern lampreys. He wondered if they might find a better analogy in the hagfish, as part of "some kind of armoured protrusible pharynx." The fossil seemed to be new to science and he needed additional opinions. Using a camera lucida to superimpose an image of the fossil onto his paper, he produced accurate drawings and sent them to Briggs in January 1982, telling him, "Tooth structure is really much more reminiscent of those of hagfish – see enclosed diagrams – increasingly this impresses me." He appended the opinions of his zoological colleagues. They had suggested that he had a larval stage or that the "teeth" might actually be gill rakers or part of the branchial basket. Even before Clarkson realized what he had found, his conversations were unknowingly revisiting earlier ideas about the conodont animal. It was as if the enigma, soon to enter its death throes, was reliving its past existence one last time.

When the drawings arrived in London, a new rumor began to develop. Now, and from Clarkson's drawings rather than from fossil or rock, the conodont animal began to emerge and acquire flesh. When the news reached Bev Halstead at the University of Reading, one of Britain's best known paleontologists, he urged Clarkson to announce the discovery in *Nature,* the premier vehicle for breaking scientific news. Halstead had a penchant for sensation, but on this occasion he was working on a volume on fossil fishes in which he would have liked to illustrate the new animal: "So please a preliminary description, illustration and name. ASAP." Clarkson was rather more circumspect: "For the moment I am quite uncertain about the spiny things inside the mouth. They look like conodonts, yes, but I would not go so far as to say they are – their story has grown a bit in the telling thereof!" Clarkson remained calm. He knew others had believed they had seen the animal, only to be greeted with ridicule and scorn. In two weeks he would be in London, there to spend a week in Briggs's company writing up the Granton "shrimps." He could wait until then to decide the matter and then publish a note on the fossil. He sent Halstead some photographs and a copy of the drawing, telling him, "The creature is clearly a cyclostome."

By the time the two men met, the seed of the idea had matured; the conodont animal was taking on definite form. Briggs teased grains of sediment away from concealed elements using weak acid, as the two compared the new fossil with natural conodont assemblages. Still they could not be absolutely sure. They needed an expert view. Fortunately, on Wednesday, March 17, 1982, Briggs was to attend a meeting of the Council of the Palaeontological Association at the Natural History Museum in London. He knew Nottingham University's conodont specialist, Dick Aldridge, would be there. An authority on conodont palaeoecology and Silurian stratigraphy, Aldridge had not researched or pursued the conodont animal or the Carboniferous rocks in which this Scottish fossil had been found. The only expertise Briggs and Clarkson required of him, however, was the ability to identify conodont fossils. When Aldridge got up that day, he had no idea that his life was about to change forever. "If there was ever a case of being in the right place at the right time, this was it," he later recalled.

After the meeting, Briggs asked Aldridge if he would take a look at a fossil. When the two met with Clarkson, Aldridge still had no idea what he was about to be shown. The rumors had certainly not reached Nottingham. Asked to look at the fossil through a binocular microscope, he soon confirmed that the amber tooth-like structures were indeed conodont-like. Looking at part and counterpart together, he could also clearly see that a conodont apparatus was preserved. Aldridge took his time to consider the possibilities. Like Clarkson, he was not one to jump to conclusions. Possibilities had to be weighed up, probabilities considered. Could this be one fossil (a worm, say) superimposed on another (a conodont apparatus)? He thought that idea most unlikely, because the elements were enclosed within the fossil impression. Was this a case of a conodont animal merely having been eaten by Clarkson's beast? With no disruption to the conodonts, and no further remains of the dead animal, this again seemed unlikely given the high quality of preservation. All the components seemed to fit together: The conodont fossils were in the right place and undisturbed and the animal was of the right scale. The realization began to dawn. He really was looking at the conodont animal!

Aldridge was numbed. That evening began to pass in a dream. Had he remained in this dream, his role in this story would have amounted to little more than a footnote. But on the train home he woke up and jotted down his observations. The next morning he wrote a two-page letter to Clarkson, copying it to Briggs, comparing the animal's apparatus with recent bedding plane assemblages described in Rod Norby's unpublished doctoral thesis at the University of Illinois in Urbana. He also included copies of a number of illustrations of schematic arrangements of conodont elements and natural assemblages. Aldridge knew far better than the other two that a rich and complex body of arcane knowledge surrounded this animal and its anatomy, but this remained implicit and unspoken in the letter; Aldridge was just giving them a few pointers. He ended, "I hope these rather garbled comments are of some help. Perhaps my enthusiasm and excitement are coming through; the more I think about it the closer I get to being happy that you've really found it. I suppose one reason is that your animal fits my prejudices much better than any previous contender, but the evidence is also looking stronger all the time. Can't wait to see it in print. Thanks for showing it to me."

Clarkson and Briggs immediately got down to writing a short account of the discovery for *Nature* titled "The conodont animal is a chordate" and asked Conway Morris and Aldridge for comments. Clarkson and Briggs also had a rather more generous offer to make Aldridge: "Euan and I have had lengthy discussions about a subsequent more detailed paper and, as I told you on the telephone, we would be delighted if you would cooperate as a joint author." He added, "To avoid any subsequent misunderstanding I should say at the outset that we have decided that the most appropriate authorship would be Briggs, Clarkson and Aldridge, in that order, and we hope that you will be happy with this. It is likely, in any case, that among conodont workers yours will be the name that will spring to mind as author of the conodont animal, once the larger paper has appeared!" The plan was to get the paper written before the summer and submit it to the journal *Palaeontology.* Aldridge was delighted. It had been pure chance that led him to the fossil, but it was the writing of his letter that changed his life so fundamentally. His expertise was indispensible, and if Briggs and Clarkson were not already convinced of this, Conway Morris told them this straight when returning comments on their proposed paper for *Nature:* "Even with my limited experience of conodonts I found the relevant parts of the discussion somewhat simplified; might I suggest that Dick Aldridge joins you in this description rather than waiting for the full blown account later." When they received Aldridge's "friendly amendments," their insecurities must have been further amplified, for Aldridge wrote in the arcane code of conodont elements – "Sc," "Sb," "Pb," "M" – and transition series elements. On that same day, Stefan Bengtson in Uppsala, Sweden, where Conway Morris was at the time, wrote a long detailed critique of their proposed paper: "If the conodonts are in situ, which seems likely although not proven, the elusive conodont animal has finally been caught, however badly battered. But is it a chordate?" After a detailed critique of this idea, he wrote, "Obviously, in a poorly preserved fossil one may to a certain extent see what one wants to see. You want to see a chordate, I want to see a chaetognath, and maybe none of us is right. But I find it to be a major weakness of your paper that you herald your fossil as a chordate without giving attention to alternative explanations. (Maybe it is a characteristic of the conodont animal that it leads its investigators to jump to conclusions too quickly. Melton & Scott certainly did so with their zeppelins, I think Simon did so with *Odontogriphus,* I think you are doing so with the present animal – and just now I have to admit that I feel enchanted by the prospect of identifying the same beast as a chaetognath!)." He continued, "I advise you not to stick out your much to[o] valuable heads with assertions on the conodont animal as a filter-feeding chordate." He then asked them to consider publishing a revised version of the paper in the paleontological journal *Lethaia,* which he edited, rather than *Nature.* "This is clearly a scoop," he added, "and it could be taken into the next available issue outside the normal waiting list." Clarkson was then a new associate editor at the journal, and Bengtson also pointed out that *Lethaia* had carried much of the recent debate on conodont biology.

When Clarkson received Bengtson's letter, he was quite taken aback and responded, "I confess that I simply had not thought of the animal as anything but a chordate, because of the apparent myomeres [muscle segments] and the ray-supported structure of the tail. But as you have shown clearly, these do not unequivocally indicate that the animal is a chordate." He told Bengtson they were still deliberating on whether to publish in *Lethaia*, believing that *Nature* would permit rapid dissemination, but Bengtson told him his journal would not be far behind and would offer a much better vehicle for the discovery. He did the hard sell.

This put Briggs and Clarkson in a difficult position. If they went with the Swedish journal it would mean a full account rather than a piece of breaking news. That would leave virtually nothing for the paper in which Aldridge was to be involved. Briggs asked Aldridge how this might be resolved, but Aldridge simply refused to be part of that discussion. It was for Briggs and Clarkson to decide. He was happy either way, and now also a little bemused because news of the animal had spread like wildfire across Sweden – Lennart Jeppsson had invited Aldridge to speak on the animal at a forthcoming meeting. This put Aldridge in a difficult position. He told Briggs, "I'll have to talk to yourself and Euan about how much or how little you are prepared for me to reveal." Aldridge was then dissolving rock Clarkson had sent him and despite early doubts managed to find conodonts and clusters. These would help to confirm that this really was the conodont animal. On May 26, 1982, Briggs and Clarkson finally decided to change their plans and invited Aldridge to accompany them in extending their five-thousand-word paper into a full account titled "The conodont animal" to be published in *Lethaia*. They also told Aldridge, recognizing that only he among them had a specialist interest in the animal, "You will be free to write a follow-up in due course. Either or both of us, hope to cooperate on this in any way (compaction of assemblages, stratigraphy, associated faunas) that seems appropriate, but leave this open to discussion." Aldridge immediately thought about getting a research student working on the topic, but Briggs asked that this be delayed; he was still hopeful that other opportunities might arise from the bed and he and Clarkson did not want to relinquish ownership. Briggs finished, in a manner reminiscent of Scott's advice to Melton, "We look forward to hearing from you about developments but counsel that you keep the specimen reasonably "close to your chest" until our joint paper is out. I guess that discoveries such as this make us all a bit paranoid!"

Clarkson sent the completed manuscript to Bengtson on June 21, 1982. They had kept to their timetable and finished it before the summer, but they had missed the October issue. It would be published in January, after a little tinkering with the finer details, and open the 1983 volume.[3]

In their description of the animal, the three men did not waste many words on all that had gone before; that was now simply irrelevant. Briggs had decided this early on, and Aldridge agreed. Melton and Scott's beast was unquestionably a conodont eater and Conway Morris's had nothing to do with conodonts at all. The new animal stood alone (figure 13.1). It was no surprise that it possessed no skeleton other than the conodont apparatus but a revelation to find that that apparatus was arranged back-to-front. From Schmidt onward, conodont workers had got this wrong. Only Jeppsson and Nicoll had dared to entertain this unorthodox thought. This put the long comb-like elements, two of which held a long, backward-pointing spine-like cusp, near the opening of the mouth. Behind these comb-like elements came the pair of stout blades, and behind these were the platform elements. Among the animal's most intriguing soft tissues was a pair of pale blue ellipses that formed dark lobes projecting toward the front of the body. Between them was a space perhaps leading to the mouth. The body of the animal also showed a line running down the center, evidence of segmentation, a tail fin, and a posterior fin.

In an instant dozens of speculative animals vanished from the minds of those who read the paper, but they were to be replaced by an animal that was frustratingly indefinite. What could it be? Briggs, Clarkson, and Aldridge narrowed the possibilities. There seemed to be just three. The eel-like shape, with possible lateral flattening toward the tail, the asymmetrical posterior and caudal fins, hints of a possible notochord, and indications of segmented muscles all suggested a chordate, but none of these features was so well preserved as to be definite. An alternative was

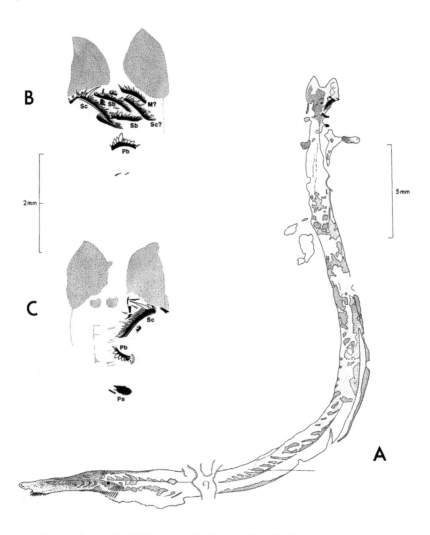

13.1. The conodont animal. This image closely resembles Clarkson's giant camera lucida drawing, 70 cm × 70 cm, of the "lamprey" he sent to Briggs. Clarkson already doubted that it really was a lamprey. The head is top right in (A), with conodont elements indicated in black and repeated structures indicated with stippling; (B) and (C) show the conodont assemblage preserved in the head region and preserved on facing rock surfaces. Reproduced with permission from E. G. Briggs, E. N. K. Clarkson, and R. J. Aldridge, *Lethaia*, 16 (1983).

the arrow worm or chaetognath as these animals have eyes positioned where the subcircular bodies appear in the conodont animal, and they possess a similar body shape. But chaetognaths do not have segmented muscles, and this new animal did not seem to possess the arrow worm's paired fins. Neither possibility was certain, and they decided to err on the side of caution. *For now,* the animal would remain in its own exclusive club, the Conodonta. It was still proving resistant. Indeed, it even proved difficult to give it a name. This relied on the identification of the platform elements, but these were poorly exposed. The authors resigned themselves to naming it *Clydagnathus? cf. cavusformis,* meaning it was probably or possibly *Clydagnathus* and like the species *Clydagnathus cavusformis.*

Concealed within this name is a little poetry, for it was originally the invention of Rhodes, Austin, and Druce. Aldridge had been Austin's student just as Austin (and Druce) had been students of Frank Rhodes. It was Whittington who suggested to Rhodes that he study conodonts. Briggs and Conway Morris, who refereed the paper, had been students of Whittington. That it should be this animal and this name was pure chance, as were Aldridge's involvement and the role of Lagerstätten. But these coincidences are not evidence for scientific ley lines, merely indications of how small the palaeontological community was and how it was organized. One further linkage now developed in the form of Stephen Jay Gould, who had previously admitted that conodonts were rather alien to him, though like all paleontologists he was intrigued by what he called – borrowing from Churchill – this "riddle wrapped in a mystery inside an enigma."[4] Gould used the discovery of the animal as material for one of his regular articles on fossils in *Natural History* magazine. In it, and claiming Clarkson as a friend – Gould had visited Clarkson on a number of occasions and the two got on well – he used this fossil as a springboard to discuss his "Wonderful Life" view of evolution, documenting the exploits of Briggs, Conway Morris, and Whittington.

Briggs and his collaborators knew that on the most critical points the fossil was more suggestive than conclusive. More material was urgently needed, but a search of other Shrimp Bed collections turned up nothing.

So Aldridge assembled a team that included his former research student, Paul Smith, and began a ground assault on the Scottish shoreline. With military force, they applied sledgehammers and crowbars to split the limestone until it would split no more. Acids etched the rock surfaces. But the animal was well and truly holed up. It did not appear. Adopting another line of attack, they tried to dissolve it out, submitting whole blocks of rock to acid immersion. Even this violent interrogation did not bring to light more than the occasional isolated element or cluster. The rock remained silent. There were no more animals.

All looked hopeless, but then things changed, and they did so remarkably. Inspired by a lecture on the animal by Clarkson, Neil Clark, an undergraduate at the University of Edinburgh, visited the site to look for himself. An inveterate fossil hunter with a nose for rarities, on this day – June 16, 1984 – accompanied by fellow collector John Hearty, he was to find the greatest of all rarities: the first in situ conodont animal and the second animal to be found. Hearty gave a "scoop" to MAPS, the newsletter of the Mid-America Paleontology Society. "As the sun beat down on us," Hearty began, as he witnessed Clark find a host of unusual fossils. "Now, by this time you will have realized that Neil was having 'one of those days' – he could do nothing wrong! . . . but Neil wasn't quite finished." It happened toward the end of the day, when his hammer blow laid open the animal. Unfortunately, this one lacked the conodont elements.[5] "Just think of the consequences if this was the only other one found!" Clark later speculated, "No conodont elements in the head region of the animal – the original would have been reinterpreted as chance association." Luckily it was not the only one. A short while later Hearty collected a block with another animal on it, which he only spotted when he got home. Clark later turned up another on a loose block. Only partially exposed, these animals were then passed to Aldridge and Briggs for further preparation.

At this time, the research site remained pretty much under the control of Clarkson and Briggs. Aldridge had been admitted onto their patch under certain conditions but also with the recognition that in some senses this research into the conodont animal would be a kind of inheritance for him since he was the only conodont specialist among them. As was the case for Melton and Scott when they believed that they held the animal, it was necessary to be sure that no one would enter the site and disrupt the research. Consequently, in April 1983, when Clarkson was contacted by a paleontologist whom Aldridge knew, and who desired samples of the Granton bed, Clarkson asked Aldridge for his opinion. Aldridge, who thought this worker completely trustworthy, told Clarkson that he could seek reassurance in "a covering letter asking him not to publish or publicize anything on the material without prior reference to you." Clarkson had already begun to have a conversation with the Nature Conservancy Council, the government agency responsible for geological conservation, about protecting the site from inappropriate exploitation, vandalism, planning developments, and so on.

News of the discovery had reached Australia by September 1982. There, Robert Nicoll, who was studying Devonian clusters from the Canning Basin, asked Clarkson for an advance copy of the paper, adding, "If I can base some of this speculation on your animal it would be of great assistance." In October, the *Scottish Journal of Geology* wanted a picture of the animal for its cover. In December, Aldridge wrote to *Nature* offering to write a news article on the animal and a "history of the speculation." The journal's Alun Anderson responded, "I'm afraid we don't allow authors to write about their own work in News and Views." Asking for a copy of the paper, he added, perhaps not as tersely as it seemed, "We will then decide whether to commission something on it from an expert in the field." They decided to do so and a précis of the paper appeared in *Nature* two months later. *New Scientist* also carried an announcement at this time under the heading, "Museum piece solves palaeontology puzzle." It ended: "The conodont animal has been found, but we still don't know what it is." Those publishing the *McGraw-Hill 1985 Yearbook of Science and Technology* wanted the story and illustrations in May 1983. Soon cards were arriving from all over the world requesting reprints of the paper. Aldridge had, beforehand, compiled a list of 205 names of individuals known to him who were to be sent the paper. When Sam Ellison received his, he admitted to Aldridge to "being skeptic on many conodont problems." It seems he remained doubtful about the new animal too.

Aldridge was not alone among the authors of the paper to also do the rounds, giving talks on the discovery of the animal. These were invariably compiled under titles like "The long tail of the conodont animal" but tended to spend little time on the enigma. As the research developed, so these talks changed, always incorporating the latest discoveries. However, there appears to have been no media frenzy; at this the team was to get much better. When, in July 1983, Aldridge read the headline "Dinosaur 'find of the century'" in the British broadsheet newspaper the *Guardian,* he took the opportunity to write to the editor to point out an equally newsworthy story that had been missed. Briggs was having some success and had been asked to write up the animal for the Field Museum of Natural History *Bulletin* – Briggs was visiting scientist at this Chicago museum – and contribute to a new book on *Problematic Fossils,* asking Aldridge to collaborate on the latter. The animal also made its way into the world's popular scientific literature, but it did so slowly. In 1984, *Géochronique* published a short article titled "Solution d'une énigme: l'animal-conodonte."

While the authors of the first paper searched for the future in real objects, others began to re-examine earlier debates in the light of the evidence of the Scottish animal. Among these was Stefan Bengtson, who was encouraged by Briggs, Clarkson, and Aldridge's tentative belief that the conodont elements were teeth. He revisited his model and dealt with the reversed apparatus and the animal's slender and mobile body. He now saw the animal as an active "macrophagous predator" (eating prey of relatively large size) in which the elements (teeth) sequentially grasped, held, directed, cut, and crushed.[6] The tentacle-covered sieve favored by some of his rivals looked, by comparison, unworkable. No filtering system would put the fine filter first, he said, but then he admitted that the asymmetry of the animal's apparatus was problematic for all existing models. His solution was to remind his readers that the conodont elements were withdrawn into nonfunctional positions in pockets. Perhaps the apparatus only took on symmetrical form when extended and in use? This kind of "dilating pharynx" was seen in other animals that swallow large prey, such as the chaetognath, so why not the conodont? This functional connection to the chaetognath also enabled Bengtson to reaffirm his view that both conodonts and chaetognaths evolved from the same stock, the chaetognath-like protoconodonts.

This view seemed further strengthened when an almost complete *Panderodus* apparatus turned up in the Ukraine. Possessing simple conical elements, it was strikingly similar to the apparatus of the chaetognath, even if the two animals were very different in size. But Polish and Ukrainian collaborators Jerzy Dzik and Daniel Drygant argued that any similarity was mere illusion. It was simply that the distinctive ancestors of the two animals – *Panderodus* and the chaetognath – had evolved to become more alike. Dzik and Drygant further weakened Bengtson's position by arguing that the fossil's hooked simplicity meant that it must have functioned rather differently from the complex conodont apparatus seen in the Scottish animal. This then broke the connection between chaetognath-like protoconodonts and that animal. There was no longer any reason to believe that the one was ancestral to the other, they argued.[7]

Sweet also felt the protoconodonts an irrelevance. He thought they were more likely Szaniawski's chaetognaths than conodont ancestors. He preferred to see the conodonts as a unique group that emerged from that great evolutionary experiment recorded in the Burgess Shale: "More than a century of trying to make conodonts something else has done little more than emphasize their uniqueness.... So a reasonable reply to the query 'What are conodonts?' ought to be 'They are conodonts,' oughtn't it?"[8] He had told Abe Zaidan, a reporter from the *Akron Beacon-Journal* who had probed Sweet about his work a quarter of a century earlier, "They were neither fish nor fowl, but conodonts." Zaidan, who thought this unhelpful, translated the arcane "conodont" into "those fascinating little whatzits." Sweet was amused but also doubtless a little frustrated that the news media could not handle the animal's name. Now he was annoyed that science could not deal with it either, for Briggs and his colleagues had called their paper "The conodont animal." This had become necessary because science had, in the absence of the animal, slipped into the bad habit of calling the tooth-like fossils conodonts. Sweet reminded the authors that these animals already possessed a name, the one Pander had given them: "The conodonts."[9]

Those outside the conodont community were rather less inclined to give the conodonts the degree of independence that comes with having phylum status. Briggs and his co-workers certainly preferred to find the animal a home in known biology, as did leading French fish paleontologist, Philippe Janvier, who would soon become one of the British team's principal antagonists. With playful mischievousness, he would soon force Briggs and his colleagues "to try harder."[10] Before the animal's discovery, Janvier believed conodonts were perhaps closest to vertebrates, a conclusion he reached from reading Gross. He thought the new animal's chevron muscle blocks and fins resembled those of chordates and suggested that the Granton animal might even belong to that exclusive chordate group known as craniates – animals with a skull. It was an idea others would adopt, re-awakening a possible relationship with jawed vertebrates,[11] but this was further than Janvier was willing to go, and even Aldridge and Briggs thought the proposition unlikely. If the conodont animal did indeed belong in this chordate world, they thought it most likely to join the jawless fishes (the Agnatha).

Already these conversations were making assumptions about the animal possessing teeth, but Robert Nicoll, who had predicted the reversed apparatus in his filtering model, now returned to that topic to elaborate. Based on fused clusters from the Devonian, Nicoll imagined the anterior bar-like elements arranged parallel to the body of the animal, set in a food groove with cusps and denticles pointing downward. They were covered in tissue, which permitted the elements to continue to grow while also supporting cilia, which captured food particles and passed them down a groove into the mouth where the other elements would gently squash or shear through the soft food. "We now have a picture of the conodont animal as one who consumed particulate matter by swimming through the water with its mouth open," Nicoll wrote. "It was . . . free to go where there was appropriate food, which may have been near the surface or at some depth. It is unlikely that the animal was exclusively a bottom feeder or that it burrowed in the sediment."[12] Far from being damaged by the new fossil, Nicoll's model seemed stronger than ever. Indeed, the vagueness with which many of the Scottish animal's features were known permitted conodont workers to continue to entertain diverse views. The animal's discovery had simply reinvigorated recent debate.

*   *   *

By now, a tiny heap of animals was accumulating in Edinburgh and Nottingham. There they remained largely out of sight. But then it was discovered that someone was interfering with their supply lines. The shift from mythological creature to a reality had placed the conodont animal on the fossil collectors' shopping list. There could be few finds more prized, and there was an increased risk that commercial collectors with an interest in profit rather than science might pillage the site in search of treasures to sell on the open market. Then, on one of his regular visits, on July 16, 1984, Clark was shocked to discover that someone had visited the site and taken away several square meters of the precious bed, leaving other rare fossils exposed to the elements. Clark told Briggs and Clarkson, who then contacted the Nature Conservancy Council. The Conservancy mustered "a lorry and a gang of tough fellows" to help remove much of what remained and take it to the Royal Museum's store at Newbattle near Dalkeith. This was on February 25 and 26, 1985. However, it was not this material, nor that previously found by Clark and Hearty, that led to the next development.

A new animal now entered stage left. No one saw it coming – not even those who had found it. It came from the Lower Silurian Brandon Bridge Dolomite in Waukesha County, near Milwaukee in Wisconsin – a rare Lagerstätte much older than that at Granton. Two amateur paleontologists, Gerald (Jerry) Gunderson and Ron Meyer, had collected it but not recognized what it was. Nor indeed had Don Mikulic and Joanne Kluessendorf, the two Illinois geologists working on the fauna. It was Briggs, his eyes now attuned to the animal's peculiar form, who spotted it in University of Wisconsin collections. Having announced the discovery with Mikulic and Kluessendorf, he was then permitted to choose his own collaborators in order to describe it. Unsurprisingly, he chose Smith and Aldridge. It had to be admitted, however, that alongside so many other wonderfully preserved fossils, this new animal was rather disappointing. Its *Panderodus* apparatus, consisting of simple cones and forming the only evidence that this was a conodont animal, was slightly detached from the remains of the soft tissues. One had to believe they belonged together. The animal did, nevertheless, preserve hints of segmentation, and this suggested that this conodont animal was not a chaetognath.[13]

However, what was more noteworthy about this discovery was the lack of sensation. The Scottish animal had acted like a switch: In an instant the sensational unknown had simply become an animal. Had this new Silurian animal been found a few years earlier, in the hands of a Scott or Conway Morris, perhaps it would have grabbed more headlines. It was nevertheless a topic of conversation for those one hundred conodont workers from twenty-nine countries who met at the European Conodont Symposium held jointly at Ronald Austin's University of Southampton and Dick Aldridge's University of Nottingham in July 1985. Aldridge was now adept at selling the importance of a fossil that could never, by itself, roar like *T. rex*. In the run up to the conference, he told the press, "The importance of the animal is not only that it helps solve one of geology's enigmas, but that it fills a 'missing link' in our evolutionary knowledge – it is quite possible that conodonts had a place in our ancestry."

This meeting was an opportunity to discuss the latest work being done on the Scottish animals and to further reflect on the meaning of those things the science had long known, such as the natural assemblages. On these latter, Briggs and Aldridge had been among the first to see new potential. Briggs had been in the States in October 1983 and met with Norby and saw some of Rhodes's specimens and Ernest Paul Du Bois's assemblages – which Briggs recognized as being outstanding. He also discovered that Norby was completely tied up with other work and unable to publish his 1976 thesis in which, according Aldridge, Norby had unscrambled the architecture of the conodont apparatus better than anyone. Briggs's solution to this problem was to suggest incorporating parts of Norby's thesis into a paper by Aldridge, Briggs, Norby, and Smith. However, he made no mention of this to Norby but understood that he would probably be amenable to such a plan. Briggs had also been chatting up other conodonts workers, such Dave Clark in Madison, and discovering other new research on assemblages. He also wondered if they might look again at Bear Gulch: "If Scott's animals were eating our creature the latter should have been preserved as well."

Aldridge had later traveled with his postdoctoral research assistant, Paul Smith, to the United States, where they examined the Illinois assemblages, collected new specimens, and discussed the possibilities of collaboration with Rod Norby with the aim of producing a refined architectural and functional model. This work was reported at the Nottingham meeting by Aldridge, Smith, Norby, and Briggs and introduced a new orthodoxy which resulted from an elegant experiment to reconstruct the three dimensional apparatus from the flattened fossils.[14] Previously, it had been assumed, explicitly or implicitly, that the variety of assemblage patterns had resulted from muscular contraction following the death of the animal (rigor mortis) and the subsequent collapse of the apparatus on decomposition of these muscles. Aldridge and his collaborators set out to discover if the various configurations could, in fact, be explained by the simple collapse of an apparatus without invoking muscular contraction. If so, it would be possible to discover the arrangement of the apparatus in life.

Briggs and graptolite specialist Henry Williams had, a few years before, developed a method that could reverse this process of collapse – at least visually.[15] It involved building models – physical models with wood and card – and photographing them from various directions. By this process, the three-dimensional object was translated into a two-dimensional photograph, so mimicking the collapse of the fossil onto the bedding plane. Aldridge and Smith built a model conodont apparatus from modeling clay, informed principally by the arrangement of the elements in the known animals. By trial and error, they arranged and rearranged the elements and photographed them from every angle. From this, they produced a single architectural model that could, by collapse in different directions, produce all the known configurations of elements seen in the fossils. The resulting arrangement of the component elements was startlingly new. It revealed an animal that was laterally compressed and therefore unlikely to be an arrow worm, and in which that the elements were not strung out in a linear array as had long been imagined. What was particularly alarming was that the arrangement challenged their original assumptions of an animal with grasping and processing teeth. Now not all the elements opposed each other. They could not have functioned like teeth. The solution to this problem was to do as Bengtson had done and presume that the elements occupied different positions at rest and in use. The team appealed to the analogy of the hagfish, suggesting that in use the elements may have been rotated by as much as ninety degrees. While celebrating the radical new insights that had arisen from so simple a technique, Aldridge and his colleagues realized that understanding the functioning of the apparatus required more work. Smith, Briggs, and Aldridge also applied this same approach to the Waukesha specimen and concluded that the apparatus was indeed like that seen in the hagfish at least in terms of its symmetry.[16]

Aldridge opened the book, *The Palaeobiology of Conodonts,* which arose from the conference, with a historical review. He did so not to revisit the myth, which seemed to belong to another era, but to empty minds of misconceptions and architectural arrangements that now had little to support them. It was an essential precursor to introducing the new. It meant the evidence of the animal need not do battle with stubborn fictions and old illusions. But he knew that to carry this argument off he would need to interrogate the past, and the objects that had produced it, with forensic precision. With eyes trained through studying the growing collection of animals, Aldridge disentangled each interpretation, rotation, re-orientation, architectural arrangement, and opposition of conodont elements. He could locate lines of influence. He noted, for example, that Schmidt had essentially reconstructed what he saw in his best specimen. Rhodes had had access to Du Bois's fine material and, influenced by one particular specimen, produced diagrammatic interpretations that reversed the direction of the component elements and included some minor repositioning. And Rhodes, through his control of the section on apparatuses in the two *Treatises,* had then turned his thinking into the orthodoxy.[17] By exposing the origins, material basis, and frailties of this thinking, Aldridge made way for the new orthodoxy he and his colleagues now presented later in this book. Like Scott before him, his possession of the animal fossils gave him a badge of authority.

Others speaking at the conference were not so fortunate. Nicoll was still making do with fused clusters, and working with Carl Rexroad re-examining Pollock's specimens from northern Indiana. Nicoll thought the rapid and preferential evolution of the platform elements at the rear of the apparatus particularly perplexing. What selective pressures could have caused this? These elements seemed to interlock or intermesh and thus must have worked in opposition. As these were fossilized together, he doubted Bengtson's suggestion that the structure was protruded when in use.[18] Nicoll imagined the conodont animal as being something like the amphioxus, which lived for up to eight years, spent two hundred days in the open ocean, and was capable of travelling over eight thousand kilometers. Amphioxus has cirri acting as a sieve and a strange ciliated wheel organ that directs food into the mouth. He imagined that the conodont's varied elements reflected similarly diverse actions in the filter-feeding animal.

Szaniawski was also at the conference. He had been undertaking comparative studies of the fine structure of conodont and chaetognath elements. This work had changed his position on conodont ancestry markedly. He no longer thought true conodonts were chaetognaths but that both groups might share a common ancestor. This did not stand in the way of the conodont animal being a chordate, he said, because there were so many uncertainties about the biological relationships of chaetognaths themselves.[19]

Clark and Hearty's fossils, found in 1984, only saw the light of day in a paper two years later. These new animals confirmed earlier interpretations, turning indistinct or uncertain features into reliable characters. Much of the tentativeness of the first paper now disappeared. Segmentation was a demonstrable fact and composed of numerous V-shaped "somites," Aldridge, Briggs, Clarkson, and Smith asserted. Embryological studies had long ago demonstrated that key structures in the vertebrate body form from these "primitive segments," or somites, including vertebrae, muscle, and the dermis of the skin. Du Bois, whose own fossil discoveries had led him long ago to postulate an animal very like those found in Scotland, imagined different kinds of conodont element being produced in this way. This did not stop Du Bois imagining the animal as a relative of those annelid worms that were then so fashionable in Illinois, but for Aldridge and his collaborators these segments seemed to confirm the chordate. They hypothesized that the animal was, like the hagfish, a jawless craniate.[20] The hagfish possesses a partial cranium or skull but no true backbone. The conodont animals simply possessed their apparatuses. The authors now took that big step and claimed that there was no longer a need to retain the conodonts as a separate phylum and drew up a shopping list of wanted diagnostic characters: the nature of the tail fins, the position of the anus, and structures of the head. The British team now began discussions with experts in cyclostomes, including Derek Yalden in Manchester and Richard Krejsa in California. Aldridge and company were looking for homologues that might connect the feeding apparatus of conodonts with those of the hagfish and lampreys. To aid discussion, they sent out proofs of the article describing their new model of the apparatus. However, Yalden told Aldridge that the model was even less convincing as the jaw apparatus of an agnathan fish than the old one.

The question of possible chordate affinities was also being independently tested by Dzik, who knew little or nothing of the newest Scottish animals. His arguments were partly based on an extraordinary Lower Devonian assemblage described by Soviet worker Tamara Mashkova in 1972. This seemed to preserve part of the three-dimensional arrangement of the apparatus. The elements were arranged symmetrically facing inward, with the long axis of each element placed vertically, the sharpest elements to the front and the most robust toward the rear. This interpretation seemed to Dzik to be corroborated by later finds and, indeed, by the first animal. But Dzik had no animals himself and thus no soft tissues from which to develop this idea, so he turned his attention to the elements. Reviewing past research into the evolution of enamel and dentine in early vertebrates, Dzik wondered if modern assumptions about bones and teeth were clouding the interpretation of the elements. Surely these materials would have evolutionary precursors. Would they have been structurally different in the past? It was this kind of reasoning that had encouraged Pander to invent his conodont fishes. Now Dzik suggested that dentine was the primitive condition of the hard parts in vertebrates and that bone evolved from it. He thought that even more primitive animals might have possessed an enamel-like substance.[21] There had long been discussion of a homology – an evolutionary connection – between dentine and enamel in vertebrates and base and crown of conodonts.

Dzik showed that these structures developed in many different ways in primitive vertebrates and argued that later kinds of dentine and enamel should not be expected in earlier animals. This thinking put Dzik ahead of the game. He was following a logical trajectory, and one rather more reliant on technological innovation than a supply of animal fossils. Dzik was confident: The conodont was a true vertebrate, and he could imagine it evolving a catching apparatus, composed of conical elements, which was then lengthened due to selective pressures. This he thought reflected a shift from a bottom-dwelling scavenger to a carnivore of open waters.

As if to consolidate a view now in the ascendancy, in 1986, Richard Jefferies of the Natural History Museum in London, an expert in the fossil ancestors of sea urchins, published *The Ancestry of the Vertebrates*, in which he envisaged a hypothetical creature, "s," which represented the first or "crown" vertebrate from which all others were descended. The creature, he said, was like the conodont animal![22] In the year Jefferies published his book he wrote to ask Aldridge for a copy of his most recent paper: "As one of the first people to advocate (though only in a speech) that conodonts should be compared with the lingual apparatus of myxinoids, I am happy to discover that I was right." Clearly the conodont was for him a foreign land.

By 1986, everyone seemed to agree that the chaetognath was off the menu and some went so far as to imagine the chordate as a mere waypoint as the animal rose to join that most exclusive of clubs, the vertebrates. The idea seemed to be gathering momentum, though Godfrey Nowlan and David Carlisle were of the opinion that craniates were a step too far. They believed the evidence suggested conodonts might be better placed with amphioxus.[23] But then, as if from nowhere, Simon Tillier and Jean-Pierre Cuif, from the Centre National de la Recherche Scientique (CNRS) in Paris, threw a group of poorly known worm-like mollusks with spiny coats into the ring. These were aplacophorans. Janvier, who had been taken with the chordate, now changed his tune; this new group looked just as convincing.[24] He wrote to Briggs to inform him of his conversion. "I have let myself be convinced by one of my colleagues," he admitted.

"Actually his arguments are striking." But having just seen Briggs and company's latest paper, in which the conodont animal now had a "very convincing" tail and "very large notochord," he was in something of quandary. He admitted, "Your new evidence make[s] me hesitate again. . . . 'Entre les deux mon coeur balance!' [Between the two my heart balances]." He then joked, "Of course, craniate affinities would be a more noble pedigree than mollusc affinities, and the Society of Vertebrate Paleontology would suddenly increase by hundreds, but, seriously, I think we are coming now very near to a definitive solution, and the chaetognath affinities (at least for true conodonts) can now be ruled out." However, he enclosed a paper, shortly to be published with Tillier, on the molluskan alternative. He told Briggs it was "intended to provoke you and raise a discussion (if not a debate), and we would like you to respond [to] it." Briggs, Aldridge, and Smith did so. Indeed, they saw it as unhelpful speculation and moved fast to close it down. Bengtson added his support, placing their rebuttal on the opening page of the next issue of *Lethaia*.[25] These bizarre and unexpected leaps by outsiders, which drew the animal into associations with the unknown, exotic, or rare, would remain a constant hazard for the conodont while the animal retained traces of its former enigmatic self. The history of the animal had, at its most superficial level, merely been a realization of that human capacity to see resemblances in things. Briggs and his colleagues were convinced that Tillier and Cuif's arguments also rested on the same kind of lookalike illusions. However, Alain Blieck of the Université de Lille wrote to Clarkson expressing his doubts about the recent interpretation of the animal as a chordate. He felt the histology of the conodont elements simply did not support this interpretation and thought Tillier's exotic slug much more plausible. He accused Briggs and his colleagues of pursuing superficial resemblances.

Janvier admitted to playing devil's advocate in the hope of forcing upon Briggs, Aldridge, and their collaborators a desire to find more convincing evidence. The British team was out in the open and now actively defending a view that the conodonts were chordates. They dug in, their defensive position shored up with new material brought in from the rear. Here they were aided by the fortuitous rescue of blocks of the Granton Shrimp Bed in 1985. When Clark worked through this material, between

May 6 and 8, 1987, he once again astounded his colleagues by finding another five conodont animals. Although remaining largely in the background, Clark's staggering talent for finding these fossils made him the most important cog in a British engine that was now trying to separate fact from fiction. As Aldridge, Briggs, Clarkson, and Smith studied these specimens, the question that concerned them was whether these new fossils possessed the items on their shopping list.

Before these new fossils could be loaded into arguments, Walt Sweet's monograph, *The Conodonta*, appeared. It was the first solo treatise on these fossils since Lindström's creative little book of 1964.[26] Like Lindström's book, Sweet's was personal, provocative, and a little experimental. It was, however, published in an entirely different era. In 1964, there were great chasms in the understanding of these fossils. The young Lindström was at the start of his career and writing the book forced him to develop a more holistic knowledge of his subject. While many of the Swedish old guard thought book writing an inappropriate activity for a scientist, his German colleagues had welcomed it. And for all its individualism, Lindström's capacity for careful study and lateral thinking had produced a number of immensely useful ideas. When Sweet's book was published in 1988, the science of conodonts was fully formed; never had it been so certain of itself. Sweet was then in the closing years of his career and had found himself collared at a conference by a very convincing woman from Oxford University Press. (Lindström had been similarly collared.)

No one could doubt Sweet's command of his field or the remarkable and forthright contributions he had made. Like Lindström's, his had been a shaping hand. But he was, unlike Lindström, willing to fight for what he believed was scientifically right. The publisher may have assumed that Sweet was going to write a summary textbook – a modern treatise, but he had never been associated with pedestrian science. He was used to thinking afresh and quite happy to take the controversial path, if he thought it was the right one to take. He had no patience with lazy or poor science. In 1988, he was still looking forward, still wishing to shape and influence. He set about writing a book that reflected how he saw the conodont world: "I was just trying desperately to put a lot of things together, and some of it was half baked and some of it was not."

Sweet would never claim to be a palaeo*biologist*, even if his work on statistical assemblages paved the way for the science to recover its biology. However, no author of a book on conodonts could ignore the animal. It was the hot topic of the moment and he could not deny the exciting new evidence provided by the Scottish animals, but he said that on many key points – points on which Briggs and company had been only too happy to draw conclusions – these objects remained "mute." Sweet willingly conceded to tentative conclusions that the fossils might preserve segmentation akin to the muscle blocks (myotomes) found in amphioxus and fish. What he found objectionable was the leap of interpretation that turned them into somites, with all that that might mean for possible vertebrate affinity. He thought it a conclusion "in no way required by the evidence." This leap, he suggested, led into a "much more subjective discussion of anatomy." The British workers had located a "head," but where was the brain? There was no evidence for advanced head structures, and Sweet preferred a less suggestive terminology. He simply could not understand why Aldridge, Briggs, and others had seen a chordate animal.

Sweet suggested that "chordates are not the only organisms that might yield such V-shaped impressions on compression." Could these impressions not belong to a flatworm (nemertine)? The modern nemertine had many of the characteristics preserved in the Scottish specimens. Sweet was not, of course, the first to suggest this relationship. And why had the British workers interpreted the lines running down the center of the body as they had? There were other possibilities here too. He felt that "Aldridge et al. allowed their interpretation of the Scottish specimens as primitive chordates to restrict their survey of anatomic possibilities" and permitted them to preferentially see "a nerve chord, notochord or dorsal aorta." Might the preserved "lines" at the front of the animal represent the nemertine's eversible proboscis, he wondered, while those preserved at the rear perhaps record the gut? If so, then that gut reached the tail of the animal. This was an invertebrate feature as it exits the body before the tail in vertebrates.

Sweet also took issue with Jefferies' mysterious "s": "marine habit, eel-shaped body, head-trunk-tail, rasping teeth, ?lensless paired eyes, ?somites." How many other animals offered this model? "I suggest that, as Jefferies intimates, the resemblance between the hypothetical animal 's' and the Scottish conodonts may be so striking because the latter were used as a general model for the former."

Sweet shared the view that the teeth possessed by hagfish merely resemble proper teeth; they bear no evolutionary relationship. They could not assist in giving the conodont chordate affinity, nor could the phosphate, the element shape, basal material, chevron impressions, or models of element function. All were unreliable. The proof, he said, was to find the notochord, gill slits, or dorsal nerve cord.

Sweet's outlook was framed by *Wonderful Life* spectacles. He had had a chance to discuss conodonts with Stephen Jay Gould on two occasions when lecturing at Harvard. The thrust of Sweet's book, which promoted the idea that conodonts were a distinctive form of life, arose from these conversations. To him, the attempts of Briggs and friends to pigeonhole the conodont in the chordates was simply illogical. He believed paleontology was moving in the opposite direction, away from blinkered classifications and presentist attitudes: "It has several times been suggested that even the venerable Chordata is possibly no more than a shaky confederation of invertebrates and vertebrates."

The key message in Sweet's arguments was not, however, about the particulars of the animal's biology but the boundaries and frailties of interpretation. In this respect his arguments read like Richard Owen's chastisement of Pander. Owen simply could not comprehend, given all the evidence to the contrary, why the Russian had thought the animal a vertebrate.

We don't know what Pander thought of Owen's opinion, but Aldridge and Briggs were certainly not the kind to take these criticisms lying down. They had been accused of producing mere "waffle."[27] When Sweet saw that Aldridge and Briggs had together written a six-page review of his book, he must have felt a little concerned. Book reviews are almost without exception written by individuals, not duos. Aldridge and Briggs were, however, gentle. Aldridge certainly knew, and had the greatest respect for, Sweet. And he was probably not that surprised that Sweet had produced a book that was as "provocative" as it was "controversial." It was, they said, food for thought.[28] But by the third page, their sleeves were rolled up: "Having been involved in presenting these candidates [the Scottish animals] for consideration and in using them to develop a hypothesis of chordate affinity . . . we have a particular interest in how they are treated in this milestone in the conodont literature."

Their assault began by countering Sweet's *Wonderful Life* view of the distant past. Now became clear the significance of Briggs's involvement in the review. It was Briggs, with Conway Morris, who had populated Gould's world with a myriad new life forms, which in turn shaped Sweet's outlook. But the tide was on the turn; Briggs admitted that some of these problematic animals were indeed finding a place in long-established phyla. This gave validity to his and Aldridge's attempts to understand the conodont animal's chordate affinity and suggested that one should not buy into Gould's vision unquestioningly.

Now the five newest animals entered the debate for the first time, for they confirmed the "V-shaped" structures that the two men were convinced were unlikely to be found in preserved nemertines. They were very like chordate muscle blocks seen in amphioxus and hagfish. Sweet had, of course, made his arguments without possessing the animals – old or new – and thus it was rather easy for Aldridge and Briggs to undermine his interpretations; the fossils simply said otherwise, they said: "The characters of the conodont animals, such as phosphatised elements, a transversely operating feeding apparatus, V-shaped somites, a laterally flattened trunk, and ray-supported posterior fins extending further along one margin than another, all bear comparison to chordates." The new specimens enhanced this view. The problem, they felt, was that Sweet had become wedded to the view that the conodonts represented a separate group and thus he looked for differences rather than similarities with other forms of life. Of course, only a few years earlier, confronted with the ambiguities of the first fossil, this had also been their refuge. In just a few years, however, the science had fundamentally changed and, they believed, it was no longer appropriate to think these old thoughts.

In a parallel review, Paul Smith saw Sweet playing the role of devil's advocate: "One does not have to be an orthodox Popperian to conclude that Sweet is here being rather mischievous and is not advancing testable hypotheses."[29] He wished Sweet had maintained a "more objective viewpoint," but that was Sweet's point: There was too little that was objective about the British animal and too much that exposed the frailties of interpretation.

Aldridge and Briggs recognized that Sweet's forthright approach had produced a useful and stimulating book with which a new generation of scientist could argue. They had, in effect, begun those arguments in this review. Those arguments would, however, soon fade from view while the book would live on. It would remain on library shelves for decades, just as Lindström's had, there to suggest to outsiders that this was the way things stood.

At the close of the 1980s, Aldridge was in Leicester and Briggs in Bristol. The decade ended with the final dispatching of the first and second contenders for conodont animal. Melton and Scott's animal was redescribed by Conway Morris in a paper that ironically, given the delayed publication of original paper, was lost in press for six years. Possessing considerably more material, Conway Morris was gentle on these earlier authors and celebrated the animal in its own right for its peculiar ecological interest and zoological strangeness. Now Conway Morris could reinvent it as the animal that fed on conodonts. The canard finally had scientific recognition. The first animal was dead.[30] Some, though, looked at the rock in which these fossils had been found and wondered how a conodont-eating animal could exist where there are, inexplicably, no signs of conodont life.

Now it was Conway Morris's turn to suffer the ignominy of error. He had admitted his mistake not long after the Scottish animal had been published, but this did nothing to protect him from a little ridicule. As mistakes go, Gould remarked, Conway Morris had "made a beauty."[31] Through scientific eyes the path of progress is littered with mistakes, though Gould thought them "not badges of dishonor." Conway Morris's interpretation of *Odontogriphus* had produced an animal of its time; had he come across it in 1989, he would have read it entirely differently. It was simply a price to be paid. But Conway Morris would, more than most, recognize how long-lived books like Gould's *Wonderful Life* could infect a scientific career. In the ephemeral press of scientific publishing, disproven interpretations are soon forgotten, but *Wonderful Life* kept Conway Morris's past alive, portraying him to readers as the author of outmoded ideas long after he had given them up. When Conway Morris later came to write his own popular book, *The Crucible of Creation*, its reviewer, Richard Fortey, detected the author's loathing of Gould.[32] He accused Conway Morris of selective amnesia and the rewriting of his own history but reflected that Gould's book perpetuated old science and retained Conway Morris as its keeper. One could understand why so many scientists doubted the validity of books. As vehicles by which progress is made, they have the unfortunate effect of producing a freeze-frame image that is then left to drift into the future, giving the illusion of still being current. The scientific book is an impossibility; almost immediately it is a history book.

Eight years after the first true conodont animal had been found, the questions it posed had not been resolved. The animal remained locked in argument. Sweet seemed to imply that Aldridge, Briggs, Smith, and Clarkson had aspirations for their animal, that they had the vertebrates in their sights. Aldridge and company, however, felt that their ideas were simply developing as new data were revealed. Indeed, Clark's constant supply of animals had progressed many of those initially tentative arguments to a point of certainty. When, in 1989, Conway Morris came to review the progress that had been made, he felt sure that the animal was a chordate and that it held tremendous potential for understanding the origins of vertebrates.[33] He also believed that more and better fossils would be found. How wonderfully ironic it would be if, after such a circuitous journey, the animal Pander had dared to imagine became a reality. Of course, there never really had been a circular journey through a hundred possible identities – that was just a myth. Those who knew the conodont – really knew the conodont – knew it had never really left the place where Pander had originally put it, that never-never land where vertebrate meets worm, and where hagfish, amphioxus, and a host of other chordates swim. The conodont animal's natural home was that most difficult and enigmatic of all palaeontological places, that place where vertebrates begin in time and in space. And this was where conodont science was now heading.

"Over the Mountains
Of the Moon,
Down the Valley of the Shadow,
Ride, boldly ride,"
The shade replied –
"If you seek for Eldorado!"

EDGAR ALLAN POE,
"Eldorado" (1849)

FOURTEEN

Over the Mountains of the Moon

IN THE MID-1980S, OUT OF SIGHT OF THOSE DEBATING THE
meaning of the first Scottish animals, the next big step was being taken
in a part of the world that had thus far proven itself completely lacking in these extraordinary fossils: South Africa. Here, along a dirt road
in the Cedarberg Mountains, some two hundred kilometers north of
Cape Town, Geological Survey officers Danie Barnardo, Jan Bredell, and
Hannes Theron came across a new borrow pit for road metal exposing
the soft and rarely seen Upper Ordovician Soom Shale.[1] They stopped
to investigate and found their curiosity rewarded with some intriguing
fossils reminiscent of graptolites. Graptolites are one of those classic
groups of extinct animals all paleontologists study at some point in their
training. Tiny, colonial – bearing a passing resemblance to corals and
bryozoans – their fossil remains are most common in shales, where they
look like minute flattened saw blades. Theron sent a specimen to Barrie
Rickards at Cambridge University in the UK, an expert on this group, to
see if they really were graptolites. Rickards said they were not.

A number of scientists at the Survey headquarters near Pretoria,
including palaeobotanist Eva Kovács-Endrödy, however, became intrigued by the similarity of these new fossils to strange spiny plants
found in much younger Devonian rocks. If this was what they were,
then these were clearly important finds. In an echo of Pander's conodont
discovery, they would push back the origins of plants by forty million
years. The oldest plants known at that time came from the late Silurian.
Theron and Kovács-Endrödy prepared a paper naming these new plants
Promissum pulchrum, meaning "beautiful promise."[2] They did not know

that rather than being harbingers of a green and pleasant Eden, these new fossils held a beautiful promise of a rather different kind.

As is the normal course with scientific publication, the paper was sent out for independent external expert opinion. The task of reviewers is to consider a paper's merits and to advise the editors on whether it should be published. On this occasion, one reviewer believed the authors had not demonstrated that these fossils really were the remains of plants but left the decision on publication to the editor. The paper was finally published alongside a reply by the critical referee and a response from the authors.

This discussion mentioned Rickards's view that the fossils were not graptolites, but Rickards was concerned that he had been quoted when he had only seen one specimen. So more material was sent to him by diplomatic pouch and he confirmed his conclusion that they were not graptolites. He thought they looked rather jaw-like and discussed them with several of his colleagues. Together they alighted on the idea that these fossils just might be conodonts. The main problem with this identification was their size; they were more than ten times bigger than the conodonts they were used to seeing. Rather than being up to two millimeters long (though usually much smaller), these were up to two centimeters long!

Rickards knew these fossils crossed the boundary of his expertise, so he contacted Dick Aldridge. Aldridge was intrigued and traveled to Cambridge the next day. It was to be another important day, for here he saw the first conodont specimens known from Africa south of the Sahara and the first complete apparatuses from the Ordovician. And if these were not in themselves major milestones, he could also confirm that they belonged to a giant animal. Preserved merely as molds, Aldridge empathized with the Survey officers who had struggled to identify them. They were, in every sense, conodonts like no others and totally unexpected.

Seeking further confirmation, in August 1987 Theron took some specimens to the Devonian Symposium held in Calgary, where he showed them to a number delegates with expertise in fossil plants and conodonts. They were overwhelming of the view that these fossils were conodonts, albeit of amazing size. Nevertheless, back in Pretoria, Kovács-Endrödy and others remained wedded to the plant and insisted that they continue with the publication of a more extended paper advancing this idea.

Theron was, however, now convinced the fossils were conodonts and withdrew as a coauthor, later joining Rickards and Aldridge in publishing a paper identifying *Promissum* as a conodont.

African conodonts, apparatuses, giants – there were many reasons for the South African geologists to welcome the news. They may have lost their landmark discovery of plants, but they had gained something quite extraordinary. Communication lines now opened between Aldridge and the Survey workers, and in 1990 he found himself traveling to South Africa to study their collections of Soom Shale fossils. At the time, all he knew was that there was a wonderful opportunity to progress science's understanding of the animal in this part of the world. Ever since the animal had been found, Briggs and Aldridge, in particular, had been searching for new resources and new ways of looking. Briggs had searched Lagerstätte collections in the United States and now Aldridge did the same in the Soom Shale collections in Pretoria. He also collected some material himself and brought it back to Leicester. It was when he was examining these new finds under his binocular microscope – playing with the lighting to get the best possible illumination – that his eye came across two indistinct oval impressions. Were these those strange, dark-lobed structures he had seen in the Scottish animals? He felt "a buzz of excitement." Had he chanced upon yet more animals and perhaps – given the relative ease with which apparatuses had been collected – the richest deposit yet? He needed to get back to South Africa – and urgently.

In 1991, with a small research grant in his back pocket, Aldridge returned South Africa to search for the animal with Theron and others. Theron had arranged for the use of a mechanical excavator which had cut a trench into the exposure at the original find site. This produced a heap of rock ready to be split, and on the second day of collecting, August Pedro was holding a complete apparatus in his hands. It had associated with it two ring-like structures. It seems that Pedro had a Neil Clark–like talent for finding these fossils, and soon more followed. The Soom Shale was that day understood anew. It too was a Konservat-Lagerstätte, and of an age hardly represented by such deposits. A new window had been opened into the deep past. It had been an important day.

With each new conodont fossil found, Aldridge gained an increasing sense of the three-dimensional structure of the hollow rings: They were deep and inwardly tapering. They did, indeed, seem to be the remains of the animal's eyes. Almost identical circular objects had been seen in the Silurian relative of the lamprey *Jamoytius*. An important animal, in the 1940s *Jamoytius* had been regarded as "undoubtedly the most primitive of the 'vertebrate' series of which we have knowledge." In the 1960s, Alexander Ritchie from the University of Sheffield visited the Scottish site where the original fossils had been found and gathered more and better specimens.[3] He interpreted the distinctive circular rings in these fossils as sclerotic cartilages surrounding the eyeball. These were so like those now found in these new conodont animals that Aldridge and Theron thought this a reasonable interpretation for them.

In total they now had about thirty apparatuses, five of which showed soft tissue preservation. They left the excavation feeling that the shale had more secrets to reveal. There were hints of more complete animals, but they needed less-weathered rock to have any hope of finding them. They remained optimistic and planned for the future – a future that had now shifted geological position. Granton had perhaps said all it was going to say. The latest animals had been reassuringly confirmatory, but none had been better than the first. In October 1991, Briggs, Aldridge, and Smith planned what they thought would be their last paper on the Granton animals.

The South African fossils had given the conodont animal project renewed momentum. In 1992, Aldridge expanded his team in order to populate those edges of the science that seemed to offer greatest potential for advancing the animal. Sarah Gabbott was recruited to research a doctoral degree on the Soom Shale Lagerstätte, and Mark Purnell, who possessed a Natural Environment Research Council fellowship, chose Leicester as the best place to work on element morphology and apparatus architecture. Following the breakup of the Nottingham team, Paul Smith also had moved. In 1990, he had become curator of the Lapworth Museum at the University of Birmingham, where he would lead his own conodont research team. From 1992, he was joined by postdoctoral researcher Ivan Sansom. In that year, then, the British effort had been considerably strengthened and reconfigured. Considered as a whole, it was now by far the largest research group ever to have worked on the animal.

How Dick Aldridge's world had changed. It seemed that luck was perpetually on his side. The first animal discovery had been delivered into his hands by chance and the South African fossils had arrived in a similar fashion, apparently drawn to him by some kind of magnetism. But in both cases, it was Aldridge's swift actions that put him firmly in the picture. There was no magnet; he was simply one of the best known conodont workers in a tiny population of such people in Britain. His involvement in the discovery of the first animal had, however, repositioned him in the field. Who else would one speak to about the biology of this animal other than Briggs, Aldridge, Smith, and Clarkson? Of these, Aldridge was the senior conodont expert. Like other successful conodont workers, Aldridge could spot an opportunity and he was not one to see it pass him by. Like Briggs, he kept his eye on the ball – he knew which directions and which resources were required to negotiate the animal's progress. The conodont animal had by then become central to his scientific being; it was not a subject to dabble in, as so many had. Now the conodont was rather more than it had been. It was knocking on the door of vertebrate ancestry and – as Alfred Romer noted long ago – therefore on the door of human conscience. No longer could it be dismissed as a fine, if arcane and exotically abstract, tool. For so long, its utilitarian value had outshone its biology, but now it seemed that even this might be reversed.

This focused dedication, backed up by considerable ambition and intellectual aspiration, meant Aldridge possessed the best fossils and the best sites and was part of a number of powerful collaborations. Luck did not come into it. The animal, and particularly its vertebrate pretensions, had turned a key. It was now possible to fund research that only a few years earlier would have seemed excessively esoteric. He and his collaborators found themselves on the mailing lists of the popular press and editors of encyclopedias; the British view of the animal was becoming the new orthodoxy – even if contested by some vertebrate paleontologists and other scientists in various parts of the world. This did not, however, leave these others entirely out in the cold, for they could participate by throwing challenges in the path of these British workers, as Janvier

14.1. Derek Briggs (*left*) and Dick Aldridge. Photos: Dick Aldridge.

delighted in doing. The animal was never a crusade, always a scientific dialogue in which the naysayers had a vital role.

To establish the vertebrate argument more widely, Aldridge, Briggs, and their collaborators knew they had to be active on all fronts, that the article in the encyclopedia, university newsletter, or popular science magazine was just as important as the basic science. They realized they could not simply "publish and be damned"; they had to fight to hold the territory they had won. But this was no blind political campaign. What they relished was the opportunity to take good science wherever it might lead. They knew that maximum exposure also resulted in new tests, challenges and opportunities, and therefore better science. But even they could not deny that they were empowered in another way, for they formed a relatively large, cohesive, intellectually diverse, support group unlike any in the world. Even they could not deny that they were a force to be reckoned with. Critics looking on, who perhaps preferred a rather different interpretation of the animal, might argue that the science they made was influenced by the vertebrate lens through which they saw their data. These critics could not help but see how "lucky" this group had become.

Scott had experienced similar empowerment when he possessed the animal, but he was more vulnerable to poaching and, being less of a conodont specialist than he claimed, he found himself somewhat isolated. Sweet and Bergström, Zeigler, Branson and Mehl, and a few others, had also shown similar determination, knowing that to win one needed both rigorous science and a willingness to see through a campaign. Not all conodont workers adopted these roles. In this small community there were really only a few combatants, and as Sweet recalls, in this close-knit world this combat was almost always chivalrous.

The British team was also large enough to internalize some of the debate, to test ideas and remove risk and uncertainty and join minds before airing results. This also gave its members extraordinary intellectual resilience. It permitted them to travel that edge where doubt lurked. They also shared an advantage over others linguistically as this science, like all science, increasingly conducted its conversations in English. Russian conodont research, which had begun the science, had long been locked behind the Iron Curtain. It found itself liberated in 1989, when the Berlin Wall fell, but then discovered another barrier stood in its way: language.

As the British conodont animal workers reorganized themselves, Theron and Pedro were hard at work in South Africa looking for new sites. They found several, the most important being some seventy kilometers to the south of the first. Aldridge and Gabbott joined Theron and his wife Elmarie there on the last couple of days of a three week collecting trip in May 1994. It was, Aldridge later recalled, an impressive spot: "The farm of Sandfontein is set in idyllic surroundings overlooked by high mountain crags. The sandstone rocks of the Table Mountain Group are weathered into stone pinnacles, in which the imaginative eye can discern the shapes of people and animals. There are few people in the area, only the local farmers and their field hands, and the skies are regularly patrolled by black eagles and jackal buzzards." In Aldridge's words, there is something of the romance of the field that has been associated with geology since its early days, but from his description we might also understand something of the work, he, Theron, and Gabbott were involved in. Fossils are not common in the Soom Shale, and excavation can soon become an unrewarding daily grind of splitting rocks. The excavation begins with its participants believing that on each next hit something remarkable will appear. Each time, anticipation is followed by disappointment or unremarkable results. After half a day, anticipation and belief become tested, and as the days go by, sights beyond the immediacy of the excavation begin to catch the eye. Soon things outside the excavation can become more attractive than those within. At Sandfontein, the rocks were blacker, less weathered, less willing to split, and less willing to give up their treasures. Inevitably, perhaps, sharing the toil of hard labor, in these pleasant surroundings, the excavators indulged in leg pulling as a way to mitigate the tedium.

The 1994 collecting trip had been a pleasant three weeks that were not in any sense unproductive, but they had not found a specimen that really stepped up the game. They began to wind down on the last morning, but then, as in the best adventures, something happened. Aldridge again: "We had dug quite an impressive trench, and I hacked out a few last slabs of rock and passed them out to be split. Inevitably, as the field season wound down, our concentration had been waning and we had been enjoying the sunshine and the banter, so we thought that Sarah Gabbott was attempting to repay Hannes for one of his jokes when she cracked open one of these final blocks and announced that she had found a conodont animal." She had! It was the animal they had been searching for. A giant, which although not complete, possessed its eyes, feeding apparatus, and about ten centimeters of its trunk.[4] Their luck had held – or rather, their sheer hard graft had finally paid off. They left knowing that the Soom had given them a hundred apparatuses, many with associated head structures. Gabbott's fossil, however, trumped them all.

Elsewhere the debate concerning the animal's vertebrate characteristics had been heating up. The decade had begun with the British proposition that the animal might indeed be a chordate, and this notion was at last finding a tiny place in reviews of the vertebrate literature. Some of those looking at primitive near-vertebrates in the field of biology held out a welcoming hand to Aldridge and his collaborators. Although related, conversations do not flow naturally between the fields of paleontology and biology; as George Simpson recognized long ago, they produce their animals using rather different materials. But the history of conodont studies also shows us that this fossil periodically attracts the attention of outsiders, and now a few biologists were beginning to take note and offer some insights on the flesh of the animal. Among them was Richard Krejsa in California, from whom Aldridge sought information about hagfishes back in 1986. Krejsa and his associates, in 1990, tried to resolve some of the outstanding issues surrounding the conodont animal – many of which had arisen in the decades before the animal had been found – by referring to living juvenile hagfishes.[5] This line of argument, however, was complicated by the fact that the hagfish were then treated by many as an encampment on the periphery of vertebrate city. Inexplicable in their form, they were regarded by some as degenerate – vertebrates that had lost many of their key vertebrate characteristics. In that idealized picture of evolution that envisaged the advancement of life, the hagfish seemed to have gone backward. By moving in next to the hagfish, the conodont animal found itself embroiled in the politics of vertebrate ancestry, a debate centered on inclusion and exclusion. The Californian team was certainly on the side of the hagfish, suggesting there was more to this animal than was commonly believed. Indeed, they argued that much existing belief relied upon old literature, old ideas, and consequently old prejudices. They felt the "teeth" of the hagfish and the conodont animal shared structural similarities, such as bubbles and tubules, even if composed of rather different materials. These bubbles occurred in the white matter in the conodont elements and had led Lindström to suggest that white matter was formed through resorption of the material making up the lamellae. Krejsa and company felt this could now be refuted. Noting that enamel and dentine grow away from each other, they thought Walter Gross's objections, to conodont elements being tooth-like, also wrong; the elements were very much like teeth. And since hagfish teeth are replaced as new ones erupt beneath them, they wondered if conodont teeth did the same.

As biologists looking in from the outside, Krejsa and his colleagues saw a field shaped by "orthodox conodontological theory." Now, with the animal found, it was difficult to understand why some of these old orthodoxies had not fallen. And while some of these arguments from California could easily be dismissed by experienced conodont workers, these biologists did the conodont workers the service of asking them to question their beliefs and offered a friendly hand of support.

Further assistance came in the form of a review of vertebrate ancestry that at first dismissed the conodonts. In 1990, Moya Smith, an expert in the structure and evolution of teeth at Guy's and St. Thomas' Hospitals in London, and Brian Hall, who was on a sabbatical there from Dalhousie University in Nova Scotia, produced a major review of the origins of the vertebrate skeleton. Nearly one hundred pages in length, this was a monumental work, and one that, during its preparation, they had discussed with Philippe Janvier and Derek Briggs. It touched more broadly on the work of Pander but denied the conodont animal its vertebrate status. For Smith and Hall, the conodont animal existed in a special place: It was a craniate (it possessed components of a skull) but it did not possess real teeth: "They simulate teeth but are not homologues of teeth" (they are not related in an evolutionary sense).[6] They thought this true of the hagfish's teeth, too, and pointed out that the platypus has horny teeth, warning against simplistic evolutionary assumptions based on form. Their dismissal of conodonts, however, confirmed a widely held view among those studying early vertebrates: "Though opinion seems to be divided as to whether conodonts are chordates or members of an invertebrate group . . . it is at least clear that they are not vertebrates."[7] This paper, and a number that followed in its wake, produced a serious challenge for "the animal that would be a vertebrate," but it gave the animal's proponents clear points of focus for future research. Objections, such as those in this paper, were always based on weaknesses and inconsistencies, which were easily identified and then attacked. It was clear from Smith and Hall's paper that ascent to vertebrate status depended on the detection of key vertebrate building blocks: cellular bone, enamel, and dentine. It was necessary to know once and for all of what the conodont elements were composed.

* * *

Fortunately, Ivan Sansom and Paul Smith in Birmingham were already on the case. While working on his doctorate degree at the University of Durham in 1990, Sansom had used a new acid-etching technique to decalcify conodont elements. He showed the scanning electron microscope images of the results to Angus Parker and Trevor Booth at the medical school at the University of Newcastle, and they "immediately identified the fossil tissues as the enamel, calcified cartilage and cellular bone which is typical of living vertebrates."[8] (Sansom showed Aldridge these images of spheroidal or globular bodies in 1991, and Aldridge thought they might also explain those curious conodont pearls.) Sansom's work at Birmingham was supervised by Paul Smith and former Nottingham PhD and established Durham academic Howard Armstrong. As with much PhD work in the sciences, there is a strong collaborative element, but, unusually, Paul Smith was keen to see that his doctoral student, Sansom, featured as lead author in their published outputs. They were now joined by Moya Smith. Sansom's task was to demonstrate that conodont elements really were directly related to vertebrate teeth and not mere imitators.[9] Moya Smith's expertise was central in this regard; above all else he needed to convince her. If she was convinced, then the opposition, who had used her review as a weapon to deny the vertebrate, would be considerably weakened.

Sansom had compared the material making up conodont elements with some of the oldest materially intact vertebrate fossils then known. These were Charles Walcott's fossil fishes from the Ordovician Harding Sandstone of Colorado – rocks and fossils, it might be remembered, that had nurtured Kirk's influential claim for the conodont fish back in 1929. Sansom and his collaborators found that conodont white matter contained spaces interconnected by irregular and radiating tubules that were identical to, and considered homologous with, those found in fossil and recent cellular bone. This, in some respects, echoed Beckmann's discovery in the late 1940s. Like Krejsa and his collaborators, Sansom and colleagues believed white matter a primary structural building material, not a later addition. They interpreted the white matter as bone. Additionally, the lamellae showed incremental growth lines, which they considered, by reference to Moya Smith's work, typical of "tissues such as enamel." And, finally, the globular structure of the material making up the basal part of the conodont had, they felt, a striking resemblance to calcified cartilage found in some vertebrates.

The news was of such significance that it made the cover of *Science,* while inside the team announced a shake up; this was "unequivocal evidence for the inclusion of conodonts within the vertebrates." Repeating Pander's original assertion, they claimed it pushed back the appearance of vertebrate hard parts by forty million years, back to the Late Cambrian. Important to this interpretation was the researchers' attachment to Jefferies's recent reorganization of the vertebrates, which made them synonymous with the craniates, a point rather buried in the footnotes. It had long been believed that vertebrate hard parts first evolved as external armor, which later developed into specialist feeding apparatuses; now they suggested the reverse was true.[10] They also contradicted the contemporary belief that dentine appeared in the fossil record before bone; these fossils appeared to contain no dentine, only cellular bone, enamel-like tissues, and cartilage.

The news spread even more rapidly than that concerning the discovery of the animal. It was reported on the front page of the *Times* under the headline "Razor-toothed fish bites into human history." It read, "Mankind's oldest ancestor was a tiny, eel-like fish with a formidable set of razor-teeth that lived 515 million years ago." And it ended, "It is still not clear what these creatures looked like. Mr Sansom said the teeth suggested carnivores. If alive today, they would give a bather a nasty nip." The paper even reflected on the discovery in its editorial. The *Independent,* a British broadsheet paper, took a more mundane line and was rather less certain of the significance of the discovery. "Limestone yields the oldest set of teeth" began "British scientists have discovered the oldest known set of teeth in the world. They have been embedded in a lump of limestone for about 500 million years. The teeth belonged to a small, ferocious marine animal that probably spent much of its time eating other members of its species. Scientists are excited about the find because it means they have identified the human race's oldest vertebrate ancestor." The *Telegraph* gave a more informed, sensation-free account in "Teething troubles": "A wriggling, toothed, eel-like animal that evolved 520

million years ago was the earliest vertebrate, scientists have reported."
The *New York Times* was even more grown up, devoting half a page to the
discovery under the heading "Vertebrates found to be much older." Here
readers were advised that "conodont" should be pronounced "KAHN-
o-dahnt," perhaps translating it into a New York accent. Doug Palmer,
writing in *New Scientist,* felt the discovery transformed the significance
of the conodonts and gave the research Sansom was undertaking great
urgency. The article carried an illustration of one of Sansom's etched
teeth. The caption read, "First vertebrate: this tooth-like fossil, only 2
millimetres in section, is the forerunner of all animals with backbones."[11]

Discover magazine, writing about the top fifty science stories of 1992,
proclaimed, "If Sansom is right, vertebrates are as old as the oldest conodonts – that is, at least 515 million years old." Here Sansom recounts the
moment of first discovery in Newcastle: "We were sitting there, and these
images appeared that were quite spectacular. . . . It was very, very clear we
were looking at enamel, bone, and cartilage." He continued: "It came as
quite a shock. . . . The structures were actually clearer in the conodonts
than in human teeth." The *Northern Echo,* a British regional paper, gave
the story the headline "Meet your great, great, great, great . . . grandad."[12]

Those pondering the origins of vertebrate life could no longer ignore
the conodont. Indeed, conodont workers of all persuasions might have
felt reason to celebrate the suddenly elevated status of their fossils. Chris
Barnes, who was immersed in events in Earth history, wondered if a
particular event in the Ordovician had marked an important moment
in the evolution of vertebrate tissue. For those already advancing on the
vertebrate from other directions, especially the Leicester team, there
could be no better piece of news; it would form a vital component in their
arguments for years to come. Briggs, who had been asked by *Science* to
write a "Perspectives" piece to accompany the article, immediately celebrated the impact of this discovery, which had thrown ideas regarding
early vertebrate evolution into disarray and in an instant expanded the
number of known vertebrate genera in the Cambrian-Ordovician from 5
to 150 (nearly all of which were conodonts). "In any event the vertebrates
can now be added to the list of major metazoan taxa that appeared during the Cambrian radiation," Briggs observed. Aldridge now felt he could
at last state, without fear of contradiction, that "conodonts are extinct

primitive soft-bodied vertebrates" and even go so far as to suggest "the soft tissue evidence now confirms their place among our earliest ancestors." Henry Gee, an editor of *Nature* and an important sounding board for the British workers, later reflected on the implications of finding this new piece in the vertebrate jigsaw. He recalled that almost since the beginning of paleontology, it had been believed that vertebrate life began with sluggish armored fishes, but now these were as nothing compared to the diversity of conodonts that joined them and in most cases were known only by their tooth-like hard parts: "Like so many geological Cheshire cats leaving their smiles to posterity." "Could later vertebrates have had instead a common ancestor among the conodonts animals?" he wondered. Reflecting on his childhood inspirations and the bizarre armored fishes that once populated his mind, he reimagined the natural history museum of that time: "The Hall of Fossil Fishes might have presented a truer picture of the Age of Fishes had the bulk of its space been devoted to conodonts, with just one small case, in a dark corner, devoted to pteraspids, sarcopterygians and their relatives." But precisely where conodonts fitted into the vertebrate family tree remained unresolved: "The mystery continued: it was a bizarre Borgesian plot made real, of future palaeontologists obliged to reconstruct human history based on several million sets of dentures, and nothing else." For an object vanishingly small, its impact was never less than remarkable. Even Gee, overcome as he was with lyrical metaphors, found himself affected.[13] The discovery was a gauntlet thrown at the feet of those who were shaping the picture of early vertebrate evolution. It demanded that science return to its microscopes and that biologists consider fossils.

The question of whether conodont elements really were vertebrate teeth was now being tackled on two fronts. While in Birmingham, Smith, Sansom, and their collaborators were revealing the fossils' material affinity, in Leicester, Mark Purnell was discovering if they could have functioned like teeth. Purnell had drawn a line in the sand at 1982 and corralled everything before that year, including all those functional models produced in the 1970s, into a period of prehistory he called "pre-animal."

The animal discoveries had provided sufficient new and quite specific data on many key aspects of apparatuses that had previously been matters of speculation. In a paper with Canadian conodont worker Peter von Bitter, Purnell attempted to reinterpret the function of the pair of robust blade-like elements that sat at the back of the apparatus. Were they covered in soft tissue? Did they roll particles of food between them? Did they bite, shear, or operate like a gate? Noticing some small nodes at one end of each blade, they reasoned that these nodes were set opposite each other, offering a rasp-like functionality in a scissor-like action. By implication this meant that the animal must have grasped "large" food particles with the anterior part of the apparatus rather than used these elements as filter devices.[14]

It had been while still working in Canada that Purnell had concluded that a fundamental objection to all previous functional models was a reliance on a test of "plausibility." Now he asked, could this be avoided? Were there other aspects of the assemblage that might more objectively point to function? At its most fundamental, the argument over function that had developed in the 1970s had revolved around whether elements were teeth or components in a filtering apparatus. Zoologists had long known that different parts of an organism grow at different rates according to the animal's needs, that growth patterns in some respects reflect mode of life. Comparative data had suggested to Purnell that a filter-feeding animal would need to preferentially grow the net – the comb-like elements near the opening of the mouth. What he found, however, was that the robust mashing elements grew more rapidly. This suggested to him that they had a molar-like function. The other elements grew at the same rate as, or slightly more slowly than, the animal as a whole. Purnell could also quash rumors that conodont teeth were, like those in the hagfish, replaced in life or lost or reabsorbed by the living animal. He deduced that the conodont animal must have been predatory and used the front of its apparatus to grasp and the rear elements to "chew."[15]

Bob Nicoll, however, was unimpressed both by the vertebrate and these food-munching teeth; he continued to defend his filtering model.[16] If it lacked fully formed jaws, Nicoll argued, how could the animal – with a muscular arrangement something like that in amphioxus – exert sufficient force to cut?

Meanwhile, in Bristol, Briggs was crafting yet another interpretive lens. He had set his postdoctoral researcher, Amanda Kear, to watch over the dead; to gather information on the decay of amphioxus (the lancelet or *Branchiostoma*). Their aim was to better understand the preservation of the conodont animals and other so-called primitive vertebrates.[17] Lancelets were considered possible "living descendents of the ancestors of vertebrates." Briggs and Kear's experiments revealed that the notochord was resilient to decay, recording its former existence in two lines like those seen in the Scottish fossils. This work permitted interpretation of these specimens to move beyond a priori assumptions about the nature of the animal and even explain some of the inadequacies in the fossils themselves. It altered expectations and clarified interpretations; it, too, moved arguments away from the test of plausibility.

These were heady days for a science that would soon be reaching the peak of its activity. In 1993, and now in possession of ten Granton specimens, the three original authors, plus Paul Smith and Neil Clark, published the final, definitive, anatomical account of the Scottish animals in the *Philosophical Transactions of the Royal Society*.[18] Aided by discoveries in South Africa, the two lobe-like head structures were now interpreted as hollow sclerotic cartilages that once supported the eyes. One specimen also seemed to preserve the animal's hearing apparatus. These were capsules that Aldridge and Briggs had once thought eyes. They had had their minds changed by the discovery of similar structures in the Mazon Creek lamprey, *Mayomyzon*, and hagfish *Myxinikela*. They also believed they could see faint traces of gill structures like those described in the Silurian relative to the lamprey, *Jamoytius*. The twin lines running down the body of the animal were now compelling evidence of a notochord, and in one heavily phosphatized specimen the area between these lines displayed a fibrous structure that just might indicate the notochord sheath. Again something similar had been seen in the Mazon Creek chordate *Gilpichthys*. Some aspects, however, remained frustratingly indefinite – a whisker away from being the proof they needed. The dorsal nerve cord was among these hazy features. But with each new specimen the beautifully preserved V-shaped segmentation could not be doubted. It formed "one of the most compelling pieces of evidence for chordate affinity." The separation of the chevrons could now be attributed to postmortem shrinkage as could be seen in Briggs and Kear's decay experiments. In one specimen, Aldridge's team even wondered if they could see evidence of original muscle fibers. Their conclusions were emphatic: "The evidence of the soft-part anatomy, together with features of element histology, show that the conodonts are vertebrates." The Birmingham team's corroborative work had given Aldridge and his colleagues a sense of certainty: "Other hypotheses that have been forwarded in recent years can now be refuted."

Sweet's tests had, in part at least, been answered. The notochord had gained solidity, and the nerve cord and gill structures tentatively interpreted. Sweet's suggestion that other invertebrate groups might preserve the trunk structures seen in the conodont animal had already been firmly denied. Again, Conway Morris and Janvier provided important supporting voices for believing the animal a chordate. Indeed, for many, including Aldridge and his colleagues, the question had since 1987 become "which chordate?" Nowlan and Carlisle, for example, had been amongst those who had placed the animal with amphioxus. However, new certainty regarding the eyes suggested that interpretation was incorrect.

The worm – the worm that had grown out of necessity in response to the discomfort caused by Pander's vertebrate – was now very dead and the reinstatement of Pander's imagined fish in full swing. However, for many involved in the debate, elevation into the hallowed realms of human sisterhood depended not simply on the fossil but also on what one could *call* a vertebrate. This had long been disputed territory, and the current view promulgated by Janvier was that the vertebrates consisted of animals with a cranium and vertebral elements; those which just possessed the cranium were then referred to as craniates and not considered vertebrates. It was, however, a messy division. What Aldridge and his collaborators required was a chink in Janvier's armor, and they found that chink by appealing to "common usage," which placed Janvier's craniates in with the vertebrates. The argument, which came from developmental biology, drew upon a shared feature present in the embryos of all vertebrates: the neural crest, a feature from which vertebrae, cranium, and other defining elements emerged. Aldridge remained convinced that the conodonts were anatomically close to the hagfish.

One important hurdle to winning this argument seemed to be to persuade Janvier, but he and Peter Forey of the Natural History Museum in London were not convinced, and certainly not by Sansom's work. They argued that the similarities between the structure of conodont elements and enamel were superficial; they looked similar but there were serious differences. They wanted dentine: "There appears to be a total absence of dentine, which is unexpected if conodonts are vertebrates. Dentine is universal in vertebrates and is thought to be the most primitive of vertebrate hard tissues." They were not alone; others also doubted Sansom and his colleagues.[19]

Most vociferous among the vertebrate paleontologists objecting to the conodont vertebrate was Sue Turner, a fossil fish specialist at Queensland Museum in Australia. She and Alain Blieck were two of the organizers of a major international symposium that both commemorated the work of Walter Gross and discussed the problem of "Palaeozoic microvertebrates" at Walliser's university in Göttingen in August 1993. For Paul Smith, Sansom, and Purnell, this was their first opportunity to present their views on the conodont animal to a traditional vertebrate paleontology community, and they soon discovered they had driven headlong into a brick wall. Paul Smith later recalled, "One lunchtime we realized that things were a bit quiet around the conference venue, only to discover that an impromptu meeting had been called to discuss how this community could counter the conodont hypothesis. It was at a reception at the end of that meeting, in Otto Walliser's house, that Sue Turner told us that 'no evidence we could ever provide' would convince her that conodonts were vertebrates." When *Nature*'s Henry Gee heard this, he christened the Birmingham-Leicester team, the "the Pommie Bastard Conodont Group," an epithet these workers then used ironically to identify themselves. Gee suggested that instead of attempting to overcome the seemingly immovable obstacle of at least some in vertebrate paleontology, they concentrate on convincing the broader and less politically organized biological community. Turner, Blieck, and a number of others at this meeting would now hold their resolve indefinitely. They would be joined by conodont workers who, while not working directing on the biology of the animal, were nevertheless unconvinced by the British animal.

✳   ✳   ✳

Sansom, with Paul Smith and Moya Smith, continued his assault on Janvier's stronghold. This time they focused on *Chirognathus*, a conodont genus with delicate elements that Branson and Mehl had described as fibrous, alluding to their frayed, wood-like structure. At one time these fossils were considered fundamentally different from other kinds of conodont, but Lindström and Ziegler had shown this structure simply resulted from a lack of white matter. Branson and Mehl had described the basal filling as bone-like and with Stauffer they marveled at the survival of fossils with such "delicate, sharp-pointed, hand-like crests with the fibrous structure." Sansom and his co-workers found that the base of these fossils did not contain the globular cartilage he had seen in the previous study but a form of dentine. This diversity of tissue was not unknown in early fish remains, and in 1994 they concluded, "It seems that both conodonts and agnathans experimented with many tissue types and tissue associations in the early history of vertebrate biomineralization." The discovery of dentine was, they believed, particularly significant. It seemed to offer "additional and conclusive evidence of the vertebrate nature of conodonts."[20]

Anne Kemp and Bob Nicoll responded to the earlier paper by searching for collagen in the conodont elements. Collagen occurs in bone, cartilage, and dentine but is not – in living animals – found in enamel. Using a stain that reacts to the presence of collagen, they investigated the organic components within the elements and found the lamellae making up the crown of the conodont element took the stain, but not the white matter or basal tissue. This suggested that the lamellae did not consist of enamel and that the white matter and basal tissues were not bone or dentine. These results flatly denied those of Sansom and his co-workers. Kemp and Nicoll continued to favor amphioxus as the most appropriate comparator organism and saw no evidence of vertebrate affinity.[21] The British, however, doubted that material this old could preserve chemically active collagen capable of responding to the stains used by biologists working on living animals.

Purnell now stepped in, supporting Sansom and his colleagues with what he believed was unequivocal evidence that conodont elements did function as teeth; food was "crushed" and "sheared" and had left the marks to prove it.[22] From studying completely preserved apparatuses, it was now possible for him to know which surfaces came together in processing food, and in yet another short paper, rapidly published in *Nature* in April 1995, he reported how he had looked for signs of wear like those found in mammal teeth. Purnell knew that such evidence did not simply support the tooth theory, it opened up entirely new possibilities for understanding the animal as a living entity. Mammal teeth, he wrote, show three types of wear: polishing, scratching, and pitting. He found these same distinctive features on his tiny conodont elements. His platform element showed pitting but no scratching and suggested a crushing mechanism. A blade element showed polishing, which suggested that area was not exposed to abrasive food. Other elements did show scratching, but to variable degrees, suggesting exposure to different kinds of food. And most interestingly, but problematically, these wear patterns were found on young conodont elements too, meaning that the elements were in use while they were still growing. The elements were, it seemed, periodically enclosed for repair but spent most of their time exposed and functioning.

With a little orchestration behind the scenes, the very next paper in this issue of *Nature* revealed Gabbott's Soom Shale animal. Remarkably, the animal had preserved in fine detail evidence of muscle-fiber and its component fibrils. Preservation of muscle around the eye suggested more advanced "encephalization" or "brain" development. This, together with other preserved features, indicated that the conodont animal might have been more advanced than the hagfish or the lamprey. Although anatomically similar to the Scottish animals, *Promissum* possessed a rather different apparatus, indicating that it belonged to a different group of conodont animals. It had, of course, come from much older rocks, and its preserved muscle tissue was now the oldest in the vertebrate fossil record. It was something like the muscle fiber found in amphioxus, but more like that found in slow agnathan fish. It suggested that the animal could cruise rather than race.[23]

All these discoveries were sufficiently in tune with the real world to attract the interest of journalists and consequently several made the national and popular scientific press in 1994 and 1995. Janvier's sometime collaborator, Peter Forey, of the Natural History Museum in London, admitted to *New Scientist,* "For many years, we as vertebrate paleontologists did not want anything to do with conodonts."[24] Now the evidence was irresistible. Janvier's resistance to the conodont craniates had been crumbling since early 1994, and in this same issue of *Nature,* faced with the weight of mounting evidence, he finally capitulated. In an article titled "Conodonts join the club," he wrote, "I think it is time for me to stop playing the Devil's advocate against this theory. There are still peculiarities in the structure of the 'conodont animal' that are at odds with what we know of classical early vertebrate fossils, but its tail, arrangement of the muscle blocks and large eyes are strikingly vertebrate-like." He was, however, a reluctant convert. Sansom and company's assertions regarding tooth structure seemed to him to be an act of "shoehorning," though he was rather more convinced by the recent evidence of dermal bone and dentine. The conodont remained problematic and, like the hagfish, he suggested that it might have evolved from a conventional early vertebrate. Like Sweet, he was puzzled by the things that had not been found. Why was there no evidence of the dense organic matter associated with gills ("no other fish lacks gills")? This was associated with another oddity: the positioning of the eyes in front of the mouth. Perhaps the front of the animal was missing, he said, and the conodont elements were associated with the pharynx and gill apparatus rather than the mouth, "like the pharyngeal denticles of jawed vertebrates"?[25] This would actually make the conodonts more vertebrate-like and explain the lack of gills: "Although the conodont animal is reconstructed as a small, frail, almost transparent animal that reflects the classical 'archetype' of the ancestral vertebrate, it may, after all, turn out to be a highly specialized close relative of the jawed vertebrates that developed suction feeding and pharyngeal denticles to grasp and shear prey."

The focus of Janvier's argument had now shifted to the positioning and functioning of the apparatus. In Leicester, Aldridge and Purnell had recruited a new doctoral student, Phil Donoghue, to take this study forward by looking for evidence in the elements themselves. Part of the problem concerned how the elements grew: How could they function and show wear and yet still grow? This, more than anything, challenged the notion that these really were teeth, particularly now that Bengtson's pockets looked increasingly untenable. Donoghue's study, which reviewed and built upon the complete literature examining the interior structure of the fossils, is remarkable testament to the seriousness with which the biology of the animal was being considered.[26] The detail of its probing was unparalleled, even in comparison to those pioneering light and scanning electron microscope studies of just a few decades before. Now these belonged to a distant past. "Without any degree of constraint over affinity this proved an unprofitable line of research, resulting in a series of esoteric accounts of hard tissue ultrastructure," Donoghue reflected. "In retrospect, it would never have been possible to reach an unequivocal conclusion regarding conodont affinity just by analyzing element morphology and internal structure alone."

The science had never put this level of resource into such a seemingly tiny question, but now rather more rested upon its conclusions. The arguments were detailed, but there was a belief that there was a possibility of discovering still more. So Donoghue examined growth across a range of element types, looking for patterns that might separate analogy from homology. The blades, he found, began by growing lamellae but almost immediately began secreting white matter, which formed the core of all denticles, the number of which was extended during growth. Later growth was concentrated in the lower portion of the element, changing its shape. The final stage saw the crowns "finished off" with lamellar tissue and growth of the white matter was halted. Other types of element showed rather more complex and varied growth patterns.

Donoghue was also interested in understanding how these growth processes had changed through the course of the creature's evolution. As to what his data meant, Donoghue permitted his interpretations to be shaped by the "current consensus," which said that conodonts were chordates. But he took issue with Sansom and colleagues' suggestion that the animal experimented with tissue structure. Instead he argued that dentine itself was so structurally variable there was no need to believe this variety resulted from experimentation. The conodonts' unique white matter remained difficult to interpret, but Donoghue thought it rather more dentine-like than like cellular bone. White matter, it seemed, had its own complex evolutionary pattern; it had not been adopted in the same way, or even at all, across the conodont group as a whole. Clearly it strengthened the element, but it was difficult to say more.

As to the problem of growth, he thought the answer lay not in considering vertebrate teeth – which cannot repair the functional enamel once erupted – but in scales. He noted that in many fish, the scales are drawn back into the dermis for additional growth before re-erupting. Conodont elements were not, then, teeth in the strictest sense, even if they functioned as such. He also postulated that, like the lamprey and hagfish, the conodont animal probably had a skeleton composed of cartilage. As cartilage is rarely preserved in the fossil record, this assertion seemed to him quite reasonable. His interpretation repeated suggestions made by Aldridge and his collaborators back in 1986, but they could not prove them then.[27]

Meanwhile, Purnell had been involved in reconstructing the Soom Shale animal's apparatus using the modeling methods Aldridge and his co-workers had exploited in the mid-1980s.[28] In 1997, along with Donoghue, Purnell presented an operational model of the animal's apparatus. This produced an arrangement of elements suggesting an analogy, in terms of its operation, to the jawless lampreys and hagfish. They imagined a hideous "eversible lingual apparatus," once held together by muscle and cartilaginous plates.[29] Its spiteful fine-toothed elements were arranged to project outward in a grasping motion, redolent of and perhaps influenced by Ridley Scott's formidable creature in *Alien* (1979) (figure 14.2).[30]

In 1995, the Leicester and Birmingham groups could celebrate a victory. The Leicester workers en masse did so in an article titled "Conodonts and the first vertebrates" in the popular science journal *Endeavour*. "Some time ago," they began, "probably during the early part of the Cambrian Period (520 million years ago), a new type of animal appeared. It was small, a few centimeters in length, and elongate; it had no hard skeleton, but a stiffening rod of cartilage along its back and V-shaped blocks of muscle along its sides; it had paired eyes, a brain and tail fins. It was the

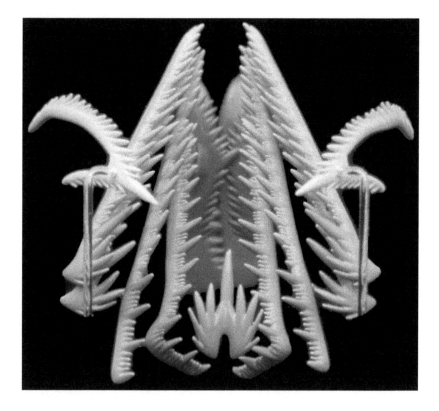

14.2. Alien jaws. Purnell and Donoghue's reconstruction of the animal's formidable apparatus seen head on. Photo: Mark Purnell, University of Leicester.

first vertebrate."[31] This notion of the ancestral vertebrate was not the conodont animal, for all that it looked very like one. Rather, it was a long-established and widely held concept, stemming originally from drawing a line from amphioxus, which was a vertebrate-like invertebrate, to the hagfish, seen by some as the most primitive living vertebrate. Recent work by British workers had shown that the conodont did not lie on this line but was more advanced – more a vertebrate – than the hagfish. Now the Leicester team explained the significance of the conodont animal and the debate that surrounded it in plain language. The textbook view was that the first vertebrates were suspension feeders living lives comparable to amphioxus and larval lampreys. According to this explanation, they only became predators a hundred million years later. But Purnell considered a predatory condition was necessary for the evolution of eyes, muscles, and skeletons. On this latter point, they too would assert that the vertebrate skeleton began not with defensive armament but with teeth. Although there were still some anatomical uncertainties about the animal, these new vertebrates nevertheless had impact for they had existed for some three hundred million years and had left an unparalleled record of their time on Earth. The article was illustrated with a photograph of conodont elements on a pinhead. One of the most striking images the science had ever produced, it at last permitted the general reader to visualize these obscure fossils in everyday terms. It soon appeared on magazine covers and websites; wherever the science needed to explain itself to a wider public (see this book's frontispiece).

Purnell translated this *Endeavour* piece into an article for children called "Armed to the Teeth."[32] Here, too, the "weird" conodont was a new and exciting clue to that mysterious vertebrate ancestor from which we sprang. "What, I hear you ask, is a conodont? It's a good question, and until a few years ago no-one really knew the answer." The conodont was made for this dinosaur-loving audience: "One of our early ancestors was a small but very vicious killer!" That killer was illustrated, though it unfortunately lacked *Alien* menace.

In another article, "The conodont controversies," Aldridge and Purnell disseminated their new vertebrate among evolutionary biologists, most of whom knew little or nothing of the animal. Gee had remarked that modern evolutionary biologists used methods that encouraged a presentist view of the past; they made their decisions based on what was now living rather than on an understanding of the extinct. It was a problem Simpson had recognized back in the 1930s and it had not gone away. But modern biology could not invent the conodont. For one and half centuries the animal had fought for its place among the vertebrates simply by virtue of its tooth-like remains, struggling against doubts that suggested it was simply too old and that its teeth were not as tooth-like as they might superficially seem. As the conodont entered this new world of vertebrate biology, it lost none of its controversy; it appeared from nowhere possessing a rich evolutionary history supported by millions of specimens, and it sought to significantly rewrite history. Of course, Aldridge and Purnell's biggest problem was a lack of certainty about

14.3. The conodont animal as it is imagined today. The animal has been depicted in a range of guises, both comic and factual. In 1993, an image appeared that had the animal possessing huge eyes, an error introduced by a picture editor who flattened a perspective drawing of the animal swimming toward the viewer. This encouraged Forey and Janvier to suggest that the animal might be a larval fish. Nevertheless, the eyes in the Scottish animal were relatively large, which Purnell explained as representing the minimum possible size for functioning vertebrate eyes. The strap-like eye muscles in Gabbott's beast suggested the animal could rotate its eye – a feature only found in vertebrates. Positioned on the front of the head, the eyes also suggested to Purnell that the animal was an active predator. New reconstruction of the animal by scientific illustrator, Debbie Maizels. ©Debbie Maizels/Simon Knell.

where the conodonts fitted in the vertebrate family tree. But this was true of many of those animals lacking a backbone yet applying for admission to the vertebrate camp. In an attempt to court this new community, Aldridge and Purnell paraded six possibilities for placing the conodont in the tree of vertebrate life. In their *Endeavour* article, they had restricted themselves to one. Of course, a great deal depended on what one might call a vertebrate, but they argued that "authors who have worked on the conodont animal specimens are united in the view that the conodonts belong within the Craniata (= Vertebrata, if the myxinoids are included in vertebrates[)]." They did not point out that these authors belonged to one extended family, but that mattered little. If other scientists were to adopt the conodont, they would do so on their own terms, just as Janvier did. Aldridge and Purnell's paper revealed that there was still much to play for. As Gee noted, "On this evidence, at least, it seems clear that conodonts represent a very early radiation of 'jawed' vertebrates, possibly distinct from the radiation of more familiar jawed vertebrates. . . . In which case, they need not tell us very much about vertebrate origins as such." But on this, he admitted, the jury was still out.[33]

These new interpretations were also finding their way into standard textbooks. The author of many, Bristol paleontologist Mike Benton, referred to the conodont as a microvertebrate: "The first fishes date from the Late Cambrian, and the commonest group then, and in the Ordovician, were the conodont animals."[34] Textbooks produce orthodoxies. Students learn and never forget that first act of learning. The conodont was a vertebrate: established in science, won by campaign, and evangelized across the media.

Of course, the battle was not entirely won, even if it had become the orthodox view. There was still the matter of tree building and negotiating that boundary where life could be said to become "vertebrate."[35] The territory that had been won for the conodont animal in this campaign continued to be defended and strengthened by those who had fought for it.[36] The new orthodoxy continued to be consolidated. At the conodont's sesquicentennial birthday celebrations in Leicester in 2006, Dick Aldridge suggested it was time to stop beating around the bush and call conodont elements "teeth." He believed the vertebrate skeleton originated in conodonts and that they were our earliest "stem ancestors." The concept of stem ancestors here meant the conodonts were the first branch on the bough of an evolutionary tree which had jawed vertebrates at its end.

Among the two hundred or so active conodont workers then on the planet, beliefs varied about what the animal was. Each mind held a slightly different animal. Nearly all believed in the chordate, though some preferred amphioxus as a model.[37] Only a tiny group had taken these thoughts into press. In 2006, the conodont controversies had shifted slightly more to consensus, not least because the mountain on which the conodont had been placed seemed unassailable – its defense was too strong. But as the heat dissipated, those who had built and defended that mountain became drawn into new projects and other animal groups. With Aldridge only a few years from retirement, was the science again to change?

Not long after this meeting, I returned some borrowed papers. Dick Aldridge looked at me with a wry smile on his face. "This will be good for your book," he said. He showed me a short paragraph in *Ordovician News* that reported that three Australian protagonists – Anne Kemp, Susan Turner, and Carole Burrow – had "initiated a polemic paper on the relationships of conodonts with vertebrates, to develop the idea that conodonts are NOT vertebrates. This is agreed in majority by the early vert experts, and perhaps also by many (if not the majority of) conodont experts. This paper will be co-authored by a group of people including Turner, S., Burrow, A., Hanke, G. F., Männik, P., Nowlan, G. S., Reif, W.-E., Rexroad, C. B., Trotter, J., Viira, V. & Young, G. C. (provisional list)."[38] Sometime later, he sent me the abstract of a paper by Alain Blieck, Susan Turner, Carole Burrow, Hans-Peter Schultze, Wolf-Ernst Reif, Carl B. Rexroad, and Godfrey Nowlan's paper explaining "why conodonts are not vertebrates." It stated, "Excessive self-citation, mis- and over-interpretation by the 'British School' in association with a blanket of publication and communication have wrongly established an acceptance by the scientific community in journals, textbooks, websites and popular articles that conodonts are vertebrates."[39] It was a manifesto for political change rather than a scientific argument. The list of signatories included a cross section of the conodont community, many of whom had never engaged in debating the animal's biology. Others had been longtime objectors and had proposed a number of alternative solutions to the enigma. They now formed a confederation that was the equal of the British school. The controversy of the animal was once again alive, and had Christian Pander somehow returned to Earth, he might well have laughed at the irony of it all. For more than 150 years there had been so much scientific effort – and so much dreaming, imagining, and arguing – and yet here we were back at the beginning, or so it seemed to some. As had been the case so many times before, two groups faced off against each other. One, possessing the evidence of real things, assumed Pander's imaginative self-assurance. The other, by contrast, felt it possessed the truth of the known world. It now adopted Owen's uncompromising stare.

In keeping with its enigmatic qualities, the conodont animal truly was as slippery as an eel. At the slightest provocation it seemed to retreat into a Kafkaesque world of disorientation and uncertainty. Aldridge's smile said a lot. He had delighted in the science but also in the debate and controversy. The animal had been an event. It had been fun. And, most important, it had lost none of its enigmatic charm. That too appealed to Aldridge. Was it possible that El Dorado was not here but somewhere in the distance? Some seemed to think so. Perhaps it is still possible to dream of solving the riddle of the conodont, to imagine that somewhere, in "some layer of subaquatic volcanic ash," one can find El Dorado.

It was so small, such a tiny, early, transitional mass, a coagulation of the unsubstantial, of the not-yet-substantial and yet substance-like, of energy, that it was scarcely possible yet – or, if it had been, was now no longer possible – to think of it as material, but rather as mean and border-line between material and immaterial.

THOMAS MANN,
The Magic Mountain 1928

The Progress of Tiny Things

THIS BOOK TELLS THE STORY OF A SCIENTIFIC JOURNEY OF TWISTS
and turns through assertions and denials, past alien monsters and incoming asteroids, through a world of unexpected discoveries and real
utility, which ultimately arrives at an animal that, rather surprisingly,
seems to say something about our own ancestry. In the course of all
this traveling, countless animals formed in scientific minds only to dissolve, replaced by new apparitions. The fossils themselves were so small,
that seeing them – really seeing them – was no easy matter. Indeed, the
millions of these things present in collections around the world today
might, if poured like so many grains of sand, fit into a few shoe boxes.
We could be forgiven, then, for not knowing the conodont. But my aim
in writing this book has not simply been to record a famous episode in
the folk history of a science. Being so tiny and evocative, these fossils
possessed a chameleon aspect. El Dorado–like, they pulled their victims
into mirror-filled rooms, ensnaring their thoughts with mirages and illusions. Vanishingly small and impossibly ambiguous, they occupied
that borderline between material and immaterial, never wholly one or
the other. If one was not careful, it was possible to see in them what one
wanted to see. If that was the case, then what they saw was no more
than a thought, an imagined thing conjured into existence. If what the
conodont workers thought they saw was then transcribed onto paper,
one has, in these representations of the fossil, a tiny lens on the minds of
those who found themselves inextricably entangled with these enigmatic
material things. It was with this thought that I began this book, and
consequently my focus has been on fossils of the mind rather than their

physical counterparts, for it was only in the scientific mind that these objects acquired their magical properties.

So I positioned myself, like an anthropologist, on the edge of this scientific community, never seeking to judge the actions or values of its members. Bronislaw Malinowski, the pioneering anthropologist, once observed, "Living in the village with no other business but to follow native life, one sees the customs, ceremonies and transactions over and over again, one has examples of their beliefs as they are actually lived through, and the full body and blood of actual native life fills out soon the skeleton of abstract constructions." I am not an anthropologist, but that field nevertheless provided me with an ethical relationship to my subjects. I viewed my actors as existing within a particular cultural world, and my interest was in how within this world these objects were produced, explained, and shared; it just so happens that this world is shaped and regulated by the practices and beliefs of science. My desire was not for the ethnographer's close observation of a moment in the life of this scientific community but to view it through the long lens of cultural history in order to understand how and why thinking changed and how ideas were passed from one generation to the next. This approach, when married with anthropological distancing and an a priori belief in the constructive nature of cultural interactions, would, I felt, permit me to understand how the conodont workers came to know their objects and produce their various animals. It would also reveal how they managed to achieve such major intellectual progress while possessing such an incomplete understanding of the objects themselves.[1]

Using this approach, I chose to treat all knowledge within this community as contingent: different in every mind, often conflicted, and frequently infected with errors and mere orthodoxies. My attitude to these undetected errors and orthodoxies, however, was to treat them as situated truths or beliefs – operationally, this is what they were. At no point was I concerned with what the animal really was, only in what these workers *believed* it to be. My data, laid out chronologically and thematically in the chapters, reveals the faceted manner in which the object – and thus the animal (the two concepts being held in a confused terminology) – acquired its identity; the animal comes into being not simply as a result of individuals hunting it down, but through glimpses of its tracks and traces. Implicitly, and often as a subplot behind the main action, the animal acquires behavioral, geographical, and historical characteristics before it acquires material form.[2]

My approach, then, has been to treat the fossils as objects of material culture, "that segment of man's physical environment which is purposely shaped by him according to a culturally dictated plan."[3] Here fossils are anthropological objects before they are mobilized as evidence in science. Paleontology is, though rarely recognized as such, a discipline deeply engaged in material culture study. Most workers saw these fossils as literally and metaphorically beautiful; they certainly had a full appreciation of the objects' materiality, but the practices of science demanded that they separate the useful data from their poetic experiences. The conodont was (and perhaps still is) one of science's great enigmas because it exists, as Thomas Mann said so eloquently of atoms in the opening quotation, "mean and border-line between material and immaterial." This is implicitly what this book explores – the dual reality of scientific objects: one material, the other immaterial; one real, the other of the mind.

TAKING KNOWLEDGE TO THE FOSSIL

To begin with, we should consider the knowledge the conodont workers took to the object, for these scientists always came with intentions. Art historian Ernst Gombrich observed, "the role which our own expectations play in the deciphering of the artists' cryptograms. We come to their works with our receivers already attuned." Studies of perception in science, art, and society suggest that the conodont workers' knowledge incorporated what might be understood as factual and accurate truths *and* all manner of things they imagined or believed to be true or probably true. Gaps in their knowledge were filled – as in the everyday world – using vague memories, possibilities, probabilities, similarities, chance encounters, and so on. We might call this "working knowledge" or, as psychologists in Gombrich's day referred to it, "a mental set," but in doing so we need to understand that this knowledge is by implication not something hard and fast. So effective is our iterative engagement with the material world that we can possess quite accurate ideas about the properties of things without knowing precisely their cause; imprecise and inaccurate beliefs can be incorporated as substitutes for truth without the world collapsing around us.[4]

The conodont workers frequently found themselves confronted by objects that forced them to think beyond their previous experiences and thus reach into their stores of expertise, which incorporated things known and (perhaps unknowingly) things *thought* to be true and relevant. Truth was contingent upon belief. For the true or untrue to have effect or agency in their scientific lives, it had to be *believed,* or at least be believable. It must for them *appear* true. The semantics of whether they considered it absolutely, hypothetically, plausibly, or probably true are not important to us. Often these truths were implicit or tacit, and they were almost always undifferentiated.[5] Truths of all kinds were performed in much the same way. The conodont workers had a tacit understanding of the possibilities of discovery, the accessibility of reality, the centrality of disinterestedness, the distancing terminology of theories, hypotheses, paradigms, and models (though rarely used it), the superiority of measurement and exact science (particularly chemistry, physics, mathematics and some biology), the necessity of open and testable data, and so on. These things were performed implicitly. They were understood as moral and ethical intentions and expectations. They were the glue that held these workers in the cohesive cultural world they knew as science. But rarely did these beliefs transport them into an abstract intellectual space divorced from other realities and other ways of knowing. Repeatedly they demonstrated that they knew much more about conodonts than other scientists, but nevertheless their expertise was contingent on a whole raft of experiences and aptitudes. They had done science firsthand and so had determined the guiding framework, the rigor applied, the fudging necessary, the things ignored, the treatment of failure, and the impact of external pressures. They knew at least some of the weaknesses and imperfections in their knowledge, but they could not know all of them.

To these banks of hard-won knowledge they also were willing to admit other kinds of "knowledge" communicated by esteemed, innovative, or charismatic colleagues, perhaps in short conference presentations. They, of course, judged the truth of such presentations against their own scientific standards and found support and challenges within them.

A third component in this knowledge making were ideas imported from the past and from other fields, whether through their education, personal research, things read, seen, or heard, or, more broadly, from lived experience. Much of this was made relevant through the selective filter of present need; this imported knowledge was often, as a result, supportive knowledge.

So what each conodont worker knew, and took to the object, was of mixed composition. And never was the knowledge they deployed in this way corralled into a discrete part of the mind away from cars, shops, novels, pets, and babies.[6] Repeatedly, workers showed the capacity to bring into their reckoning all manner of external influences, though only some of these were articulated in the papers themselves. Being an object-based interpretive science, which is reliant upon connoisseurship skills, paleontology has come to implicitly value this intellectual eclecticism.[7]

Each individual, then, possessed a constellation of knowledge about the fossil that had been derived from diverse sources.[8] At its core was firsthand knowledge, which defined each individual's territory and scientific identity. It was from the security of this heartland of deep experience that individual conodont workers felt most able to develop and defend new ideas about these fossils, as it permitted them to both evaluate and avoid risks, and maneuver around difficulties. This core experience – often nebulous, undigested, and unarticulated – also held within it latent knowledge that had the potential to be configured and used as called upon when problems presented themselves. Paleontological combatants, such as Stephen Jay Gould, knew only too well that great advantage could be gained in an argument if one could drag opponents out of this comfort zone and into an alien territory where they would feel at sea. Gould calculated that he possessed greater knowledge in these other fields, which often drew upon episodes in science, science history, and the classical arts, in which opponents would be too embarrassed to show their ignorance. One way of overcoming these difficulties, however – particularly for experienced conodont workers forced to migrate into new intellectual territories as their research developed – was to seek collaborations with other specialists and/or undergo a degree of self-education.

The variable, eclectic, and contingent manner in which knowledge was acquired and then applied within the conodont research community made individualism a core resource. Among the conodont workers it appears that only Lennart Jeppsson came under the spell of Björn Kurtén, the cave geologist, and only Maurits Lindström held conversations with algal biologist Adolf von Stosch. As a result they could bring unique perspectives into a debate. A quick survey of the thoughts of senior conodont workers on the nature of the animal, conducted in 2007, for example, revealed a surprising array of thinking, despite recent breakthrough discoveries. The imagined animal present in these minds possessed no singularity of form, even if it seemed to coalesce around a number of discrete possibilities. Each worker based part of his or her understanding on the same animal specimens, though most of them had no firsthand experience of the fossils themselves. Each individual viewed this evidence through the filter of his or her own experiences. Far from being problematic, these varied beliefs and performances gave conodont research an intellectual richness, suppleness, and adaptability. The lack of a singularity of view – of a narrowly conceived paradigm for the animal – meant that the animal's elusive identity remained unimportant. It could be worked around or held constantly in negotiation. And anyway, resolving its identity was never the conodont workers' central occupation.

The other implication of this eclecticism was the frequent appearance of what conodont workers referred to as "luck." Few stopped to consider that a socially connected, opportunistic, and intellectually focused and eclectic culture might have little need for mere chance.[9] This feeling, however, demonstrates that members of this community felt that surprising progress was being made; in other words, this socially negotiated, culturally conforming, yet freely intellectualizing community was effective.

These aspects of science culture, frequently discussed by historians and philosophers of science, empowered individuals to mobilize objects as evidence. The history of the conodont animal, however, reveals a huge number of such mobilizations: Countless kinds of animal and plant were put forward as the owners of Pander's mysterious tiny teeth. In a rigorous scientific field, populated with talented individuals, how could this happen? Surely the fossils themselves – the material evidence of the thing – would prevent this?

THE INTANGIBILITY OF FOSSILS

In 1988, Walt Sweet observed that workers possessed a large number of isolated tooth-like conodont fossils representing some four thousand species, together with five hundred assemblages and five conodont animals.[10] He stated that these alone were the facts; "all the rest is really speculation or, if you will, interpretation." Sweet believed that anything other than these few material facts was contestable, and although he referred to "species" – which he knew resulted from acts of interpretation – this was simply his concise way of saying "four thousand precisely known types of material object." Lennart Jeppsson gave a nice illustration of this disconnection between material reality (specimens) and immaterial belief (species): "One specimen = one species; two specimens = two species; 100 specimens = one species." His point here is that once a conodont worker had sufficient material, he or she became a connoisseur of species' variation and was then able to distinguish differences in an informed manner.

American paleontologist David Raup once admitted that "the road to good scholarship is paved with imagined patterns." One of those imagined patterns, I suggest, is the fossil as it appears in science. Fossils preserve traces of former life and are composed of minerals that in most cases are not those that existed in the living animal or plant. Yet in common usage we simplify this understanding; we think of the fossil as a trace of life, disregarding the complex processes (geological and intellectual) by which it has come to be considered as such. There is, for example, an unarticulated and ambiguous boundary between the fossilized animal or plant and the materials of which it is composed or a part.[11] This ambiguity is apparent in the work of Karl von Zittel and Josef Victor Rohon in the late nineteenth century. They reinterpreted structures long thought to record the anatomy of the animal as mere artifacts of fossilization. Things once considered reminiscent of life were made irrelevant; they were now just rock or mineral. This interpretive act should indicate to us that the fossils that existed within the minds of the conodont workers were different from those preserved in stone. Indeed, the fossil that participates in science is only ever a conceptual or immaterial one: a fossil imagined and believed and not the one that has been hewn, boiled, or dissolved out of the rock. To make use of the material fossil, science must interpret it and in doing so decide what characteristics define it, its material and taxonomic boundaries, its significances, and so on. It is this value-laden fossil – a conceptual fossil, a representation of the reality – that then enters the mind and the science and is partially captured in word and image. This notion has its origins more than two millennia ago, and it is not my intention to attempt to engage with general philosophy. I merely wish to explain the broad basis on which I have explored the practices of this particular group of scientists. A few sound bites might at least set the scene. Ian Parker, for example, noted, "We *must* separate the world from our knowledge of it. We live in an *Umwelt*, beyond which there are currently unimagined material possibilities. We must assume that the world is richer than we know." Goethe remarked in 1823, "My thinking is not separate from objects; that the elements of the object, the perceptions of the object, flow into my thinking and are fully permeated by it." Edmond Husserl wrote that "the objects that surround us function less 'as they are' than 'as they mean,' and objects only mean for someone. . . . To see implies seeing meaningfully." Bertrand Russell observed, "It is not correct to say that I am believing the actual event; what I am believing is something now in my mind . . . since the event is not occurring but the believing is. . . . What is believed . . . is not the actual fact that makes the belief true, but a present event related to the fact. This present event, which is what is believed, I shall call the 'content' of belief." Finally, Alfred Schütz: "Even the thing perceived in everyday life is more than a simple sense presentation. It is a thought object, a construct of a highly complicated nature."[12]

The material object from which our conceptual fossil is produced remains simply that: material and mute but bearing witness to its origins and the context from which it has come. Science can take its immaterial (or conceptual) fossils wherever it likes, and in the case of the conodont fossils this has been to imagine them in a wide variety of ways. The silent, material fossil – the real fossil – remains in a drawer or on a microscope slide. Although situated in the tame environment of a collection, this object sits apart from science, always belonging to a reality beyond science. Being real, it is there to question science and to be the subject of new investigations as new knowledge and technologies permit; it alone – it must be believed – has the potential to reveal the ultimate truth.[13] Science's great task is to unlock that truth or, rather, make its truth the same.

To suggest this reading, rather than adopt the realism the scientists themselves deploy, is not to weaken our conception of the science. It simply means that science's inaccuracies and mistaken beliefs remain separated from reality by an invisible and impenetrable barrier; the factuality of material objects is never dependent upon the vagaries of scientific belief.[14] History shows that real fossils have never been affected by thought; they have proven immune to designer gods and successive creations, untouched by the Flood and all those variants of evolution that were believed before and after Darwin. The evidential or immaterial or conceptual object *appeals* to the truth of the material fossil, but its connection to its material twin is detached and fluid – it lives in another world. The object in our thoughts seems material, definite, and fixed, but it is in fact intangible, contingent, and transient.[15]

This is a useful analytical frame for thinking about the relationship between the fossil and interpretations that produced its enigmatic qualities; it is not an argument against scientific realism. Each approach is culturally situated. The conodont workers needed to approach their subject as realists and believe that material truth was accessible. This was an essential component in the performances necessary for separating truth from non-truth. But these performances were affected by spatial and temporal structures and disruptions that had broad implications for how the material fossil was perceived. We shall now turn to these.

THE SPATIAL DYNAMICS OF INTERPRETATION

The plurality of knowledge in the conodont community offered some protection from erroneous ideas and gave the field some consistency of thinking. While bizarre solutions to the problem of the animal's identity would still surface with surprising regularity, these most often came from outsiders caught up in the mythology. Nevertheless, the conodont research community remained tiny and potentially vulnerable. It was this that permitted Walter Gross to silence a generation when he demonstrated, apparently conclusively, that the conodont fossils could not be teeth. Given his authority in the more fully developed field of fish paleontology, and his exceptional conodont material, along with the novelty, rigor, and logic of his study, other conodont workers were rendered powerless to contest the change he imposed on their worldview. For more than a decade, the nature of the animal and the functioning of the tiny tooth-like objects were hardly discussed. This mind shift led to, and was further helped by, the erasing of a deeply embedded technical language drawn from the study of teeth, which had been used to describe the conodont fossils for more than a century. To retain these terms was now considered unscientific. Then, in the 1970s, paleontology acquired a new intellectual liberalism that in turn produced arguments suggesting that the fossils could be teeth after all. With further research the idea gained support, and at a conference in 2006, Dick Aldridge told his audience to stop beating around the bush and start referring to the fossils as "teeth."

It is difficult to conceive of Gross's assertion as paradigmatic because the identity of animal itself was not central to conodont science. The assertion was a "truth" that had to be accommodated but could also simply be ignored. It was possible to do this because the culture producing conodont science manifested itself in ongoing social relations rather than concrete institutions. "When we look at how people experience and negotiate authenticity through objects," Sian Jones notes, "it is the networks of relationships between people, places and things that appear to be central, not the things in themselves."[16]

This social adaptability meant Gross's inconvenient truth was negotiated in exactly the same way workers dealt with changes to their discipline in general and inconsistencies introduced by geographical variations in practice. The doing of conodont science, like the doing of paleontology, was never constant or uniform. In the sparsely populated United States of the 1930s, conodont science developed differently in Chicago (and Illinois), Washington D.C., and Missouri. And each of these centers developed an engagement with the fossil that was different from that present in Göttingen. At each center particular individuals played a critical shaping role: Croneis in Chicago, Ulrich in Washington, Branson in Missouri, and Stadtmüller in Göttingen. In the first half of the twentieth century, the United States supported two kinds of paleontology, both interested in the conodont fossils but distributed unevenly across the nation. And in what was still a small and well-connected paleontological community, it was possible for the likes of Ted Branson to assume the role of geological Baron of Missouri and thus exert a degree of control over the local scientific culture and the participants engaged within it.[17] Each group tackled these fossils according to intellectual resources, preferences, biases, and ambitions available locally; each formed a distinct "interpretive community."[18] Repeatedly conodont science produced these important, geographically specific cultures that had a profound effect on the development of the science: the University of Iowa in the 1940s and 1950s, Ohio State University in the 1960s, the University of Marburg from the 1950s to the 1970s, and the "British school," centered on universities in the English Midlands in the 1980s and 1990s.

In Europe, national cultures shaped the engagement. Here language configured separation and collaboration. British conodont veteran Ronald Austin, for example, recalls being required to take French as a subsidiary subject at university, and this determined his collaborations on the Continent in later life. Austin found meeting colleagues in Eastern Europe, then behind the Iron Curtain, extraordinarily expensive and effectively discouraged. In the Soviet Union, a much smaller and largely self-contained conodont research group had developed. With the rise of English as a global scientific language, particularly after 1990, non-English speakers became increasingly disadvantaged. Thus language alone ensured that at no point was conodont studies ever an even playing field, and in the 1960s there were frequent appeals for translations of key works. Few workers, for example, could read Pander's book in its original German and had instead to rely upon an American translation, which in subtle ways incorporated mid-twentieth-century thinking.

These geographical inconsistencies within the field were accompanied by changing ideas about the nature of paleontological science in general, particularly after World War II. The drive then was to make the study of fossils more rigorous, sophisticated, and biological. Young workers who entered the science at this time were more than ready for this injection of new ideals; they wanted nothing less than a new paleontology. In the early 1970s, when geology as a whole was embracing big theory, paleontologists again reinvented their science, liberating practitioners to engage more fully in pure reason. In the 1980s, the emergence of planetary thinking took this reasoning to new heights and the imaginings of conodont workers became wonderfully bold and grand. Paleontology had always been a science that opened up rich possibilities for the imagination, but it took time for the imagination to find an appropriate role in legitimate interpretation.

CONFLICT AND CHIVALRY

These spatial and temporal disruptions of conodont science were further affected by the science's treatment of the individual. I have already said that individualism manifested itself in a constellation of beliefs that made the science both adaptable and capable of living with unknowns, difficulties, and errors. This individualism was actively encouraged when a student began his or her doctorate studies. At this moment the student was allocated a geological resource (a set of field sites within a geographically defined area), a subject (a particular group of fossils and rocks), a period of geological time (such as the Upper Devonian), and particular questions and methodologies. This level of management was necessary to prevent students coming into competition with one another. The aim was to create an environment in which a student could thrive and thus pass the examination without having to confront personal rivalries; it served to keep things rational and objective.[19] However, it sometimes happened that two students, from different universities, found themselves on the same patch. Frank Rhodes, for example, wanted his student, Ronald Austin, to study some important rock sequences in Yorkshire, but Austin found that territory occupied and moved his research to a field site near Bristol. In this geographical, paleontological, and geological space – uncontaminated by earlier work and free from rivalries – Austin forged his scientific identity. Like the names of most conodont workers, his became inextricably connected to a period of geological time, a specific place and a style of study. To say a worker's name was to implicitly communicate these things.

This territorial division, which is pervasive in paleontology for these very good reasons, configures scientific performances in particular way as each individual can argue they possess their own unique and situated knowledge – their core knowledge. And while this territorialization ensures that the science maximizes its intellectual reach, it also ensures that truth is argued not from within a shared material resource, but with each worker possessing his or her own particular resources. At first sight this appears anti-competitive in what is an agonistic field. This is, however, an effective means to develop and test conflicting views. For example, while it was possible for several groups to hold different opinions based on the fossil animals that appeared in Scotland in the 1980s, only one of those groups actually possessed these specimens. This meant that interpretations of the group in possession were tested against interpretations based on material of another kind. The test was, in effect, to see if ideas triangulated across a range of objects and sites. There is, then, in this division of labor, a peculiar balance between the agonistic and communal necessities of the science that was achieved through a tight control over its material things: both the individual possession of particular fossils and rocks and the science's more general views on the morality and ethics of that possession. If Sweet recalls that almost everyone debated their points in a civilized and friendly fashion, it is because so much was in place to protect an individual's science and thus his or her identity. It prevented arguments becoming personal. It was only when identities were challenged, such as when a scientist's taxonomic children came under assault, that real animosities broke out.

Some workers, however, simply avoided arguments. They never saw the defense of a point of view as central to their contribution, preferring instead the creative moment of invention. Klaus Müller believed that he and Lindström were not fighters. They would publish but not crusade. A few, however, and most notably Aldridge and his various collaborators, played the game more completely. They constructed a military campaign to change minds within the conodont research community and beyond by winning a succession of scientific arguments. They possessed unique material, produced radical interpretations, and saw far-reaching significance in their discoveries. These factors permitted them to develop (relatively) large and powerful research teams. It gave them access to grants and made them the authors of choice for writing the new orthodoxy in encyclopedias, yearbooks, and magazines. But these workers were not simply interested in imposing a view; they valued, and grew strength from, the battle. They believed that debate was vital and that the tests of opponents would ultimately produce better science. Indeed, the doubting of others shaped the course their research was to take. They strengthened their arguments by recruiting talented individuals, which expanded and diversified the mental processing and permitted obscure topics to be explored and new methods to be deployed. These research groups acquired communal strength, permitting them to experiment with risk and make rapid advances in their research programs. However, those on the outside became only too aware of how uneven things had become; a whole army of mainly British workers was now developing a shared view while those who contested it were lone workers lacking the same level of resource. Eventually, this led to the formation of a political confederation of opponents holding no single shared view other than a belief that the British school was wrong on certain key points.

FASHION AND FORGETTING

Taken as a whole this unevenness in this global research culture – the geographical clusters, the individualism, the linguistic barriers, the research teams, the access to particular technologies and fossils – made progress uneven. During the 1950s and early 1960s, for example, German workers advanced an understanding of Devonian conodonts more rapidly than their colleagues in the United States. They benefited from a shared mission, individual competition, local inspiration, new technologies, better rock sequences, and raw talent and ambition. These things catalyzed research in a Germany recovering from the war.

But concerted effort of this kind was rare, and the conodont research community remained sufficiently small for individuals to continue to have considerable impact. Polish worker Hubert Szaniawski's finely argued evidence for arrow worms, for example, altered the visual apparatus of a number of his contemporaries for a period of time, influencing the direction their research took. The actions of a single individual could inject adrenalin into a particular strand of research, causing it to preferentially advance.

These shifts in interpretation inevitably made some earlier ideas re-dundant. Occasionally, this involved outright rejection – as happened with all the failed explanations for the conodont animal – but often ideas simply slipped into the backwaters without a fight.

Repeatedly the conodont workers started out along a particular path that seemed to offer progress but before the end was reached – before the once-imagined definitive result had been delivered – those on the path were drawn into new things. This occurred because at the moment when particular questions were asked and the map of progress was drawn up, the features on that map were composed of things already within view. Consider, for example, the 1950s advocates for a new ecological approach to study. They drew upon a few pioneering studies that showed the po-tential for this kind of research. But as the conodont ecologists voyaged forward, they soon hit seas disrupted by new ecological, global, and plan-etary theories. They looked again at their map and saw that the features marked upon it were not those imagined in 1955. They needed a new ship and new map, and a new direction in which to voyage. It was not simply that the expenditure of effort in this field had entered a phase of diminishing returns, but that changes around that field of research had diminished its value.

In this way, once-fashionable ideas about the fossil drifted out of sight, joined by those ideas rejected by new knowledge or ideas that were never promoted or defended or had only lasted while their proponents remained alive. While certain classic studies by Christian Pander and a few others persisted beyond the active involvement of the individual, other intellectual positions disappeared with their authors. Each new generation possessed little sentimentality about their predecessors and were never encouraged to historicize the past. All things, old and new, were for them contemporary, to be retained or rejected as that moment seemed to demand.

THE PROGRESS OF TINY THINGS

So what picture of progress can we paint for these enigmatic tiny things? An examination of any textbook on conodonts, or indeed the newsletter that has kept these researchers connected for the past forty years, seems to suggest that knowledge of the conodont fossil emerged through acts of specialization: from questions of biology and taxonomy to the use of these fossils in stratigraphic, evolutionary, ecological, and geographical studies, incorporating new research on fossil associations, animal physiognomy, chemistry, structure, and so on. While there is a degree of truth in this interpretation, we need to understand that these specializations are merely categories through which the field expressed itself and filters through which the individual pursued his or her creativity. These were not discrete subfields operating in relative isolation from one another. Those who manned the scanning electron microscopes in the early 1970s were already established in stratigraphy, taxonomy, and, in varying degrees, animal biology. Intellectually, individuals transgressed those categories into which the science seemed to be divided. The emergence of the animal may have been aided by this focusing of attention, but the conceptual – immaterial – fossils that occupied the conodont worker's mind were never divided in this way; these areas of focus were merely facets in the thinking of each mind.

In this book, my focus has been on the emergence of the animal and, indeed, on showing that it begins to take on aspects of its biological form in activities that had no interest in the animal. It is easy from a cursory view to imagine the history of conodont research as an enigma planted in the soil of ignorance that gave rise to a branching and ever-specializing engagement with the fossil. The structure of this book might, indeed, give grounds for this thinking. But this is an artifact of narrative (applied here and necessary in the communications of the workers themselves) and of imposing a long historical lens. If, however, we fold up our telescope and consider conodont science as a succession of cultural moments, then the manner in which the animal was pieced together seems to echo the way in which the animal itself evolved. Like the animal, the conodont workers frequently innovated from the conservative heartland of stratigraphic study around which conodont science was built. This core of practice was something all workers understood and shared, and they knew that it, more than anything else, had given the fossil (and therefore their work) significance. New strands of research would emerge from this conservative core as creative flourishes before entering a period of decline for reasons I have discussed above. One might say that circumstances and resources at one moment permitted a fire to burn, but as with the burning of all things, in time those resources (whether desires, ambitions, individuals, ideas, and so on) diminish. Each of these flourishes seemed to reveal a small aspect of the animal. When the animal was eventually found, it was already in many respects known, and much of this knowing could not have come from the animal specimens themselves.

When the animal finally appeared, its significance was acquired in the reverse of a statement I made earlier in this piece: Now the object had to become scientific before it could become anthropological. Only then could it be clothed in a great history of knowing and mythologizing and united with all manner of memories and so much wondering. Those who possessed the fossils also possessed the scientific authority – at least for a time – and they would state that the world now possessed this singular animal, an animal that was real and material, that could be seen and held. This became a scientific animal objectified in publication. It was, however, as I have already explained, simply one mental response. This animal was never the only animal possessed by the science – the animal that occupied the minds of individual conodont workers always existed in plural. These latter animals had only ever swam and evolved in a sea of thoughts, perhaps coalescing around a number of possibilities but nevertheless "a transitional mass, a coagulation of the unsubstantial, of the not-yet-substantial and yet substance-like ... mean and border-line between material and immaterial." The ambiguities of the fossil, like the speed of a being in rapid movement, prevented the animal from being fully observed or physically or mentally possessed. Workers had to content themselves with incomplete sightings. Another way to think about this is to return to Gombrich's observations on the role of illusion in art. For what we seem to be observing in this enigma is the boundary between artistic and scientific interpretation. Gombrich noted "how much the artist of the Western tradition came to rely upon the power of indefinite forms": "What we called 'mental set' may be precisely that state of readiness to start projecting, to thrust out the tentacles of phantom colors and phantom images which always flicker around our perceptions. And what we call 'reading' an image may perhaps be better described as testing it for its potentialities, trying out what fits." As the storm clouds gather around the British animal, some conodont workers might argue that no fossil has thus far been capable of protecting conodont workers from this artistic engagement, of lifting them out of a world of ambiguity and illusion, out of the thrusting tentacles of phantoms like those that, to return to the metaphor of El Dorado, "seemed to flee before the Spaniards, and to call on them unceasingly."[20]

Notes

1. THE ROAD TO EL DORADO

1. C. H. Pander, *Monographie der fossilen Fische des silurischen Systems der russisch-baltischen Gouvernements* (St. Petersburg: Akademie der Wissenschaften, 1856), 6.

2. Baer to W. v. Ditmar, 10 July 1816, in B. E. Raikov, *Christian Heinrich Pander: Ein bedeutender Biologe und Evolutionist*, trans. W. E. von Hertzenberg and P. H. von Bitter (Frankfurt: Verlag Waldemar Kramer, 1984), 18. See also G. A. Wells, "Goethe and evolution," *J. History of Ideas* 28 (1967): 537–50; C. H. Pander, "Beiträge zur Entwicklungsgeschichte des Hühnchens im Ei" (Würzburg, 1817), quoted by E. S. Russell, *Form and Function: A Contribution to the History of Animal Morphology*, Project Gutenberg, 2007, 133, http://www.gutenberg.org/; Professor Ignaz Döllinger (1770–1841).

3. Raikov, *Pander*, 49; S. Schmitt, "From eggs to fossils: Epigenesis and transformation of species in Pander's biology," *Int. J. Dev. Biol.* 49 (2005): 1–8; Robert J. Richards, *The Romantic Conception of Life: Science and Philosophy in the Age of Goethe* (Chicago: Chicago University Press, 2002), 483.

4. Raikov, *Pander*, 78. The Geological Society of London's copy of C. H. Pander, *Beiträge zur Geognosie des russischen Reiches: Die Umgebungen von St. Petersburg* (St. Petersburg, 1830) was presented by Murchison when he was president.

5. S. J. Knell, *The Culture of English Geology, 1815–1851: A Science Revealed Through Its Collecting* (Basingstoke, UK: Ashgate, 2000); James A. Secord, *Controversy in Victorian Geology: The Cambrian-Silurian Dispute* (Princeton: Princeton University Press, 1986); R. Murchison, *The Silurian System* (London: John Murray, 1839), 657.

6. "Barabbas," Murchison on visit to Pander in 1841, in M. Collie and J. Diemar, *Murchison's Wanderings in Russia* (London: British Geological Survey, 2004), 137–38. For "painstaking" Charles Bunbury, see John C. Thackray, *To See the Fellows Fight* (London: British Society for the History of Science Monograph 12, 2003), 200.

7. Stephen Jay Gould, *Wonderful Life: The Burgess Shale and the Nature of History* (Harmondsworth, UK: Penguin, 1989), 65; "I. Trutnew" possibly Russian artist Ivan Petrovich Trutnev (1827–1912)?

8. Pander, *Monographie*, 5, trans. extract by Wladimir Ayvazoglou, in Wilbert H. Hass, "Morphology of conodonts," *J. Paleont.* 15 (1941): 71–81, and reprinted in Wilbert H. Hass, "Conodonts," in Raymond. C. Moore (ed.), *Treatise on Invertebrate Paleontology*, Part W Miscellanea (1959; reprint, Lawrence: GSA/Uni-

versity of Kansas Press, 1962), W3–W69. As non-German speakers had no direct access to Pander's description, this became an important surrogate.

9. R. I. Murchison, *Siluria,* 2nd ed. (London: J. Murray, 1867), 236, 355–56. Should Pander's eye disease predate Helmersen's letter, then 1848 or 1849 seem likely. R. O. Fay, *Catalogue of Conodonts,* Paleont. Contr. 3 (Lawrence: University of Kansas, 1952), 4, incorrectly believed the 2 November 1853 mss. date of a paper by Karl Eduard von Eichwald (1795–1876) to record the first mention of the conodont.

10. Pander in J. Barrande, "Sur une découverte de fossils faite dans la partie inférieure du terrain silurien de Russie," *Bulletin de la Société Géologique de France,* ser. 2, no. 8 (1851) : 251–59, 254.

11. R. I. Murchison, *Siluria,* 1st ed. (London: J. Murray,1854), 323.

12. K. E. Eichwald, "Beitrag zur geographischen Verbreitung der fossilen Theire Rußlands Alte Periode," *Bull. Soc. Imper. Natur. Moscou* 30, no. 2 (1857): 305–54, 338–39.

13. R. Owen, *Palaeontology,* 2nd ed. (Edinburgh: Adam and Charles Black, 1861).

14. Owen's opinion appeared in a letter, 1858, in the 1867 edition of Murchison's *Siluria,* Appendix E, later versions in Owen's *Palaeontology,* 1860 and 1861 editions.

15. Huxley to W. Macleay, Sydney, 9 November 1851, in L. Huxley, *Life and Letters of Thomas Henry Huxley,* vol. 1 (London: Macmillan, 1900), 94.

16. J. Harley, "On the Ludlow Bone-Bed and its crustacean remains," *QJGS* 17 (1861): 542–52; Murchison, *Siluria* (1867),148.

17. Alexander von Volborth (1800–1876). Letter from Volborth to Harley, 12 May 1861, reprinted in Harley, "Ludlow Bone-Bed," 551–52. This is a partial translation of A. Volborth, "Vorkommen von Conodonten in England und Schweden,"

Neues Jahrbuch fur Mineralogie, Geognosie, Geologie und Petrarefakten-kunde 11B (1861): 464–65.

18. Murchison, *Siluria* (1867), 356.

19. Charles Moore, "On Triassic beds near Frome, and their organic remains," *Rep.* BAAS 1857 (1858): 93–94; Charles Moore (1815–1881). The fissures contained fossils dissolved from overlying rocks, including the Carboniferous Limestone. It is unlikely that Moore's conodonts came from the Triassic.

20. Charles Moore, "Report on mineral veins in Carboniferous Limestone and their organic contents," *Rep.* BAAS 1869 (1870): 360–80.

21. R. A. Davis, "Science in the hinterland: The Cincinnati school of paleontology," GSA Annual Meeting Abstract, 2001; D. S. Brandt, "Lagerstätte and luck: The role of the type Cincinnatian in shaping paleontological research in North America (1838–1961)," GSA Annual Meeting Abstract, 2003.

22. E. O. Ulrich, "Observations on fossil annelids and description of some new forms," *Cincinnati Soc. Nat. Hist. J.* 1 (1878): 87–91, 87.

23. J. S. Newberry, *Descriptions of Fossil Fishes* (Ohio: Geological Survey Report, 1875), 41–46; C. A. White, *Biographical Memoir of John Strong Newberry* (Washington, D.C., 1908).

24. G. B. Grinnell, "Notice of a new genus of annelids from the Lower Silurian," *Am. Jour. Sci.,* ser. 3, 14, no. 81 (1877): 229–30.

25. Ulrich, "Observations," 88.

26. Hugh Miller, *The Old Red Sandstone* (Edinburgh: John Johnstone, 1841), 33–34; H. Woodward, "George Jennings Hinde," *Geol. Mag.* 6 (1918): 233–40. Through marriage to Edith Octavia Clark, there is a family connection to the Woodward geological dynasty claiming Henry, of the British Museum, as his "cousin." M. O'Connell, "George Jennings Hinde,"

Science 48 (1918): 588–90; G. W. Land-pugh, "Anniversary address of the president," *QJGS* 75 (1919): lvii–lix.

27. John Smith (1846–1930), incorrectly named by Hinde. R. B. Wilson, *John Smith of Dalry, Geologist, Antiquarian and Natural Historian* (Ayr: Ayrshire Archaeological and Natural History Society, 1995), 23–29; John Young, "Notes on the fossils found in a thin bed of impure Carboniferous Limestone at Glencart, near Dalry, Ayrshire," *Proc. Nat. Hist. Soc. Glasgow* 5 (1884): 234–40, 235–36; Anon., "The twenty-ninth annual general meeting, 24 September 1878," *Proc. Nat. Hist. Soc. Glasgow* 4 (1881): 1–5, 3; John Smith, "Conodonts from the Carboniferous Limestone strata of the west of Scotland," *Trans. Nat. Hist. Soc. Glasgow* 5 (1900): 336–46. Thanks to Mike Taylor for information on Smith.

28. John Smith, "The conodonts of the Carboniferous rocks of the Clyde drainage area," in J. B. Murdoch (ed.), *The Geology and Palaeontology of the Clyde Drainage Area* (Glasgow: Geological Society of Glasgow, 1904), 510; John Young, "On a group of fossil organisms termed conodonts," *Proc. Nat. Hist. Soc. Glasgow* 4 (1881): 5–7, 6.

29. G. J. Hinde, "On conodonts from the Chazy and Cincinnati Group of the Cambro-Silurian, and from the Hamilton and Genesee-Shale Divisions of the Devonian, in Canada and the United States," *QJGS* 35 (1879): 351–69.

30. G. J. Hinde, "On annelid jaws from the Cambro-Silurian, Silurian, and Devonian Formations in Canada and from the Lower Carboniferous in Scotland," *QJGS* 35 (1879): 370–87.

31. U. P. James, "On conodonts and fossil annelid jaws," *J. Cincinnati Soc. Nat. Hist.* 7, no. 3 (1884): 143–49; J. M. Clarke, "On the higher Devonian faunas of Ontario County, New York," *U.S. Geol. Surv. Bull.* 3, no. 16 (1885): 35–120; A. W. Grabau, "Geology and palaeontology of Eighteen

Mile Creek and the lake shore sections of Erie County, New York," *Buffalo Soc. Nat. Sci. Bull.* 6 (1898–99).

32. Now known as *Branchiostoma*; Freidrich Rolle, "Fische," in Adolf Kenngott (ed.), *Handwörterbuch der Mineralogie, Geologie und Palaeontologie*, vol. 1 (Breslau, 1882), 408; K. A. Zittel and J. V. Rohon, "Über conodonten," *Bayer Akad. Wiss. München Math-Phys. K.l. Sitzungsber* 16 (1886): 108–36, 111.

33. E. H. Ehlers, *Die Borstenwurmer* (Leipzig: W. Engelmann, 1864–68).

34. A. Geikie, *Text-book of Geology* (London: Macmillan, 1903), 942; J. V. Rohon, "Uber unter-Silurische fische," *Bull. Scientifique Publié par L'Académie Imperiale des Sciences de Saint Pétersbourg, Acad. St. Petersb.* 33 (1890): 269.

2. A BEACON IN THE BLACKNESS

1. Roger M. Olien and Diana Davids Olien, *Oil and Ideology: The Cultural Creation of the American Petroleum Industry* (Chapel Hill: University of North Carolina Press, 2000), 120, 141, 189.

2. P. V. Roundy, "Introduction, the microfauna in Mississippian formations of San Saba County, Texas," *U.S. Geol. Surv. Prof. Pap.* 146 (1926); Hugh S. Torrens, *The Practice of Geology, 1750–1850* (Farnham, UK: Ashgate, 2002); S. J. Knell, "The road to Smith: How the Geological Society came to possess English geology," in C. Lewis and S. J. Knell (eds.), *The Making of the Geological Society of London* (London: Geological Society, 2009), 1–47, 6–7.

3. C. Croneis, "Micropaleontology – past and present," *Bull. AAPG* 25 (1941): 1208–55; Roundy, "San Saba," 1. On professionalization, see also R. S. Bassler, "Development of invertebrate paleontology in America," *Bull. Geol. Soc. Am.* 44 (1933): 265–86; Carl O. Dunbar, "Symposium on fifty years of paleontology: A half century of paleontology," *J. Paleont.* 33 (1959): 909–14; S. Powers, "History of

the American Association of Petroleum Geologists," *Bull. AAPG* 13 (1929): 153–70; R. H. Dott, "Founding of the AAPG," *Bull. AAPG* 37 (1953): 1117–21; H. T. Morley, "A history of the American Association of Petroleum Geologists," *Bull. AAPG* 50 (1966): 669–820.

4. Laurence L. Sloss, Twenhofel Award acceptance speech, Denver, 10 June 1980, http://www.earth.northwestern. edu/twenhofel.html/; J. D. Fischer, *The Seventy Years of the Department of Geology University of Chicago, 1892–1961* (Chicago: University of Chicago Press, 1963), 43; Carey Gardiner Croneis (1901–1972); Croneis, "Micropaleontology,"1238.

5. Roundy, "San Saba," 2.

6. Hass, "Conodonts," in Moore, *Treatise*, W3–W69, W4.

7. C. L. Cooper, "Conodonts from the Arkansas Novaculite, Woodford Formation, Ohio Shale and Sunbury Shale," *J. Paleont.* 5 (1931): 143–51.

8. God comment by August Foerste in Curt Teichert, "From Karpinsky to Schindewolf – memories of some great paleontologists," *J. Paleont.* 50 (1976): 1–12, 11; E. O. Ulrich, "Revision of the Paleozoic systems," *Bull. Geol. Soc. Am.* 22 (1911): 281–680, 477; E. O. Ulrich, "Correlation by displacements of the strand-line and the function and proper use of fossils in correlation," *Bull. Geol. Soc. Am.* 27 (1916): 451–90; R. S. Bassler, "The Waverlyan period of Tennessee," *U.S. Nat. Mus. Proc.* 41, no. 1851 (1911): 209–24; K. E. Caster, "Memorial to Ray S. Bassler (1878–1961)," *Proc. Geol. Soc. Am.* (1965): 167–73.

9. E. O. Ulrich, "The Chattanoogan series with special reference to the Ohio shale problem," *Am. J. Sci.* 34 (1912): 157–83; E. M. Kindle, "Unconformity at the base of the Chattanooga Shale in Kentucky," *Am. J. Sci.* 33 (1912): 120–36, 128.

10. Ulrich, "Chattanoogan series"; E. M. Kindle, "The stratigraphic relations of the Devonian shales of northern Ohio,"

Am. J. Sci. 34 (1912): 187–213; Girty to Kindle, 1 November 1912, Smithsonian Institution Archives (hereafter cited as SIA), Record Unit 7329, George Girty Papers (hereafter cited as Girty Papers).

11. Kindle to David White, n.d., Girty Papers.

12. E. O. Ulrich, "Kinderhookian age of the Chattanoogan series (abstract)," *Bull. Geol. Soc. Am.* 26 (1914): 97–99.

13. Ulrich, "Revision," 290.

14. W. L. Bryant, "The Genesee conodonts," *Buffalo Soc. Nat. Sci. Bull.* 13 (1921): 1–59, 9, 23. See also Roundy, "San Saba," 9, 13.

15. Raymond R. Hibbard to Bassler, receiving conodonts from Bassler (9 July 1924), developing conodont collections and library (17 November 1933), collecting (29 August 1937), acid preparation of scolecodonts ("worm jaws") (29 September 1939), and so on; Bassler to Hibbard, in which Bassler says with Branson's activity he has given up conodonts (12 November 1933), all in SIA, Record Unit 7234, Box 3, Folder 5, Ray S. Bassler Papers (hereafter cited as Ray S. Bassler Papers).

16. E. O. Ulrich and R. S. Bassler, "A classification of the toothlike fossils, conodonts, with description of American Devonian and Mississippian species," *Proc. U.S. National Mus.* 68 (1926): 1–63; R. S. Bassler, "Classification and stratigraphic use of the conodonts," *Bull. Geol. Soc. Am.* 36 (1925): 218–20. Also R. S. Bassler, "The stratigraphic use of conodonts (abstract)," *Washington Academy Sci. J.* 16 (1926): 72–73.

17. Grace. B. Holmes, "A bibliography of the conodonts with descriptions of early Mississippian species," *Proc. U.S. National Mus.* 72 (1928). Holmes was directed by Ulrich and Bassler and thus serves to back up their conclusions.

18. Clarice B. Strachan, "Biographical sketches of recently elected honorary members: Edward Oscar Ulrich," *Bull.*

AAPG 20 (1936): 1265–68. But note that Roundy had already drawn up his own "Bibliography of conodont and Paleozoic annelid jaw literature in USGS Library, Reston," in 1925 in the knowledge that these fossils would prove important to the oil industry.

19. C. R. Stauffer, "Conodonts from the Decorah shale," *J. Paleont.* 4 (1930): 121–28; C. R. Stauffer, "Decorah Shale conodonts from Kansas," *J. Paleont.* 6 (1932): 257–64; C. R. Stauffer and H. J. Plummer, "Texas Pennsylvanian conodonts and their stratigraphic relations," *Univ. Texas Bull.* 3201 (1932): 13–50. Helen Plummer (neé Skewes) was married to Frederick Byron Plummer.

20. F. H. Gunnell, "Conodonts from the Fort Scott limestone of Missouri," *J. Paleont.* 5 (1931): 244–52.

21. F. H. Gunnell, "Conodonts and fish remains from the Cherokee, Kansas City, and Wabaunsee groups of Missouri and Kansas," *J. Paleont.* 7 (1933): 261–97.

22. C. L. Cooper, "Arkansas"; C. L. Cooper, "New conodonts from the Woodford Formation of Oklahoma," *J. Paleont.* 5 (1931): 230–43.

23. Carl Rexroad told me he was known as "Ted"; Rexroad received a cowboy belt from Branson when a child. C. R. Longwell, "Edwin Bayer Branson (1877–1950)," *Bull. AAPG* 35 (1951): 1706–10; S. P. Ellison, "Memorial for Maurice Goldsmith Mehl (1877–1966)," *Proc. GSA* (1966): 219–24; C. C. Branson, "Maurice G. Mehl (1887–1966)," *Oklahoma Geol. Notes* 26 (1966): 139–40.

24. E. B. Branson and M. G. Mehl, *Conodont Studies* (Columbia: University of Missouri Studies 8, 1933–34), 8. I have not listed the individual papers below.

25. Their paranoia was not helped when Stauffer showed them old yet unworn conodonts – "delicate, sharp-pointed, hand-like crests with the fibrous structure" – mixed with a younger fauna that was

considerably more worn! C. R. Stauffer, "Conodonts of the Glenwood Beds," *Bull. Geol. Soc. Am.* 46 (1935): 125–68.

26. C. R. Stauffer, "The conodont fauna of the Decorah shale (Ordovician)," *J. Paleont.* 9 (1935): 596–620.

27. C. L. Cooper, "Review of Conodont studies," *J. Geol.* 43 (1935): 443–45.

28. J. W. Huddle, "Marine fossils from the top of the New Albany shale of Indiana," *Am. J. Sci.* 25 (1933): 303–14; J. W. Huddle, "Conodonts from the New Albany shale of Indiana," *Bull. Am. Paleont.* 21 (1934).

29. C. L. Cooper, "Conodonts from the Upper and Middle Arkansas Novaculite, Mississippian, at Caddo Gap, Arkansas," *J. Paleont.* 9 (1935): 307–15.

30. E. B. Branson and M. G. Mehl, "The conodont genus *Icriodus* and its stratigraphic distribution," *J. Paleont.* 12 (1938): 156–66.

31. M. M. Knechtel and W. H. Hass, "Kinderhook conodonts from Little Rocky Mountains, northern Montana," *J. Paleont.* 12 (1938): 518–20, 520; E. B. Branson and M. G. Mehl, "New and little known Carboniferous conodont genera," *J. Paleont.* 15 (1941): 97–106; Wilbert H. Hass (1906–1959); C. C. Branson, c. 1962; *Geol. Soc. Am. Proc. for 1960* (1962): 104–106; C. R. Stauffer, "Conodonts of the Olentangy shale," *J. Paleont.* 12 (1938): 411–43.

32. C. L. Cooper, "Conodonts from a Bushberg-Hannibal horizon in Oklahoma," *J. Paleont.* 13 (1939): 379–422; W. H. Hass, "Corrections to the Kinderhook conodont fauna, Little Rocky Mountains, Montana," *J. Paleont.* 17 (1943): 307–309.

33. Branson and Mehl, "New and little known," 97.

34. C. C. Branson, "Conodonts in the Permian," *Science* 75 (1932): 337–38; C. C. Branson, "Origin of phosphate in the Phosphoria Formation (abstract)," *Bull. Geol. Soc. Am.* 43 (1932): 284. This discovery was widely publicized at the time. E. B.

Branson and M. G. Mehl, "The recognition and interpretation of mixed conodont faunas," *Denison University Bull. Jour. Sci. Lab.* 35 (1941): 195–209. Ellison makes much of the importance of this work in his obituaries of Mehl. E. B. Branson and M. G. Mehl, "A record of typical American conodont genera in various parts of Europe," *Denison University Bull. Jour. Sci. Lab.* 35 (1941): 189–94.

35. W. M. Furnish, "Conodonts from the Prairie du Chien (Lower Ordovician) beds of the upper Mississippi Valley," *J. Paleont.*12 (1938): 318–40, 323; W. M. Furnish, E. J. Barragy, and A. K. Miller, "Ordovician fossils from upper part of type section of Deadwood Formation, South Dakota," *Bull.* AAPG 20 (1936): 1329–41; William Madison Furnish (b. 1912); Arthur K. Miller (b. 1902).

36. S. P. Ellison Jr., "Revision of the Pennsylvanian conodonts," *J. Paleont.* 15 (1941): 107–43.

37. E. B. Branson and M. G. Mehl, "Conodonts," in H. W. Shimer and R. R. Shrock (eds.), *Index of Fossils of North America* (New York: Wiley, 1944), 235–46.

38. S. P. Ellison Jr., "Conodonts as Paleozoic guide fossils," *Bull.* A A P G 30 (1946): 93–110.

39. W. H. Hass, "Conodont zones in Upper Devonian and Lower Mississippian formations of Ohio," *J. Paleont.* 21 (1947): 131–41.

40. Gil Klapper, pers. comm.. 16 October 2005; W. H. Hass, "Conodonts of the Barnett Formation of Texas," *U.S. Geol. Surv. Prof. Pap.* 243-F (1953): 69–94; Hass, "Chattanooga"; W. H. Hass, "Conodonts from the Chappel Limestone of Texas," *U.S. Geol. Surv. Prof. Pap.* 294-J (1959): 365–99. The latter paper included a new utilitarian classification of the conodonts that he said he first proposed in 1941.

41. Branson, "Ellison."

42. Croneis, "Micropaleontology," 1233.

43. A. N. Dusenbury, "Brooks Fleming Ellis (1897–1976)," *Micropaleontology* 22 (1976): 4, 377–78. Also "The Micropaleontology Press," http://micropress.org/history.html/.

44. Taken from the Boy Scout Geology Merit Badge Cooper helped develop. Houston Geology Society, "Petroleum Geology and the Development of the Boy Scout Geology Merit Badge," 2004, http://www.hgs.org/en/articles/print-view.asp?48/.

45. N. D. Newell, "Towards a more ample invertebrate paleontology," in B. Kummel (ed.), "Status of Invertebrate Paleontology," *Bull. Mus. of Comparative Zool.* 112 (1954): 93–97; T. J. M. Schopf, *Models in Paleobiology* (San Francisco: Freeman, Cooper, 1972), 10.

46. G. A. Cooper, "The science of paleontology," *J. Paleont.* 32 (1958): 1010–18; Sweet recalls Cooper as a gloomy pessimist. Sweet, pers. comm., 16 July 2010.

3. THE ANIMAL WITH THREE HEADS

1. Carl Branson to James Steele Williams, 6 November 1948, SIA, Record Unit 7328, Box 1, Folder 3, Carl Branson Folder, James Steele Williams Papers (hereafter cited as Williams Papers); Croneis, "Micropaleontology," 1242 (see ch. 2, n. 3); H. G. Schenck, "The biostratigraphy aspect of micropaleontology," *J. Paleont.* 2 (1928): 158–65.

2. Stauffer and Plummer, "Texas Pennsylvanian," 16 (see ch. 2, n. 19); a view also given by Bryant, "Genesee" (see ch. 2, n. 14); R. S. Bassler, "Bibliographic Index of American Ordovician and Silurian Fossils, *Bull. U.S. National Mus.* 92 (1915): 1426.

3. Bryant, "Genesee," 3, 6, 12, 24.

4. J. M. Macfarlane, *The Quantity and Sources of Our Petroleum Supplies: A Review and a Criticism* (Philadelphia: Noel Printing, 1931), 227 admits to copying Hinde.

5. J. M. Macfarlane, *Evolution and Distribution of Fishes* (Burlington, N.J.: Enterprise, 1923); J. M. Macfarlane, *Fishes the Source of Petroleum* (New York: Macmillan, 1923). For more recent review, R. Jenner, "Foiling vertebrate inversion with the humble nemertean," *Paleont. Assoc. Newsletter* 58 (2005): 32–39; R. Jenner, "Meeting a nemertean nemesis," *Paleont. Assoc. Newsletter* 59 (2005): 37–43.

6. Anonymous review of Macfarlane, *Quantity and Sources,* in *Nature* 130 (1932): 832.

7. Roundy to Cooper, 11 September 1929, Girty Papers.

8. Branson to James Steele Williams, 26 October 1933, handwritten addendum, Box 1, Folder 4, Edwin Branson Folder, Williams Papers.

9. C. L. Cooper, "Actinopterygian jaws from the Mississippian black shales of the Mississippi Valley," *J. Paleont.* 10 (1936): 92–94.

10. S. R Kirk, "Conodonts associated with the Ordovician fish fauna of Colorado – A preliminary note," *Am. J. Sci.* 18 (1929): 493–96; C. D. Walcott, "Preliminary notes on the discovery of a vertebrate fauna in Silurian (Ordovician) strata," *Bull. Geol. Soc. Am.* 3 (1892): 153–72.

11. Branson and Mehl, *Conodont Studies,* 5 (see ch. 2, n. 24).

12. Gunnell, "Cherokee," 263 (see ch. 2, n. 21); F. H. Gunnell, "Conodonts in relation to petroleum," *American Midland Naturalist* 13 (1932): 324–25.

13. Macfarlane, *Evolution and Distribution,* 260; Stauffer and Plummer, "Texas Pennsylvanian," 22; Stauffer, "Conodonts from the Decorah shale," 258.

14. W. Eichenberg, "Conodonten aus dem Culm des Harzes," *Palaeont. Z.* 12 (1930): 177–82.

15. H. Schmidt, "Condonten-Funde in ursprunglichen Zusammenhang," *Palaeont. Z.* 16 (1934): 76–85.

16. Harold William Scott (1906–1998), PhD completed in 1935.

17. "The life of Harold W. and Joann Scott, Urbana, Ill.," University of Illinois Archives, Box 6, Harold Scott Papers (hereafter cited as Scott Papers).

18. H. W. Scott, "The zoological relationships of the conodonts," *J. Paleont.* 8 (1934): 448–55, 450.

19. Croneis and Scott published three abstracts in *Bull. Geol. Soc. Am.* 44 (1933): 207–208. Stauffer was also expert in this field and is recorded in the same volume.

20. D. J. Jones, "Conodont assemblages from the Nowata shale," master's thesis, (University of Oklahoma, 1935) and reported under the same title in *J. Paleont.* 9 (1935): 364; R. L. Denham, "Conodonts," *J. Paleont.* 18 (1944): 216–18.

21. F. B. Loomis, "Are conodonts gastropods?" *J. Paleont.* 10 (1936): 663–64; H. A. Pilsbry, "Are conodonts molluscan teeth?" *Nautilus* 50 (1937): 101.

22. On anti-German sentiments in the early 1930s, G. C. Cadée, "The history of taphonomy," in S. K. Donovan (ed.), *The Processes of Fossilization* (London: Belhaven, 1991), 3–21; Stauffer, "Conodont fauna of the Decorah," 599; E. B. Branson and M. G. Mehl, "Conodont assemblages (abstract)," *Geol. Soc. Am. Proc. for 1937* (1938): 270; E. B. Branson and M. G. Mehl, "Geological affinities and taxonomy of conodonts," *Geol. Soc. Am. Proc. for 1935* (1936): 436.

23. F. Demanet, "Filtering appendices on the branchial arches of *Coelacanthus lepturus* Agassiz," *Geol. Mag.* 76 (1939): 215–19.

24. E. D. Currie, C. Duncan, and H. M. Muirwood, "The fauna of Skipsey's Marine Band," *Trans. Geol. Soc. Glasgow* 19 (1937): 413–51.

25. J. S. Cullison, "Dutchtown fauna of southeastern Missouri," *J. Paleont.* 12 (1938): 219–28. The fossil was named *Archeognathus.* It was confirmed as a

fish by A. K. Miller, J. S. Cullison, and W. Youngquist, "Lower Ordovician fish-remains from Missouri," *Am. J. Sci.* 245 (1947): 31–34, and denied by Lindström in 1964. An attempt to reestablish significance in G. Klapper and S. M. Bergström, "The enigmatic Middle Cambrian fossil *Archeognathus* and its relations to conodonts and vertebrates," *J. Paleont.* 58 (1984): 949–76. Bergström (pers. comm., 2011) notes: "I do not think any informed conodont worker would now regard the conodont nature of *Archeognathus* as controversial."

26. Jones to Scott, 11 May 1937, General Correspondence, 1937, Scott Papers; D. J. Jones, *The Conodont Fauna of the Seminole Formation of Oklahoma* (Chicago: University of Chicago Press, 1941), announced in an abstract in *Oil and Gas Journal* 37 (1939): 74; Gertrude I. Burnley, "The conodonts of the shale overlying the Lexington Coal Bed of Lafayette County and Jackson County," master's thesis (University of Missouri–Columbia, 1938).

4. ANOTHER FINE MESS

1. Stauffer, "Olentangy," 414 (see ch. 2, n. 33); Furnish, "Prairie du Chien," 324 (see ch. 2, n. 35); Frank H. T. Rhodes, "The zoological affinities of conodonts," *Biol. Rev. Cambridge. Philos. Soc.* 29 (1954): 419–52, 428, noted that "cancelled denticles in reflected light appear peglike" and probably dropped out due to preferential attack by acid.

2. Hass, "Morphology of conodonts."

3. See chapter 2. See also W. H. Hass and Marie L. Lindberg, "Orientation of crystal units of conodonts," *J. Paleont.* 20 (1946): 501–504.

4. S. P. Ellison Jr., "The composition of conodonts," *J. Paleont.* 18 (1944): 133–40.

5. H. W. Scott, "Conodont assemblages from the Heath Formation, Montana," *J. Paleont.* 16 (1942): 293–300.

6. E. P. Du Bois, "Evidence on the nature of conodonts," *J. Paleont.* 17 (1943): 155–59.

7. F. W. Clarke, "The data of geochemistry," *Bull. Geol. Soc. Am.* 770 (1924): 527–32.

8. Branson and Mehl, "Conodonts," 236 (see ch. 2, n .37); S. P. Ellison Jr., "Ecology of conodonts," *NRC Div. Geol. Geogr. Ann. Rep. App. K, Report of the Committee on Marine Ecology as related to Paleontology,* 1943–44 (1944): 1–4.

9. J. W. Huddle, "Historical introduction to the problem of conodont taxonomy," *Geol. et Palaeont.* SB 1 (1972): 3–16.

10. R. Brinkmann, *Abriß der historischen Geologie* (Stuttgart: Ferdinand Enke, 1948); H. Schmidt, "Nachträge zur Deutung der Conodonten," *Decheniana* 104 (1950): 11–19, 11.

11. H. Beckmann, "Conodonten aus dem Iberger Kalk (Ober-Devon) des Bergischen Landes und ihr Feinbau," *Senckenbergiana* 30 (1949): 153–68, 163.

12. Schmidt, "Nachträge."

13. Rhodes's doctorate, "British Lower Palaeozoic conodont faunas," 1950. On his skepticism, F. H. T. Rhodes, "Recognition, interpretation, and taxonomic position of conodont assemblages," in Raymond C. Moore (ed.), *Treatise on Invertebrate Paleontology,* Part W Miscellanea (Lawrence: GSA/University of Kansas Press, 1962), W70–W83, W76; Gould, *Wonderful Life,* 83 (see ch. 1, n. 7); F. H. T. Rhodes, "Some British Lower Palaeozoic conodont faunas," *Phil. Trans. Roy. Soc.* B 237 (1953): 261–334.

14. F. H. T. Rhodes, "A classification of Pennsylvanian conodont assemblages," *J. Paleont.* 26 (1952): 886–901. The terms "form species" and "form genera" refer to an anatomical component treated as a species. However, H. W. Shimer and R. R. Shrock, *Index Fossils of North America* (New York: Wiley, 1944), 1, raised the issue conflicting meanings muddying the

water. Picked up by K. J. Müller, "Taxonomy, nomenclature, orientation, and
stratigraphic evaluation of conodonts,"
J. Paleont. 30 (1956): 1324–40, who suggests an alternative.

15. R. J. Aldridge, "Conodont palaeobiology: A historical review," in R. J.
Aldridge (ed.), *Palaeobiology of Conodonts*
(Chichester, UK: Horwood, British Mi-
cropalaeontological Society, 1987), 11–34.

16. Rhodes, "Classification," 890.

17. Fay, *Catalogue of Conodonts*, 5;
Teichert, "From Karpinsky," 10 (see ch. 2,
n. 8); *Houston Chronicle*, 23 January 1972;
Houston Post, 23 January 1972; *Proceedings
of the Philosophical Society of Texas*, 1972;
Who's Who in the South and Southwest,
vol. 11.

5. OUTLAWS

1. Scott, "Conodont assemblages,"
294 (see ch. 4, n. 5).

2. C. Croneis and J. McCormack,
"Fossil Holothuroidea," *J. Paleont.* 6
(1932): 136.

3. C. Croneis, "Utilitarian classification for fragmentary fossils," *J. Geol.* 46
(1938): 975–84; C. Croneis, "A military classification for fossil fragments," *Science* 89
(1939): 314–15, and echoed by Scott (above)
referring particularly to conodonts.

4. Croneis, "Micropaleontology,"
1245. On the commission's decline and
recovery, see Dunbar, "Symposium on
fifty years," 913–14; G. Deflandre and M.
Deflandre Regaud, "Proposed new system
of nomenclature for fragments of fossil
invertebrates found in sedimentary rocks:
Rejection of proposal," *Bull. Zool. Nomen.*
4 (1948): 274, 294; R. C. Moore and P. C.
Sylvester-Bradley, "Proposed insertion
in the 'Règles' of provisions recognizing
'Parataxa' as a special category for the
classification and nomenclature of discrete fragments or of life-stages of animals
which are inadequate for identification
of whole-animal taxa, with proposals

of procedure for the nomenclature of
'Parataxa,'" *Bull. Zool. Nomen.* 15 (1957):
5–13, 6.

5. Sinclair to Scott, 9 February 1952,
Box 2, General Correspondence 1952,
Scott Papers; H. W. Scott, "Siliceous
sponge spicules from the Lower Pennsyl-
vanian of Montana," *American Midland
Naturalist* 29 (1943): 732–60.

6. G. W. Sinclair, "The naming of conodont assemblages," *J. Paleont.* 27 (1953):
489–90.

7. The ICZN warned against such
practices and in 1943 remarked that "the
tendency to enter into public polemics
over matters which educated and refined
professional gentlemen might so easily
settle in refined and diplomatic correspondence is distinctly unfavorable to a
settlement." Extract at http://www.mbl
.edu/BiologicalBulletin/KEYS/INVERTS/
opinions.html/.

8. F. H. T. Rhodes, "Nomenclature
of conodont assemblages," *J. Paleont.* 27
(1953): 610–12.

9. F. W. Lange, "Polychaete annelids
from the Devonian of Parana, Brazil," *Bull.
Am. Paleont.* 33 (1949): 1–102.

10. Hinde (1880) quoted in ibid., 49.

11. Rhodes, "Nomenclature," 611.

12. P. C. Sylvester-Bradley, "Formgenera in paleontology," *J. Paleont.* 28
(1954): 333–36.

13. Müller, "Taxonomy, nomencla-
ture," 1328 (see ch. 4, n. 14); M. Lindström,
"Conodonts from the lowermost Ordovician strata of south-central Sweden," *Geol.
Foren. Stockholm Forhandl.* 76 (1954):
517–603, 541.

14. Moore to Bassler, 27 January 1956,
Ray S. Bassler Papers.

15. R. C. Buchanan, *"To bring together,
correlate, and preserve": A History of the
Kansas Geological Survey, 1864–1989*, Kansas Geological Survey Bulletin 227:7.

16. D. F. Merriam, "Raymond Cecil
Moore: A great 20th century geological

synthesizer," GSA *Today* 13, no. 8 (2003): 13–18.

17. Moore to Bassler, 8 April 1926, Ray S. Bassler Papers; R. C. Moore, "The use of fragmentary crinoidal remains in stratigraphic paleontology," *J. Sci. Lab. Denison Univ.* 33 (1939): 165–250; R. S. Bassler and M. W. Moodey, *Bibliographic and Faunal Index of Paleozoic Pelmatozoan Echinoderms,* GSA Special Paper 45 (1943).

18. Moore and Sylvester-Bradley, "Proposed insertion," 10–11. Subsequent debate takes place in the *Bulletin of Zoological Nomenclature,* often in sequential correspondence, and is not referenced in full. The University of Leicester Library's volumes of this journal formerly belonged to Sylvester-Bradley and contain his annotations. Moore was not a commissioner. Sylvester-Bradley was probably the most active member of the commission.

19. Moore to Scott, Rhodes, and "Carl Mueller," 1 June 1956, Box 2, General Correspondence 1956, Scott Papers; R. C. Moore and P. C. Sylvester-Bradley, "First supplemental application: Application for a ruling of the International Commission directing that the classification and nomenclature of discrete conodonts be in terms of 'Parataxa,'" *Bull. Zool. Nomen.* 15 (1957): 15–34.

20. W. J. Arkell, "Proposed Declaration that a generic or specific name based solely upon the 'aptychus' of an ammonite (Class Cephalopoda, Order Ammonoidea) be excluded from availability under the Règles," *Bull. Zool. Nomen.* 9 (1954): 266–69; R. C. Moore and P. C. Sylvester-Bradley, "Second supplemental application: Application for a ruling of the International Commission directing that the classification and nomenclature of ammonoid aptychi (Class Cephalopoda) be in terms of 'Parataxa,'" *Bull. Zool. Nomen.* 15 (1957): 35–69.

21. Documents 1/72 and 1/73, Colloquium on Zoological Nomenclature,

London, 1958 (ZN [L] 18), ICZN Archive, London.

22. Report of proceedings of the colloquium on zoological nomenclature, London, 14 July 1958 (ZN [L] 44), ICZN Archive, London.

23. F. H. T. Rhodes and K. J. Müller, "The conodont genus *Prioniodus* and related forms," *J. Paleont.* 30 (1956): 695–99.

24. Hass, "Conodonts," in Moore, *Treatise,* W3–W69; R. C. Moore, "Conodont classification and nomenclature," in Raymond C. Moore (ed.), *Treatise on Invertebrate Paleontology,* Part W Miscellanea (Lawrence: GSA/University of Kansas Press, 1962), W92–W98.

25. Rhodes to Scott, 6 June 1961, which includes Rhodes's "Comments on R.C. Moore's article on 'Conodont Classification and Nomenclature,'" Scott Papers.

26. R. C. Moore, C. G. Lalicker, and A. G. Fischer, *Invertebrate Fossils* (New York: McGraw-Hill, 1952).

6. SPRING

1. Klapper, pers. comm., 16 October 2005; Sweet interview, 23 May 2007.

2. Harry Kreisler interview with Frank Rhodes, 1999, Conversations with History, Institute of International Studies, University of California–Berkeley; Rhodes, "Zoological affinities," 419 (see ch. 4, n. 1); F. H. T. Rhodes and P. Wingard, "Chemical composition, microstructure and affinities of the Neurodontiformes," *J. Paleont.* 31 (1957): 448–54; Hass, "Conodonts," in Moore, *Treatise,* W25; M. Lindström, *Conodonts* (Amsterdam: Elsevier, 1964), 23.

3. Austin interview, 8 May 2006.

4. Müller interview, 23 April 2007; his translation.

5. SIA, Record Unit 7318, G, introduction, Arthur Cooper Papers, http://siarchives.si.edu/findingaids/FARU7318.htm/.

6. Sweet, pers. comm., 16 July 2010; Hibbard to Bassler, 29 September 1939,

Ray S. Bassler Papers. On Furnish, Strothmann, and the Missouri workers, Gil Klapper, pers. comm., 15 September 2005; Branson and Mehl, "Conodonts,"236 (see ch. 2, n. 37); S. P. Ellison Jr. and R. W. Graves Jr., "Lower Pennsylvanian (Dimple Limestone) conodonts of the Marathon region, Texas," *Univ. Missouri School of Mines and Metallurgy Bull., Teach.,* ser. 14, no. 3 (1941): 1–21; Rhodes, "Some British," 268 (see ch. 4, n. 13). Moore, Lalicker, and Fischer, *Invertebrate Fossils* (see ch. 5, n. 26) list this as a standard conodont technique in 1952.

7. A lack of awareness of the potential richness of limestones is implicit in Moore, Lalicker, and Fischer, *Invertebrate Fossils,* 733.

8. Walter C. Sweet, *The Conodonta* (Oxford: Clarendon Press, 1988). The residues of acid preparation – mineral fragments and fossils – were separated using the heavy liquid bromoform. Differing densities cause conodonts to sink and common minerals like quartz and calcite to float. It was another mass processing technique. Bromoform was soon discovered to be carcinogenic and other techniques would be deployed.

9. G. Bischoff and W. Ziegler, "Die Conodontenchronologie des Mitteldevons und des tiefsten Oberdevons," *Abh. Hess. Landesamtes Borden-forsch.* 22 (1957): 11; H. Beckmann, "Zur Anwendug von Essigsäure in der Mikropaläontologie," *Palaeont. Z.* 26 (1952): 138–39.

10. This limestone was something of an enigma: It occurred in the relatively young glacial drift of northern Germany; its true source remained unknown.

11. H.-P. Schultze, "Walter R. Gross, a palaeontologist in the turmoil of 20th century Europe," *Modern Geology* 20 (1996): 209–33.

12. W. R. Gross, "Zur Conodonten-Frage," *Senckenbergiana Lethaea* 35 (1954): 73–85.

13. W. R. Gross, "Uber die Basis der Conodonten," *Palaeont. Z.* 31 (1957): 78–91; Lindström, "Lowermost Ordovician," 537 (see ch. 5, n. 13); M. Lindström, "Om conodonter," *Svensk Faunistisk Revy* 3 (1955): 1–4; G. A. Stewart and W. C. Sweet, "Conodonts from the Middle Devonian bone beds of central and west-central Ohio," *J. Paleont.* 30 (1956): 261–73, 262.

14. W. R. Gross, "Uber die Basis bei den Gattungen Palmatolepis und Polygnatthus (Conodontida)," *Palaeont. Z.* 34 (1960): 40–58.

15. Müller, "Taxonomy, nomenclature," (see ch. 4, n. 14).

16. Lindström, "Lowermost," 539; Müller, "Taxonomy, nomenclature," 1330.

17. Beckmann, "Conodonten," 167 (see ch. 4, n. 11); Bischoff and Ziegler, "Conodontenchronologie," 7; Hass, "Conodonts," in Moore, *Treatise,* W40.

18. K. Weddige, "Willi Ziegler," *Pander Society Newsletter* 35 (2003): 1; Bischoff and Ziegler, "Conodontenchronologie," 11; D. Sannemann, "Oberdevonische Conodonten (to II a)," *Senckenbergiana Lethaea* 36 (1955): 123–56.

19. G. Bischoff, "Die Conodonten-Stratigraphie des rheno-herzynischen Unterkarbons mit Berücksichtigung der Wocklumeria-Stufe und der Devon/Karbon-Grenze," *Abh. Hess. Landesamtes Bodenforsch.* 19 (1957): 1–64, 7.

20. Walliser interview, 25 April 2007, and Müller interview.

21. K. J. Müller and D. Walossek, "Morphology, ontogeny and life habit of *Agnostus pisiformis* from the Upper Cambrian of Sweden," *Fossils and Strata* 19 (1987): 1–124. This interpretation remains contested, and most trilobite workers believe Müller wrong.

22. Lindström interview, 22 March 2007.

23. The discovery was first made by A. K. Ghosh and A. Bose, "Occurrence of microfossils in the Salt Pseudomorph

beds of the Punjab Salt Range," *Nature* 160 (1947): 796. A number of other papers by this duo extended the belief that vascular plants existed in the Cambrian. There had been a similar, but similarly disbelieved, discovery in Sweden in 1937: W. C. Darrah, "Spores of Cambrian plants," *Science* 86 (1937): 154–55.

24. Remarks made by Ziegler in 1996 on the award of the Pander Society medal to Maurits Lindström, *Pander Society Newsletter* (1997).

25. S. C. Matthew, "Conodonts," *Nature* 206 (1965): 646.

26. Sweet interview.

7. DIARY OF A FOSSIL FRUIT FLY

1. D. M. Raup, and R. E. Crick, "Evolution of single characters in the Jurassic ammonite *Kosmoceras*," *Paleobiology* 7 (1981): 200–215.

2. E. J. Larson, *Evolution: The Remarkable History of a Scientific Theory* (New York: Modern Library, 2004), 224; G. G. Simpson, *Concession to the Improbable: An Unconventional Autobiography* (New Haven: Yale University Press, 1978), 114. Simpson, *Concession*, 115, believed Otto Schindewolf's 1936 synthesis "entirely unacceptable." See also G. G. Simpson, *The Major Features of Evolution* (New York: Columbia University Press, 1953).

3. On implications of utilitarian stratigraphy in the United States, Simpson, *Concession*, 114, and contributions by Rhodes, 41, and Newell, 64, to P. C. Sylvester-Bradley, ed., *The Species Concept in Palaeontology* (London: Systematics Association, 1956); N. L. Thomas, "The use of evolutionary changes in geologic correlation," *J. Paleont.* 1 (1927): 135–39, 135; R. W. Harris and R. V. Hollingsworth, "New Pennsylvanian conodonts from Oklahoma," *Am. J. Sci.* 25 (1933): 193–204; Branson and Mehl, "New and little known," 103 (see ch. 2, n. 58).

4. Ellison, "Conodonts" (see ch. 2, n. 38); C. B. Rexroad, "The conodont homeomorphs *Taphrognathus* and *Streptognathodus*," *J. Paleont.* 32 (1958): 1158–59.

5. G. G. Simpson, *Tempo and Mode in Evolution* (New York: Columbia University Press, 1944), xv. See also Simpson, *Concession*, 12, 128.

6. Simpson, *Concession*, 115–16; G. G. Simpson, "Types in modern taxonomy," *Am. J. Sci.* 238 (1940): 413–31: Dunbar, "Symposium on fifty years," 911.

7. Sylvester-Bradley, *Species Concept*, 4, and contributions by I. Parker (9) and Frank Rhodes (37).

8. Austin interview; A. J. Scott and C. W. Collinson, "Intraspecific variability in conodonts: *Palmatolepis glabra*," *J. Paleont.* 33 (1959): 550–65; Klapper, pers. comm., 16 October 2005.

9. J. Helms, "Die 'nodocostata-Gruppe' der Gattung *Polygnathus*," *Geologie* 10, no. 6 (1961): 674–711, 674–75; Müller interview; K. J. Müller, "Zur Kenntnis der Conodonten-Fauna des europäischen Devons, 1; Die Gattung *Palmatolepis*," *Abh. der Senckenbergischen Naturfors. Gesellschaft* 494 (1956).

10. T. H. Morgan, *Evolution and Genetics* (Princeton: Princeton University Press, 1925), 140–41.

11. Walliser and Müller interviews; W. Ziegler, "Conodontenfeinstratigraphische Untersuchungen an der Grenze Mittledevon/Oberdevon und in der Adorfstufe," *Notizbl. Hess. L,-Amt Bodenforsch* 87 (1958): 7–77; W. Ziegler, "Phylogenetische Entwicklung stratigraphisch wichtiger Conodonten-Gattungen in der Manticoceras-Stufe (Oberdevon, Deutschland)," *Neues Jahrb. Geol. und Paläontol. Abh* 114 (1962): 142–68; W. Ziegler, "Taxonomie und Phylogenie Oberdevonischer conodonten und ihre stratigraphische Bedeutung," *Abh. Hess. Landes. Bodenforsch.* 38 (1962): 1–166, 6.

12. J. Helms, "Zur 'Phylogenese' und Taxionomie von *Palmatolepis* (Conodontida, Oberdevon)," *Geologie* 12 (1963): 449–85.

13. Gould, *Wonderful Life*, 60.

14. Klapper, pers. comm., 16 October 2005; G. Klapper and W. M. Furnish, "Conodont zonation of the early Upper Devonian in eastern Iowa," *Iowa Acad. Sci. Proc.* 69 (1963): 400–410.

15. B. F. Glenister and G. Klapper, "Upper Devonian conodonts from the Canning Basin, Western Australia," *J. Paleont.* 40 (1966): 777–842; W. Ziegler, "Conodont stratigraphy of the European Devonian," in W. C. Sweet and S. M. Bergström (eds.), *Symposium on Conodont Stratigraphy,* GSA Memoir 127, 227–84, 265.

16. Klapper, pers. comm., 16 October 2005.

17. *Pander Society Newsletter* 7 (January 1974); R. J. Aldridge and P. von Bitter, "The Pander Society (1967–2007): A brief history at forty," *Paleontographica Americana* 62 (2009): 11–21.

18. F. H. T. Rhodes, "Conodont research: Programs, progress and priorities," in F. H. T. Rhodes (ed.), *Conodont Paleozoology,* GSA Special Paper 141, 277–86, 277; Soviet group established in December 1966, *Pander Society Newsletter* 4 (1970): 3; *Pander Society Newsletter* 7 (1974).

19. Subcommission on Devonian Stratigraphy formed in 1973.

20. Hass, "Conodonts," in Moore, *Treatise,* W42; K. J. Müller, "Taxonomy, evolution, and ecology of conodonts," in Raymond. C. Moore (ed.), *Treatise on Invertebrate Paleontology,* Part W Miscellanea (Lawrence: GSA/University of Kansas Press, 1962), W83–W91; K. J. Müller, "Wert and Grenzen der Condonten-Stratigraphie," *Geol. Rundschau* 49 (1960): 83–92.

21. Müller, "Taxonomy, nomenclature," 1335 (see ch. 4, n. 14).

22. K. J. Müller, "Kambrische Conodonten," *Zeitschr. Deutsch. Geol. Gesell.* 111 (1959): 434–85; K. J. Müller, "Cambrian conodont faunas," in W. C. Sweet and S. M. Bergström (eds.), *Symposium on Conodont Biostratigraphy,* GSA Memoir 127, 5–20; K. J. Müller and I. Hinz, "Upper Cambrian conodonts from Sweden," *Fossils and Strata* 28 (1991): 4; Bergström (pers. comm.): "The history of *Westergaardodina* goes all the way back to 1893 and 1903 when Wiman illustrated this fossil as a 'ganz rätselhafter Organismus' (totally enigmatic organism)."

23. D. Ager, *Principles of Paleoecology* (New York: McGraw-Hill, 1963), 97.

24. K. J. Müller, "Supplement to systematics of conodonts," in Raymond. C. Moore (ed.), *Treatise on Invertebrate Paleontology,* Part W Miscellanea (Lawrence: GSA/University of Kansas Press, 1962), W246–49; W. C. Sweet and S. M. Bergström, "Conodonts from the Pratt Ferry Formation (Middle Ordovician) of Alabama," *J. Paleont.* 36 (1962): 1214–52, 1250; Lindström, *Conodonts* 31 (see ch. 6, n. 5); Allison R. (Pete) Palmer.

25. Müller, "Cambrian conodont faunas," 6; V. Poulsen, "Early Cambrian distacodontid conodonts from Bornholm," *Biol. Medd. Dan. Vid. Selsk.* 23 (1966): 1–10.

26. W. Youngquist, "Triassic conodonts from southeastern Idaho, *J. Paleont.* 26 (1952): 650–55; W. C. Sweet et al., "Conodont biostratigraphy of the Triassic," in W. C. Sweet and S. M. Bergström (eds.), *Symposium on Conodont Biostratigraphy,* GSA Memoir 127, 441–70, 443; U. Tatge, "Conodonten aus dem germanischen Muschelkalk," *Palaeont. Z.* 30 (1956), 108–27. However, Müller was also publishing information on Triassic conodonts at this time and claimed priority of discovery.

27. C. R. Stauffer, "Conodonts from the Devonian and associated clays of

Minnesota," *J. Paleont.* 14 (1940): 417–35; R. Huckriede, "Die Conodonten der mediterranen Trias und ihr stratigraphischer Wert," *Palaeont. Z.* 32 (1958): 141–75; K. Diebel, "Conodonten in der Oberkreide von Kamerun," *Geologie* 5 (1956): 424–50; Huckriede, "Conodonten," 165.

28. Hass, "Conodonts," in Moore, *Treatise,* W39.

29. Lindström, *Conodonts,* 9, 65, 124.

30. K. J. Müller, "Some remarks on the youngest conodonts," *Proceedings of the Second West African Micropaleontological Colloquium, Ibadan, 1965* (Leiden: E. J. Brill, 1966), 137–41; L. C. Mosher, "Are there post-Triassic conodonts?" *J. Paleont.* 41 (1967): 1554–55.

31. S. Nohda and T. Steoguchi, "An occurrence of Jurassic conodonts from Japan," *Mem. Coll. Sci. Kyoto Univ.,* ser. B, 33 (1967): 227–37.

32. K. J. Müller and L. C. Mosher, "Post-Triassic conodonts," in W. C. Sweet and S. M. Bergström (eds.), *Symposium on Conodont Biostratigraphy,* GSA Memoir 127, 467–70. Mosher spent ten months under Müller's guidance, but Müller himself did not really work on this problem (Müller interview). L. C. Mosher, "Evolution of Triassic platform conodonts," *J. Paleont.* 42 (1968): 947–54.

33. D. Raup and S. M. Stanley, *Principles of Paleontology* (San Francisco: W. H. Freeman, 1978), x.

34. *Harvard Gazette,* 20 May 2002; *Guardian,* 22 May 2002.

35. N. Eldredge and S. J. Gould, "Punctuated equilibria: An alternative to phyletic gradualism," in T. J. M. Schopf (ed.), *Models of Paleobiology* (San Francisco: Freeman, Cooper, 1972), 82–115; N. Eldredge, "The allopatric model of phylogeny in Paleozoic invertebrates," *Evolution* 25 (1971): 156–67; S. J. Gould, "Evolutionary palaeontology and the science of form," *Earth-Science Reviews* 6 (1970): 77–119;

N. Eldredge and S. J. Gould, "Morphological transformation, the fossil record, and the mechanisms of evolution: A debate. Part II the reply," in T. Dobzhansky et al. (eds.), *Evolutionary Biology,* vol. 7 (New York: Plenum, 1975), 303–308; M. K. Hecht, N. Eldredge, and S. J. Gould, "Morphological transformation, the fossil record, and the mechanisms of evolution: A debate," *Evolutionary Biology* 7 (1974): 295–308.

36. S. J. Gould and N. Eldredge, "Punctuated equilibria: The tempo and mode of evolution reconsidered," *Paleobiology* 3 (1977): 115–51, 117, 125.

37. A. B. Shaw, "Adam and Eve, paleontology, and the non-objective arts," *J. Paleont.* 43 (1969): 1085–98, 1094–95; Eldredge and Gould, "Punctuated equilibria: An alternative," 92.

38. Gould and Eldredge, "Punctuated equilibria: The tempo," 124–25; G. Klapper and D. B. Johnson, "Sequence in conodont genus *Polygnathus* in Lower Devonian at Lone Mountain, Nevada," *Geol. et Paleont.* 9 (1975): 65–83; Klapper, pers. comm., 16 October 2005.

39. N. Eldredge and S. J. Gould, "Evolutionary models and biostratigraphic strategies," in E. G. Kauffman and J. E. Hazel (eds.), *Concepts and Methods of Biostratigraphy* (Stroudsburg, Pa.: Dowden, Hutchinson and Ross, 1977), 25–40, 31–33; Eldredge, "Allopatric model."

40. Eldredge and Gould, "Punctuated equilibria: The tempo," 31.

8. FEARS OF CIVIL WAR

1. At the time, Rhodes had no idea if the rumors of Triassic conodonts were actually true. Rhodes, "Zoological affinities." Rhodes's reimagined assemblages were published in Moore's first *Treatise* just eleven pages after Hass's reprinting of Scott's.

2. Müller, "Taxonomy, nomenclature."

3. H. Schmidt and K. J. Müller, "Weitere Funde von Conodonten-Gruppen aus dem oberen Karbon des Sauerlandes," *Palaeont. Z.* 38 (1964): 105–35, 106.

4. Ibid., 108, 133.

5. Lindström, *Conodonts*, 124; C. L. Cooper, "Conodont assemblage from the lower Kinderhook black shales (abstract)," *GSA Bull.* 56 (1945): 1153 simply states, "Five pairs of denticulated bars are recognized, with one complete *Hibbardella* unpaired." M. Lindström, "Conodonts from the Crug limestone (Ordovician, Wales)," *Micropaleontology* 5 (1959): 427–52, 431; Lindström, *Conodonts*, 124 (see ch. 6, n. 2); H. R. Lane, "Symmetry in conodont element-pairs," *J. Paleont.* 42 (1968): 1258–63; A. Voges, "Conodonten aus dem Unterkarbon I und II (Gattendorfia- und Pericyclus-Stufe) des Sauerlandes," *Palaeont. Z.* 33 (1959): 266–314; W. C. Sweet and S. M. Bergström, "Conodonts from the Pratt Ferry Formation (Middle Ordovician) of Alabama," *J. Paleont.* 36 (1962): 1214–52.

6. Walliser interview; Huckriede, "Die Conodonten"; O. H. Walliser, "Conodonten des Silurs," *Abh. Hess. Landesamtes Bodenforsch.* 41 (1964): 1–106.

7. Sweet and Bergström, "Pratt Ferry"; Sweet interview: "And, by golly, we hit it right, we got it right on the nose as far as its age is concerned. We were so pleased with that because it was a good European or Scandinavian fauna."

8. Lindström, *Conodonts*, 77ff, 129.

9. C. B. Rexroad and R. S. Nicoll, "A Silurian conodont with tetanus?" *J. Paleont.* 26 (1964): 771–73.

10. Sweet, pers. comm., 16 July 2010; Sweet interview. The details of their work together is explored more fully in the next chapter

11. S. M. Bergström and W. C. Sweet, "Conodonts from the Lexington Limestone (Middle Ordovician) of Kentucky and its lateral equivalents in Ohio and Indiana," *Bull. Am. Paleont.* 50, no. 229 (1966): 271–424, 280.

12. Sweet interview.

13. G. F. Webers, T. J. M. Schopf, and W. C. Sweet, "Multielement Ordovician conodont species (abstract)," *GSA Program for 1965*, 180–81; G. F. Webers, *The Middle and Upper Ordovician Conodont Faunas of Minnesota*, Minnesota Geological Survey Special Publication SP-4 (1966).

14. T. J. M. Schopf, "Conodonts of the Trenton Group (Ordovician) in New York, southern Ontario, and Quebec," *New York State Mus. Sci. Serv. Bull.* 405 (1966): 105.

15. Bergström and Sweet, "Lexington Limestone."

16. Sweet, pers. comm., 16 July 2010.

17. Sweet, *Conodonta*, 38, 6.

18. Bergström and Sweet, "Lexington Limestone," 302.

19. Sweet, pers. comm., 16 July 2010.

20. C. R. Barnes, "A questionable natural conodont assemblage from Middle Ordovician limestone, Ottawa, Canada," *J. Paleont.* 41 (1967): 1557–60; R. L. Austin and F. H. T. Rhodes, "A conodont assemblage from the Carboniferous of the Avon Gorge, Bristol," *Palaeont.* 12 (1969): 400–405.

21. F.-G. Lange, "Conodonten-gruppenfunde aus Kalken des Oberdevon," *Geol. Palaeontol.* 2 (1968): 37–57. Also Bergström, pers. comm.

22. C. A. Pollock, "Fused Silurian conodont clusters from Indiana," *J. Paleont.* 43 (1969): 929–35, submitted for publication in April 1968; T. Mashkova, "*Ozarkodina steinhornensis* (Ziegler) apparatus, its conodonts and biozone," *Geol. Palaeontol.* 1 (1972): 81–90; Walliser interview.

23. Lane, "Symmetry."

24. Sweet and Bergström later claimed that this partly resulted from misidentification. At the time, sorting and breakage were considered the main impediments.

The term "multi-element species" was introduced to distinguish these new, hopefully biological, species.

25. Sweet interview.

26. J. J. Kohut, "Determination, statistical analysis, and interpretation of recurrent conodont groups in Middle and Upper Ordovician strata of the Cincinnati Region (Ohio, Kentucky, and Indiana)," *J. Paleont.* 43 (1969): 392–412, 412; J. J. Kohut and W. C. Sweet "The American Upper Ordovician standard. X. Upper Maysville and Richmond conodonts from the Cincinnati region of Ohio, Kentucky and Indiana," *J Paleont.* 42 (1968): 1457–77. Although both papers were submitted on the same day in 1967, Kohut's single-authored account of his methods did not appear until March 1969. Also *Pander Society Newsletter* 2 (6 July 1968).

27. Huddle reporting in *Pander Society Newsletter* 2:10.

28. Huddle, "Historical introduction," 8 (see ch. 4, n. 9) 1972. A large number of authors followed Kohut's lead. For example, in September 1969, Ronald Austin was working on the "application of information analysis techniques to conodonts" and Willi Ziegler was working on a statistical analysis of Devonian conodonts. *Pander Society Newsletter* 3. Sweet, *Conodonta*, 38, says these authors used slightly different clustering techniques.

29. W. C. Sweet and S. M. Bergström (eds.), *Symposium on Conodont Biostratigraphy*, GSA Memoir 127 (1970). This publication was not distributed until 1971.

30. Sweet interview.

31. Rhodes, "Conodont research," 285; also *Pander Society Newsletter* 4:9.

32. Jeppsson interview; Huddle in *Pander Society Newsletter* 5:10; W. C. Sweet and S. M. Bergström, "Multielement taxonomy and Ordovician conodonts," *Geol. Palaeontol.* 1 (1972): 29–42, 32 on strength of interpretations and size of collections.

33. M. Lindström and W. Ziegler, "Marburg symposium on conodont taxonomy, 1971," *Geol. Palaeontol.* 1 (1972): 1–2, 1.

34. D. L. Clark, "Early Permian crisis and its bearing on Permo-Triassic conodont taxonomy," *Geol. Palaeontol.* 1 (1972): 147–58, 147; Lindström and Ziegler, "Marburg," 1.

35. Rhodes to "Colleagues," 4 January 1972, Conodont File, Scott Papers.

36. R. Melville, "Further proposed amendments to the International Code of Zoological Nomenclature," *Bull. Zool. Nomen.* 36:11–14, and Melville's continuing debate in this journal.

37. From Aldridge's correspondence: Melville to Aldridge, 1 and 30 July 1980, Aldridge to Lennart Jeppsson, 25 August 1980, and Melville to Lennart Jeppsson, 1 September 1980, all in Aldridge Files, Geology Department, University of Leicester, UK (hereafter cited as Aldridge Files); R. W. Huddleston, "Comments on the proposed amendments to the International Code of Zoological Nomenclature concerning paranomenclature," *Bull. Zool. Nomen.* 37 (1980): 143–44, and later correspondence in that journal.

38. Pander Society quoted by Melville in *Bull. Zool. Nomen.* 38 (1981): 43, 46; Aldridge and von Bitter, "Pander Society," 3.

39. Sweet to Melville, 30 September 1980, Aldridge Files.

40. Melville to Sweet, 10 October 1980, Aldridge Files.

41. Melville in *Bull. Zool. Nomen.* 38 (1981): 42, also 238.

42. W. C. Sweet and P. C. J. Donoghue, "Conodonts: Past, present and future," *J. Paleont.* 75 (2001): 1174–84.

9. THE PROMISED LAND

1. Bischoff and Ziegler, "Conodontenchronologie," 10 (see ch. 6, n. 9); Müller, "Taxonomy, evolution" (see ch. 7, n. 20); M. Lindström, "Conodont provincialism and paleoecology – a few

concepts," in C. R. Barnes (ed.), *Conodont Paleoecology,* 3–9, 3, (see also vii); Lindström interview; Müller, "Zur Kenntnis," 1334 (see ch. 7, n. 9); K. J. Müller and E. M. Müller, "Early Upper Devonian (Independence) conodonts from Iowa, Part 1," *J. Paleont.* 31 (1957): 1069–1108, 1077; W. Youngquist, R. W. Hawley, and A. K. Miller, " Phosphoria conodonts from southeastern Idaho," *J. Paleont.* 25 (1951): 356–64; Huckriede, "Conodonten" (see ch. 7, n. 27); Müller, "Taxonomy, nomenclature" (see ch. 4, n. 14); Lindström, "Lowermost Ordovician" (see ch. 5, n. 13); Lindström, *Conodonts* (see ch. 6, n. 2), 66–97; M. Lindström, "Two Ordovician conodont faunas found with zonal graptolites," *Geol. Foren. Stockholm Forhandl.* 79 (1957): 161–78.

2. H. S. Ladd and J. W. Hedgpeth (eds.), *Treatise on Marine Ecology and Paleoecology,* GSA Memoir 67; S. P. Ellison Jr., "Economic applications of paleoecology," in Alan M. Bateman (ed.), *Economic Geology: Fiftieth Anniversary Volume* (Urbana, Ill.: Economic Geology Publishing, 1955), 867–84, 868.

3. P. E. Cloud Jr., "Paleoecology – retrospect and prospect," *J. Paleont.* 33 (1959): 926–62. On theorizers, Ager, *Principles,* vii (see ch. 7, n. 23).

4. D. L. Clark, "Paleoecology," in R. A. Robins et al. (eds.), *Treatise on Invertebrate Paleontology,* Part W, Supplement 2 Conodonta (Lawrence: GSA/University of Kansas,1981).

5. R. C. Moore, "Modern methods of paleoecology," *Bull.* AAPG 41 (1957): 1775–1801. See also Ellison, "Economic applications," 867.

6. C. B. Rexroad, "Conodonts from the Chester Series in the type area of southwestern Illinois," *Illinois Geol. Survey Rep. Inv.* 199 (1957); Rexroad, "Conodont homeomorphs" (see ch. 7, n. 4); C. B. Rexroad and M. K. Jarrell, "Correlation by conodonts of Golconda Group (Ches-

terian) in Illinois Basin," *Bull.* AAPG 45 (1961): 2012–17.

7. Lindström, "Two Ordovician."

8. Sweet, pers. comm., 21 April 2005; W. C. Sweet, C. A. Turco, E. Warner, and L. C. Wilkie, "The American Upper Ordovician Standard 1. Eden conodonts from the Cincinnati region of Ohio and Kentucky," *J. Paleont.* 33 (1959): 1029–68.

9. Sweet and Bergström, "Pratt Ferry," 1214 (see ch. 8, n. 5); S. M. Bergström, "Conodont biostratigraphy of the Middle and Upper Ordovician of Europe and Eastern North America," in W. C. Sweet and S. M. Bergström (eds.), *Symposium on Conodont Biostratigraphy,* GSA Memoir 127, 83–161.

10. Bergström and Sweet, "Lexington Limestone" (see ch. 8, n. 11). Ziegler reviewed this paper in 1966.

11. Bergström, "Conodont biostratigraphy," 88. Also contributions in this same volume by Sweet, Ethington, and Barnes (165). Kohut and Sweet, "American Upper Ordovician," 1460, 1464, 1467 (see ch. 8, n. 26); Webers, *Minnesota,* 18 (see ch. 8, n. 13).

12. P. C. Sylvester-Bradley, "Dynamic factors in animal palaeogeography," in F. A. Middlemiss, P. F. Rawson, and G. Newall (eds.), *Faunal Provinces in Space and Time,* Special Issue 4 (Liverpool: Seel House Press, 1971), 1–18, 16. On agreeing, see Middlemiss and Rawson in same volume (200).

13. Schopf, *Models,* 10 (see ch. 2, n. 45).

14. Glenister and Klapper, "Upper Devonian," 786 (see ch. 7, n. 15).

15. E. C. Druce, " Devonian and Carboniferous conodonts from the Bonaparte Gulf basin, northern Australia," *Aust. Bur. Miner. Resour. Geol. Geophys. Bull.* 98 (1969): 1–242; *Pander Society Newsletter* 34 (2002): 1. Druce ceased working with conodonts in 1979 and became an international figure in the stamp-collecting world.

16. G. Seddon, " Frasnian conodonts from the Sadler Ridge-Bugle Gap area,

Canning basin, Western Australia," *J. Geol. Soc. Aust.* 16 (1970): 723–53.

17. *Pander Society Newsletter* 4 (1970): 10.

18. The back reef proved to be generally without conodonts.

19. E. C. Druce, "Upper Paleozoic conodont distribution (abstract)," GSA *Abstracts with Programs* 2 (1970): 386.

20. G. Seddon, "Devonian biofacies in the Canning Basin, Western Australia (abstract)," *Abstr.* GSA *Proc. 4th Ann. Mtg. N. Cent Sec.* (1970): 404–405; E. Druce, "Upper Paleozoic and Triassic conodont distribution and the recognition of biofacies," in F. H. T. Rhodes (ed.), *Conodont Paleozoology,* GSA Special Paper 141, 191–237, 210.

21. G. Seddon and W. C. Sweet, "An ecologic model for conodonts," *J. Paleont.* 45 (1971): 869–80.

22. Druce, "Upper Paleozoic and Triassic," 211.

23. W. Ziegler, M. Lindström, and R. McTavish, "Monochloracetic acids and conodonts – a warning," *Nature* 230 (1971): 584–85. This method was based on the use of monochloracetic acid rather than plain acetic acid. See also chapter 12.

24. G. K. Merrill, "Facies relationships in Pennsylvanian conodont faunas (abstract)," *Texas J. Sci.* 14 (1962): 418; G. K. Merrill, *Allegheny (Pennsylvanian) Conodonts (Abstract),* GSA Special Paper 115, 147–48 (1967); C. L. Cooper, "Role of microfossils in interregional Pennsylvanian correlations," *J. Geol.* 55 (1947): 261–70, 270; G. K. Merrill, "Pennsylvanian conodont paleoecology," in F. H. T. Rhodes (ed.), *Conodont Paleozoology,* GSA Special Paper 141, 239–74 (244 cites master's student D. A. Drake as first discovering the relationship). Merrill's two "genera" were *Cavusgnathus* (dominant in shales) and, a "plexus" of similar forms that could be treated as a single biological entity, *Idiognathodus-Streptognathodus.*

25. F. H. T. Rhodes and D. L. Dineley, "Devonian conodont faunas from southwest England," *J. Paleont.* 31 (1957): 353–69, 356; R. L. Ethington, "Conodonts of the Ordovician galena formation," *J. Paleont.* 33 (1959): 257–92, 271; Druce, "Pennsylvanian conodont," 194.

26. Harris quoted in J. McPhee, *In Suspect Terrain* (New York: Farrar, Straus and Giroux, 1983), 122–23.

27. J. W. Valentine, "Plate tectonics and shallow marine diversity and endemism, an actualistic model," *System. Zool.* 20 (1971): 253–64; J. W. Valentine and E. M. Moores, "Plate tectonic regulation of faunal diversity and sea-level: A model, *Nature* 228 (1970): 657–59.

28. C. R. Barnes, C. B. Rexroad, and J. F. Miller, "Lower Paleozoic conodont provincialism," in F. H. T. Rhodes (ed.), *Conodont Paleozoology,* GSA Special Paper 141, 157–90.

29. L. E. Fåhræus, "Conodontophorid ecology and evolution related to global tectonics," in C. R. Barnes (ed.), *Conodont Paleoecology,* Geological Association of Canada Special Paper 15, 11–26, 12.

30. C. R. Barnes and L. E. Fåhræus, "Provinces, communities and the nektobenthic habit of Ordovician conodontophorids," *Lethaia* 8 (1975): 133–49.

31. C. R. Barnes (ed.), *Conodont Paleoecology,* Geological Association of Canada Special Paper 15, vii.

32. L. Jeppsson, "Autecology of Late Silurian conodonts," in C. R. Barnes (ed.), *Conodont Paleoecology,* Geological Association of Canada Special Paper 15, 105–18. The inspiration was B. Kurtén, "On the variation and population dynamics of fossil and recent mammal populations," *Acta Zoologica Fennica* 6 (1953): 122; Jeppsson interview.

33. R. J. Aldridge, "Comparison of macrofossil communities and conodont distribution in the British Silurian," in C. R. Barnes (ed.), *Conodont Paleoecology,*

Geological Society of Canada Special Paper 15, 91–104, 92.

34. K. Weddige and W. Ziegler, "The significance of *Icriodus: Polygnathus* ratios in limestones from the type Eifelian," in C. R. Barnes (ed.), *Conodont Paleoecology,* Geological Association of Canada Special Paper 15, 187–99.

35. L. E. Fåhræus and C. R. Barnes, "Conodonts as indicators of palaeogeographic regimes," *Nature* 258 (1975): 515–18.

36. M. A. Buzas, in *Paleobiology* 3 (1977): 330–32.

37. D. Raup and S. M. Stanley, *Principles of Paleontology,* 2nd ed. (San Francisco: W. H. Freeman, 1978), x; J. W. Hedgpeth, "Review: Structure and classification of paleocommunities," *Paleobiology* 3 (1977): 110–14.

38. G. Klapper and J. E. Barrick, "Conodont ecology; pelagic versus benthic," *Lethaia* 11 (1978): 15–23.

39. S. M. L. Pohler and C. R. Barnes, "Conceptual models in conodont paleoecology," *Courier Forsch.-Inst. Senckenberg* 118 (1990): 409–40; G. Klapper and J. G. Johnson, "Endemism and dispersal of Devonian conodonts," *J. Paleont.* 54 (1980): 400–455; C. A. Sandberg and R. Dreesen, "Late Devonian icriodontid biofacies models and alternate shallow-water conodont zonation," in D. L. Clark (ed.), *Conodont Biofacies and Provincialism,* GSA Special Paper 196, 143–69.

10. THE WITNESS

1. McPhee, *Suspect Terrain,* 6–7, 45–46 (see ch. 9, n. 26); Sweet, pers. comm., 21 April 2005.

2. A. G. Epstein, J. B. Epstein, and L. D. Harris, "Conodont color alteration – an index to organic metamorphism," *U.S. Geol. Surv. Prof. Pap.* 995 (1977): 4; A. G. Epstein, J. B. Epstein, and L. D. Harris, "Incipient metamorphism, structural anomalies, and oil and gas potential in the Appalachian basin determined from conodont color," *GSA Abstracts with Programs* 6 (1974): 723–24; A. G. Epstein, J. B. Epstein, and L. D. Harris, "Conodont color alteration – an index to diagenesis of organic matter," AAPG and SEPM, *Ann. Mtg. Abstracts* 2 (1975): 21–22.

3. W. Alvarez, *T.rex and the Crater of Doom* (Princeton: Princeton University Press, 1997); L. Alvarez et al., "Extraterrestrial cause for the Cretaceous-Tertiary extinction," *Science* 208 (1980): 1095–1108.

4. D. M. Raup, *The Nemesis Affair* (New York: Norton, 1986), 64, 112; W. Glen (ed.), *The Mass-Extinction Debate* (Palo Alto, Calif.: Stanford University Press, 1994).

5. Walliser interview; O. Schindewolf, "Neokatastrophismus?" *Deutsch Geologische Gesellschaft Zeitschrift Jahrgang* 114 (1962): 430–45; N. D. Newell, "Paleontological gaps and geochronology," *J. Paleont.* 36 (1962): 592–610; N. D. Newell, "Crisis in the history of life," *Scientific American* 208, no. 2 (1963): 1–16; F. H. T. Rhodes, "Permo-Triassic extinction," in W. B. Harland et al. (eds.), *The Fossil Record* (London: Geological Society of London, 1967), 57–76; D. J. McLaren, "Time, life and boundaries," *J. Paleont.* 44 (1970): 801–15.

6. O. H. Walliser, "Pleading for a natural D/C boundary," *Cour. Forsch.-Inst. Senckenberg* 67 (1984): 241–46.

7. O. H. Walliser, "Natural boundaries and commission boundaries in the Devonian," *Cour. Forsch.-Inst. Senckenberg* 75 (1985): 401–408. This was part of his Moscow presentation.

8. McLaren, "Time," 812–13; D. J. McLaren, "Bolides and biostratigraphy," *GSA Bull.* 94 (1987): 313–24; R. S. Dietz, "Astroblemes," *Sci. Am.* (August 1961): 50–58.

9. McLaren, "Time," 812; D. J. McLaren, "Frasnian-Famennian extinctions," in L. T. Silver and P. H. Schultz

(eds.), *Geological Implications of Impacts of Large Asteroids and Comets on the Earth*, GSA Special Paper 190, 477–89, 482.

10. Müller and Hinz, "Upper Cambrian," 4; J. F. Miller, "Conodont fauna of the Notch Peak Limestone (Cambro-Ordovician) House Range, Utah," *J. Paleont.* 43 (1969): 413–39; D. L. Clark and R. A. Robison, "Oldest conodonts in North American," *J. Paleont.* 43 (1969): 1044; Clark, "Early Permian Crisis."

11. D. L. Clark, "Extinction of conodonts," *J. Paleont.* 57 (1983): 652–61, was stimulated by the debate surrounding Leigh Van Valen's evolutionary law. L. Van Valen, "A new evolutionary law," *Evolutionary Theory* 1 (1973): 1–30; D. L. Clark, "Conodonts: The final fifty million years," in R. J. Aldridge (ed.), *Conodont Palaeobiology* (Chichester, UK: Horwood, 1987), 165–73.

12. Fåhræus, "Conodontophorid" (see ch. 9, n. 29).

13. O. H. Walliser, "International Palaeontological Association General Assembly, Paris, 10th July 1980," *Lethaia* 13 (1980): 288; O. H. Walliser (ed.), *Global Events and Event Stratigraphy in the Phanerozoic* (Berlin: Springer-Verlag, 1996), 7.

14. A body set up by the IUGS (International Union of Geological Sciences) and UNESCO in 1972; O. H. Walliser, "Global events and evolution: First IPA research programme," *Lethaia* 15 (1982): 198.

15. McLaren, "Bolides"; A. Boucot, "Does evolution take place in an ecological vacuum?" *J. Paleont.* 57 (1983): 1–30; O. H. Walliser, "Geologic processes and global events," *Terra Cognita* 4 (1984): 17–20; J. J. Sepkoski Jr., "Mass extinctions in the Phanerozoic oceans: A review," in J. J. Sepkoski (ed.), *Geological Implications of Impacts of Large Asteroids and Comets on the Earth*, GSA Special Paper 190, 283–90; Glen, *Mass-Extinction*, 50–53.

16. Walliser, "Patterns," 16; P. E. Playford et al., "Iridium anomaly in the Upper Devonian of the Canning Basin, Western Australia," *Science* 226 (1984): 437–39.

17. Walliser, "Natural," 402, 405; C. A. Sandberg, J. R. Morrow, and W. Ziegler, "Late Devonian sea-level changes, catastrophic events and mass extinctions," in C. Koeberl and K. G. MacLeod (eds.), *Catastrophic Events and Mass Extinctions: Impacts and Beyond*, GSA Special Paper 356, 473–87, 475.

18. Walliser, "Natural," 406.

19. C. W. Sandberg, W. Ziegler, and R. Dreesen, "Abrupt conodont biofacies changes redate and delimit the Frasnian (Late Devonian) extinction even in Euramerica (abstract)," *Terra Cognita* 7 (1987): 209–10.

20. W. Ziegler and H. R. Lane, "Cycles in conodont evolution and Devonian to mid-Carboniferous," in R. J. Aldridge (ed.), *Conodont Palaeobiology* (Chichester, UK: Horwood, 1987), 147–63.

21. C. W. Sandberg et al., "Late Frasnian mass extinction: Conodont event stratigraphy, global changes and possible causes," *Cour. Forsch.-Inst. Senckenberg* 102 (1988): 263–307; C. W. Sandberg, W. Ziegler, and R. Dreesen, "Late Frasnian mass extinction: Associated sea-level changes reflected by conodont faunas and biofacies," *Cour. Forsch.-Inst. Senckenberg* 102 (1988): 253–54.

22. O. H. Walliser et al., "On the Upper Kellwasser Horizon (boundary Frasnian/Famennian)," *Cour. Forsch.-Inst. Senckenberg* 110 (1989): 247–56.

23. L. Jeppsson, "Aspects of Late Silurian conodonts," *Fossils and Strata* 6 (1974): 1–54.

24. For example, L. Jeppsson, D. Fredholm, and B. Mattiasson, "Acetic acid and phosphate fossils – a warning," *J. Paleont.* 59 (1985): 952–56.

25. L. Jeppsson, "Sudden appearances of Silurian conodont lineages – provincialism or special biofacies?" in D. L. Clark (ed.), *Conodont Biofacies and Provincialism*, GSA Special Paper 196, 103–12.

26. A. G. Fischer, "Climatic oscillations in the biosphere," in M. H. Nitecki (ed.), *Biotic Crises in Ecological and Evolutionary Time* (San Diego: Academic Press, 1981), 103–31, 105, 127; A. G. Fischer and M. A. Arthur, "Secular Variations in the Pelagic Realm," in H. E. Cook and P. Enos (eds.), *Deep-Water Carbonate Environments* (Tulsa: SEPM Special Publication 25, 1977), 19–50.

27. P. Wilde and W. B. N. Berry, "Destabilization of the oceanic density structure and its significance to marine "extinction' events," *Palaeogeogr., Palaeoclimat., Palaeoecol.* 48 (1984): 143–62.

28. L. Jeppsson, "Lithological and conodont distributional evidence for episodes of anomalous oceanic conditions during the Silurian," in R. J. Aldridge (ed.), *Conodont Palaeobiology* (Chichester, UK: Horwood, 1987), 129–45.

29. L. Jeppsson, "An oceanic model for lithological and faunal changes tested on the Silurian record," *J. Geol. Soc. Lond.* 147 (1990): 663–74; R. J. Aldridge, L. Jeppsson, and K. J. Dorning, "Early Silurian oceanic episodes and events," *J. Geol. Soc. Lond.* 150 (1993): 501–13.

30. Jeppsson's many publications record the progressive elaboration of his model. He gives his best summary in L. Jeppsson, "Silurian oceanic events: Summary of general characteristics," in E. Landing, and M. E. Johnson (eds.), *Silurian Cycles,* New York State Museum Bulletin 491, 239–57 (1998). See also L. Jeppsson, "The anatomy of the Mid-Early Silurian Ireviken Event and a scenario for P-S events," in C. E. Brett and G. C. Baird (eds.), *Paleontological Events* (New York: Columbia University Press 1997), 451–92.

11. THE BEAST OF BEAR GULCH

1. Lindström, *Conodonts,* 117.

2. H. W. Scott, "Discoveries bearing on the nature of the conodont animal," *Micropaleontology* 51 (1969): 420–26. On disappointing the audience, *Pander Society Letter* 2 (July 1968): 11.

3. Scott in 1969 reflecting on his recent discoveries in an unpublished lecture, "Concerning Devonian conodont assemblages from Germany," Box 1, Unpublished MSS Folder, Scott Papers.

4. Scott, "Discoveries," 423.

5. Scott, "Concerning"; F.-G. Lange, "Conodonten-Gruppenfunde aus Kalken des tieferen Oberdevon," *Geol. et Palaeontol.* 2 (1968): 37–57; Scott, "A conodont-acanthodian association," c. 1969, Unpublished MSS Folder, Scott Papers.

6. W. G. Melton Jr., "The Bear Gulch Limestone and the first conodont bearing animals," 21st Annual Field Conference, Montana Geological Society, September 1972, 65–68; H. W. Scott, "New specimens of Conodontochordata, Cycloidea and associated animals from the Bear Gulch Formation, Montana," n.d., Box 1, Unpublished MSS Folder, Scott Papers.

7. Melton, "Bear Gulch," 66.

8. Ibid.; *Pander Society Letter* 3 (September 1969): 9; Scott to Norbert Cygen, 8 September 1969, Box 1, Correspondence Re: Conodont Animal, Scott Papers. The Correspondence Re: Conodont Animal archive folder holds an almost complete record of the correspondence surrounding the discovery of the Bear Gulch animals. I have mentioned key actors in the main text and maintain the narrative timings in these copious letters for the most part. For this reason I do not give specific reference to each letter here. All unreferenced remarks and story elements come from this correspondence. S. J. Gould, "Nature's great era of experiments," *Natural History* July (1983): 12–22, 12. See also *Pander Society Letter* 3 (1970):7 and W. G. Melton Jr., "The Bear Gulch fauna from central Montana," *Proceedings of the North American Paleontological Convention,* vol.1 (Lawrence, Kans.: Allen Press, 1970), 1202–1207.

9. E. L. Yochelson, "Introduction," in *Proceedings of the North American Paleontological Convention,* vol.1 (Lawrence, Kans.: Allen Press, 1970). Thanks to Derek Briggs and Dick Aldridge for this important source.

10. Austin (pers. comm.) recalled this rumor, though he did not subscribe to it himself. Rhodes, pers. comm., 29 October 2010.

11. William G. Melton died on December 25, 1991. He was sixty-eight. He was the preparator of vertebrate paleontology under Claude W. Hibbard from March 1957 to August 1966 and worked with him for many summers in Meade, Kansas. *Geoscience News,* University of Michigan alumni newsletter, December 1992, http://www.geo.lsa .umich.edu/geonews/archive/9212.pdf/.

12. *Pander Society Letter* 2 (October 1969).

13. E. S. Richardson, "The conodont animal," *Earth Science* 6 (1969): 256–57.

14. Rhodes expressed his skepticism of the Melton and Scott animal in the book in which it was published, and again with Austin in the *Treatise* published in 1981, though he did so with such delicacy that readers, including Scott, may have been uncertain precisely where he stood on the matter. Certainly, he did not conclusively support their interpretation. Frank Rhodes, pers. comm., 29 October 2010.

15. R. J. Riedl, "Gnathostomulisa from America," *Science* 163 (1969): 445–52; C. J. Durden, J. Rogers, E. L. Yochelson, and R. J. Riedl, "Gnathostomulida: Is there a fossil record? (correspondence)," *Science* 164 (1969): 855–56; O. Wetzel, "Die in organischer substanz erhaltenen mikrofossilien des baltischen Kriedefeuersteins," *Palaeontogr. Abt. A. Paleozoool.-Stratigr.* 78 (1933): 1–110.

16. John R. Horner (b. 15 June 1946) would become a distinguished palaeontologist who demonstrated the sociability of some dinosaurs and later provided technical advice for the *Jurassic Park* films.

17. *Pander Society Letter* 4 (August 1970).

18. W. G. Melton and H. W. Scott, "Conodont-bearing animals from the Bear Gulch Limestone, Montana," in F. H. T. Rhodes (ed.), *Conodont Paleozoology,* GSA Special Paper 141; L B. Halstead, *The Pattern of Vertebrate Evolution* (San Francisco: Freeman 1969).

19. *Pander Society Letter* 4 (August 1970).

20. Scott, "New specimens."

12. THE INVENTION OF LIFE

1. K. Fahlbusch, "Bildung von Calciumphosphat bei fossilen Algen," *Naturwissenschaften* 50 (1973): 517–18; K. Fahlbusch, "Die stellung der Conodontida im biologischen system," *Palaeontographica* A 123 (1964): 137–201; Lindström, *Conodonts,* 121; H. Beckmann et al., "Sind Conodontent Reste fossiler Algen?" *N. Jb. Geol. Paläont. Abh.* 7 (1965): 385–99. On Nease, *Pander Society Newsletter* 3 (1969).

2. Lindström, *Conodonts,* 123–30.

3. H. Pietzner et al., "Zur chemischen Zusammensetzung and Mikromorphologie der Conodonten," *Palaeontographica* 128 (1968): 115–52. The first fossils to be studied in this way were brachiopods and bivalves, for which, Barnes, Rexroad, and Miller, "Lower Paleozoic," 3; R. W. Pierce and R. L. Langenheim Jr., "Ultrastructure in *Palmatolepis* sp. and *Polygnathus* sp.," *GSA Bull.* 80 (1969): 1397–1400.

4. K. J. Müller, "Bürstenbildung bei Conodonten," *Palaeont. Z.* 43 (1969): 64–71; K. J. Müller and Y. Nogami, "Über den feinbau der Conodonten," *Mem. Fac. Sci., Kyoto Univ., Geol. and Min.* 38 (1971): 1–87; K. J. Müller, "Micromorphology of elements: Internal structure," in R. A. Robison (ed.), *Treatise of Invertebrate Paleontology,* pt. W, suppl. 2, Conodonta (Boulder, Colo./Lawrence: GSA and Univ. Kansas Press, 1981), W20–41; P. C. J. Donoghue,

"Growth and patterning in the conodont skeleton," *Phil. Trans. R. Soc. Lond.*, ser. B, 353 (1998): 633–66.

5. C. R. Barnes, D. B. Sass, and E. A. Monroe, "Preliminary studies of the ultrastructure of selected Ordovician conodonts," *R. Ont. Mus. Life Sci. Contrib.* 76 (1970): 1–24; C. R. Barnes, D. B. Sass, and M. L. S. Poplawski, "Conodont ultra-structure," *R. Ont. Mus. Life Sci. Contrib.* 90 (1973): 1–36.

6. F. H. T. Rhodes (ed.), *Conodont Paleozoology,* GSA Special Paper 141, vii.

7. M. Lindström, "On the affinities of conodonts," in F. H. T. Rhodes (ed.), *Conodont Paleozoology,* GSA Special Paper 141, 85–102; M. Lindström, "The conodont apparatus as a food-gathering mechanism," *Palaeontology* 17 (1974): 729–44, 731–32.

8. S. Rietschel, "Zur Deutung der Conodonten,' *Natur und Museum* 103 (1973): 409–18; F. R. Schram, "Pseudocoelomates and a nemertine from the Illinois Pennsylvanian," *J. Paleont.* 47 (1973): 985–89.

9. Lindström, "On the affinities."

10. S. Conway Morris, "A new Cambrian lophophorate from the Burgess Shale of British Columbia," *Palaeontology* 19 (1976): 199–222.

11. This was *Protohertzina* discovered by Missarzhevskij, see S. Bengtson, "The structure of some Middle Cambrian conodonts, and the early evolution of conodont structure and function," *Lethaia* 9 (1976): 185–206, 185.

12. Gould, *Wonderful Life,* 149; B. F. Glenister et al., "Conodont pearls?" *Science* 193 (1976): 571–73; D. McConnell et al., "Nautiloid uroliths composed of phosphatic hydrogel," *Science* 199 (1978): 208–209. Donoghue later suggested these pearls belonged to an extinct group of bryozoans.

13. J. Priddle, "The function of conodonts," *Geol. Mag.* 111 (1974): 255–57; Priddle to Scott, 24 January 1974, Scott

to Priddle, 4 February 1974, Box 1, Correspondence Re: Conodont Animal, Scott Papers.

14. G. C. O. Bischoff, "On the nature of the conodont animal," *Geol. & Palaeont.* 7 (1973): 147–74.

15. J. Hofker, "Eine mögliche Tiergruppe, welche die Trägerin der sogenannten Conodonten war," *Palaeont. Z.* 48 (1974): 29–35.

16. Bengtson, "Structure."

17. S. Bengtson, "Conodonts: The need for a functional model," *Lethaia* 13 (1980): 320 admits to this attraction.

18. K. J. Müller and D. Andres, "Eine conodontengruppe von *Prooneotodus tenuis* (Müller, 1959) in natürlichen Zusammenhang aus dem Oberen Kambrium von Schweden," *Palaeont. Z.* 50 (1976): 193–200; P. Carls, "Could conodonts be lost and replaced?" *N. Jb. Geol. Paläont. Abh.* 155 (1977): 18–64.

19. E. Landing, "'Prooneotodus' tenuis (Müller, 1959) apparatuses from the Taconic allochthon, eastern New York: Construction, taphonomy and the proto-conodont 'supertooth' model," *J. Paleont.* 51 (1977): 1072–84.

20. R. S. Nicoll, "Conodont apparatuses in an Upper Devonian palaeoniscid fish from the Canning Basin, Western Australtia," BMR *J. Austral. Geol. and Geophys.* 2 (1977): 217–28.

21. V. H. Hitchings and A. T. S. Ramsay, "Conodont assemblages: A new functional model," *Paleogeog., Palaeoclimat., Palaeoecol.* 24 (1978): 137–49.

22. L. Jeppsson, "Conodont element function," *Lethaia* 12 (1979): 153–71.

23. S. Conway Morris, "Conodont function: Fallacies of the tooth model," *Lethaia* 13 (1980): 107–108; L. Jeppsson, "Function of the conodont elements," *Lethaia* 13 (1980): 228; Bengtson, "Conodonts."

24. H. Szaniawski, "Chaetognath grasping spines recognized among Cambrian

protoconodonts," *J. Paleont.* 56 (1982): 806–10; H. Szaniawski, "Structure of protoconodont elements," *Fossils and Strata* 15 (1983): 21–27; Sweet, *Conodonta*, 174.

25. R. Buchsbaum, *Animals without Backbones* (Harmondsworth, UK: Penguin, 1951), 1:199–200.

26. K. J. Müller, "Zoological affinities of conodonts," in R. A. Robison (ed.), *Treatise of Invertebrate Paleontology*, pt. W, suppl. 2, Conodonta (Boulder, Colo./ Lawrence: GSA and Univ. Kansas Press, 1981), W78–W82.

13. EL DORADO

1. This chapter has, in addition to published resources, drawn upon an unpublished book-length account Dick Aldridge wrote of his scientific research, in part published in R. J. Aldridge and D. E. G. Briggs, "The discovery of conodont soft tissue anatomy and its importance for understanding the early history of vertebrates," in D. Sepkoski and M. Ruse (eds.), *The Paleobiological Revolution* (Chicago: University of Chicago Press, 2009), 73–88. This chapter differs considerably from that account by attempting to locate a broader overview of debate in a longer history of discovery. Derek Briggs and Euan Clarkson read a late draft of this chapter and offered some important correctives to factual accuracy. Walt Sweet and Neil Clark provided me with recollections. My last task was to interrogate the extensive files of correspondence that Dick Aldridge had gathered together from various actors in the drama. For the sake of brevity, I have not given precise references to these resources below.

2. D. E. G. Briggs, "The search for paleontology's most elusive entity: The conodont animal," *Field Mus. Nat. Hist. Bull.* 55 (1984): 11–18; D. E. G. Briggs, E. N. K. Clarkson, and R. J. Aldridge, "Conodont," *McGraw-Hill Yearbook of Science and Technology 1985* (New York: McGraw-Hill,
1984), 132–35; D. E. G. Briggs and E. N. K. Clarkson, "The Lower Carboniferous Granton 'shrimp bed,' Edinburgh," in D. E. G. Briggs and P. D. Lane (eds.), *Trilobites and Other Early Arthropods,* Palaeontological Association Special Papers in Palaeontology 30, 616–77 (1983); D. E. G. Briggs, N. D. L. Clark, and E. N. K. Clarkson, "The Granton 'shrimp-bed,' Edinburg – a Lower Carboniferous Konservat-Lagerstätte," *Trans. Roy. Soc. Edinb. Earth Sci.* 82 (1991): 65–85; D. Tait, "Notice of a shrimp-bearing limestone in the Calciferous Sandstone Series at Granton, near Edinburgh," *Trans. Edinb. Geol. Soc.* 11 (1924): 131–15.

3. D. E. G. Briggs, E. N. K. Clarkson, and R. J. Aldridge, "The conodont animal," *Lethaia* 16 (1983): 1–14.

4. Gould, "Nature's great era," 12.

5. J. J. Hearty, "A day at Granton Harbour – conodont II," *MAPS [Mid-America Paleontological Society] Digest,* November 1984.

6. S. Bengtson, "A functional model for the conodont apparatus," *Lethaia* 16 (1983): 38; S. Bengtson, "The early history of the Conodonta," *Fossils and Strata* 15 (1983): 5–19.

7. J. Dzik and D. Drygant, "The apparatus of panderodontid conodonta," *Lethaia* 19 (1986): 133–41.

8. W. Sweet, "Conodonts: Those fascinating little whatzits," *J. Paleont.* 59 (1985): 485–94.

9. R. J. Aldridge and D. E. G. Briggs, "Conodonts," in A. Hoffman and M. H. Nitecki (eds.), *Problematic Fossil Taxa* (Oxford: Oxford University Press, 1986), 227.

10. P. Janvier, "Conodont affinity: A reply," *Lethaia* 21 (1988): 27; P. Janvier, "'L'animal-conodonte' enfin demasqué?" *Recherche* 14, no. 145 (1983): 832–33.

11. J. K. Rigby Jr., "Conodonts and the early evolution of the vertebrates," *GSA Abstracts with Programs* 15 (1983): 671.

12. R. S. Nicoll, "Multielement composition of the conodont species *Polygnathus xylus xylus* Stauffer, 1940 and *Ozarkodina brevis* (Bischoff & Ziegler, 1957) from the Upper Devonian of the Canning Basin, Western Australia," *BMR J. Austral. Geol. Geophys.* 9 (1985): 133–47, 146.

13. D. G. Mikulic, D. E. G. Briggs, and J. Kluessendorf, "A Silurian soft-bodied biota," *Science* 228 (1985): 715–17; D. G. Mikulic, D. E. G. Briggs, and J. Kluessendorf, "A new exceptionally preserved biota from the Lower Silurian of Wisconsin, USA," *Phil. Trans. Roy. Soc. Lond.*, ser. B, 311 (1985): 78–85.

14. R. J. Aldridge, M. P. Smith, R. D. Norby, and D. E. G. Briggs, "The architecture and function of Carboniferous polygnathacean conodont apparatuses," in R. J. Aldridge (ed.), *Palaeobiology of Conodonts* (Chichester, UK: Ellis Horwood, 1987), 63–75.

15. D. E. G. Briggs and S. H. Williams, "The restoration of flattened fossils," *Lethaia* 14 (1981): 157–64.

16. M. P. Smith, D. E. G. Briggs, and R. J. Aldridge, "A conodont animal from the lower Silurian of Wisconsin, USA, and the apparatus architecture of panderodontid conodonts," in Aldridge, *Palaeobiology of Conodonts*, 91–104.

17. Aldridge, "Conodont palaeobiology" (see ch. 4, n. 15).

18. R. S. Nicoll, "Form and function of the Pa element in the conodont animal," and R. S. Nicoll and C. B. Rexroad, "Reexamination of Silurian conodont clusters from Northern Indiana," both in Aldridge, *Palaeobiology of Conodonts*, 77–90.

19. H. Szaniawski, "Preliminary structural comparisons of protoconodont, paraconodont and euconodont elements," in Aldridge, *Palaeobiology of Conodonts*, 35–47.

20. R. J. Aldridge, D. E. G. Briggs, E. N. K. Clarkson, and M. P. Smith, "The affinities of conodonts – new evidence from the Carboniferous of Edinburgh, Scotland," *Lethaia* 19 (1986): 279–91; M. Benton, "Conodonts classified at last," *Nature* 325 (1987): 482–83.

21. Mashkova, "*Ozarkodina*" (see ch. 8, n. 22); J. Dzik, "Chordate affinities of the conodonts," in Hoffman and Nitchki, *Problematic Fossil Taxa*, 240–54.

22. R. P. S. Jefferies, *The Ancestry of the Vertebrates* (Cambridge: Cambridge University Press, 1986).

23. G. S. Nowlan and D. B. Carlisle, "The cephalochordate affinities of conodonts," *Can. Paleont. Biostrat. Seminar, Prog. Abstracts* (1987): 7.

24. Janvier, "Conodont affinity: A reply."

25. S. Tillier and J. P. Cuif, "L'animale-conodonte est-il un mollusque Aplacophore," *Cr. Hebd. Séanc. Acad. Sci., Paris* 303 (1986): 627–32; S. Tillier and P. Janvier, "Le retour de l'animal-conodonte," *Recherche* 17 (1986): 1574–1575; D. E. G. Briggs, R. J. Aldridge, and M. P. Smith, "Conodonts are not aplacophoran mollusks," *Lethaia* 20 (1987): 381–82.

26. Sweet, *Conodonta* (see ch. 6, n. 8). A book had been produced in Chinese but had limited impact.

27. See also H. Gee, "Four legs to stand on . . . ," *Nature* 342 (1989): 738–39.

28. R. J. Aldridge and D. E. G. Briggs, "Sweet talk," *Paleobiology* 16 (1990): 241–46.

29. M. P. Smith, "The Conodonta – Palaeobiology and evolutionary history of a major Palaeozoic chordate group," *Geol. Mag.* 127 (1990): 365–69.

30. S. Conway Morris, "*Typhloesus wellsi* (Melton and Scott, 1973), a bizarre metazoan from the Carboniferous of Montana, USA," *Phil. Trans. Roy. Soc. Lond.*, ser. B, 327 (1990): 595–624, is the most comprehensive account. The author presented on the animal in the United States in 1979. That account was not published until 1985.

31. Gould, *Wonderful Life,* 148; D. E. G. Briggs, and S. Conway Morris, "Problematica from the Middle Cambrian Burgess Shale of British Columbia," in Hoffman and Nitecki, *Problematic Fossil Taxa,* 167–83. Bengtson and others published on this animal in 2006.

32. R. Fortey, "Shock lobsters," review of Conway Morris's *Crucible of Creation, Lond. Rev. Books* 20, no. 19 (1998).

33. S. Conway Morris, "Conodont palaeobiology: Recent progress and unsolved problems," *Terra Nova* 1 (1989): 135–50, 141.

14. OVER THE MOUNTAINS OF THE MOON

1. This account of the Soom Shale investigation draws upon Dick Aldridge's unpublished account and correspondence files (see ch. 13, n. 1). I am also grateful, via Dick, to Hannes Theron for recounting details of the discovery.

2. J. N. Theron and E. Kovacs-Endrody, "Preliminary note and description of the earliest known vascular plant, or an ancestor of vascular plants, in the flora of the Lower. Silurian Cedarberg Formation, Table Mountain Group, South Africa," *S. African J. Sci.* 82 (1986): 102–105.

3. R. J. Aldridge and J. N. Theron, "Conodonts with preserved soft tissue from a new Ordovician Konservat-Lagerstatte," *J. Micropalaeo.* 12 (1993): 113–17; A. Ritchie, "New evidence on *Jamoytius kerwoodi* White, an important ostracoderm from the Silurian of Lanarkshire, Scotland," *Palaeontology* 11 (1968): 21–39.

4. R. J. Aldridge, J. N. Theron, and S. E. Gabbott, "The Soom Shale: A unique Ordovician fossil horizon in South Africa," *Geology Today* 10 (1994): 218–21.

5. R. J. Krejsa, P. Bringas, and H. C. Slavkin, "A neontological interpretation of conodont elements based on agnathan cyclostome tooth structure, function and development," *Lethaia* 23 (1990): 369–78; R. J. Krejsa, P. Bringas, and H. C. Slavkin,

"The cyclostome model: An interpretation of conodont element structure and function based on cyclostome tooth morphology, function and life history," *Cour. Forsch.-Inst. Senckenberg* 118 (1990): 473–92.

6. M. M. Smith and B. K. Hall, "Development and evolutionary origins of vertebrate skeletogenic and odontogenic tissues," *Biol. Rev.* 65 (1990): 277–373; R. J. Aldridge et al., "The anatomy of conodonts," *Phil. Trans. Roy. Soc.,* ser. B, 340 (1993): 405–21.

7. D. K. Elliot, A. R. M. Blieck, and P.-Y. Gagnier, "Ordovician vertebrates," in C. R. Barnes, and S. H. Williams (eds.), *Advances in Ordovician Geology,* Geological Survey of Canada Paper 90-9, 93–106 (1991).

8. D. Palmer, "Early vertebrates given new teeth," *New Scientist,* 29 August 1982, 16.

9. I. J. Sansom et al., "Presence of earliest vertebrate hard tissues in conodonts," *Science* 256 (1992): 1308–11.

10. M. Smith, I. J. Sansom and P. Smith, "'Teeth' before armour: The earliest vertebrate mineralized tissues," *Modern Geology* 20 (1996): 303–19, a paper presented in 1993 and updated during delayed publication.

11. N. Nuttal, "Razor-toothed fish bites into human history," *Times,* 10 June 1992; "Limestone yields the oldest set of teeth," *Independent,* 10 June 1992; J. H., "Teething troubles," *Telegraph,* 22 June 1992; Palmer, "Early vertebrates."

12. *Discover,* January 1993, 68; *Northern Echo,* 11 June 1992.

13. C. R. Barnes, R. Fortey, and S. H. Williams, "The patterns of global bioevents during the Ordovician Period," in O. H. Walliser (ed.), *Global Events and Event Stratigraphy in the Phanerozoic* (Berlin: Springer-Verlag, 1996), 142; D. E. G. Briggs, "Conodonts – a major extinct group added to the vertebrates," *Science* 256 (1992): 1285–86; Aldridge, Theron, and Gabbott, "Soom Shale," 220–21; H. Gee,

Deep Time: Cladistics, the Revolution in Evolution (London: Fourth Estate, 2000); H. Gee, *Before the Backbone* (London: Chapman & Hall, 1996).

14. M. A. Purnell, "Skeletal ontogeny and feeding mechanisms in conodonts," *Lethaia* 27 (1994): 129–38; M. A. Purnell and P. H. von Bitter, "Blade-shaped conodont elements functioned as cutting teeth," *Nature* 359 (1992): 629–31.

15. M. A. Purnell, "Feeding mechanisms in conodonts and the function of the earliest vertebrate hard tissues," *Geology* 21 (1993): 375–77; R. J. Aldridge and M. A. Purnell, "The conodont controversies," *Trends in Ecology and Evolution* 11 (1996): 463–68.

16. R. S. Nicoll, "Conodont element morphology, apparatus reconstructions and element function: A new interpretation of conodont biology with taxonomic implications," *Cour. Forsch. Inst. Senckenberg* 182 (1995): 247–62. See also O. H. Walliser, "Architecture of the polygnathid conodont apparatus," *Cour. Forsch. Inst. Senckenberg* 168 (1994): 31–36.

17. D. E. G. Briggs and A. J. Kear, "Decay of the lancelet *Branchiostoma lanceolatum* (Cephalochordata): Implications for the interpretation of soft-tissue preservation in conodonts and other primitive chordates," *Lethaia* 26 (1994): 275–87.

18. Aldridge et al., "Anatomy of conodonts."

19. P. Forey and P. Janvier, "Agnathans and the origin of jawed vertebrates," *Nature* 361 (1993): 129–34. See also Aldridge and Purnell, "Conodont controversies," 464, for further debate.

20. I. J. Sansom, M. P. Smith, and M. M. Smith, "Dentine in conodonts," *Nature* 368 (1994): 591.

21. A. Kemp and R. S. Nicoll, "Protochordate affinities of conodonts," *Cour. Forsch. Inst. Senckenberg* 182 (1995): 235–45; A. Kemp and R. S. Nicoll, "A histochemical analysis of biological residues in conodont elements," *Modern Geology* 20 (1996): 287–302. Also H.-P. Schultze, "Conodont histology: An indicator of the vertebrate relationship," *Modern Geology* 20 (1996): 275–85.

22. M. A. Purnell, "Microwear in conodont elements and macrophagy in the first vertebrates," *Nature* 374 (1995): 798–800.

23. S. E. Gabbott, R. J. Aldridge, and J. N. Theron, "A giant conodont with preserved muscle tissue from the Upper Ordovician of South Africa," *Nature* 374 (1995): 800–803.

24. R. Monastersky, "Fossil enigma bares teeth, tells its tale," *Science News* 147 (1995): 261.

25. P. Janvier, "Conodonts join the club," *Nature* 374 (1995): 761–72; J. Mallet, "Ventilation and origin of jawed vertebrates: A new mouth," *Zool. J. Linn. Soc.* 117 (1996): 329–404; P. Janvier, "The dawn of the vertebrates: Characters versus common ascent in the rise of current vertebrate phylogenies," *Palaeontology* 39 (1996): 259–87.

26. Donoghue, "Growth and patterning."

27. P. C. J. Donoghue and M. A. Purnell, "Growth, function and the conodont fossil record," *Geology* 27 (1999): 251–54.

28. R. J. Aldridge et al., "The apparatus architecture and function of *Promissum pulchrum* Kovács-Endrödy (Conodonta, Upper Ordovician), and the prioniodontid plan," *Phil. Trans. Roy. Soc. Lond.*, ser. B, 347 (1995): 275–91.

29. M. A. Purnell and P. C. J. Donoghue, "Architecture and functional morphology of the skeletal apparatus of ozarkodinid conodonts," *Phil. Trans. Roy. Soc. Lond.*, ser. B, 352 (1997): 1545–64.

30. P. C. J. Donoghue and M. A. Purnell, "Mammal-like occlusion in conodonts," *Paleobiology* 25 (1999): 58–74; M. A. Purnell, P. C. J. Donoghue, and R. J. Aldridge, "Orientation and anatomical nota-

tion in conodonts," *J. Paleont.* 74 (2000): 113–22; M. A. Purnell, "Feeding in extinct jawless heterostracan fishes and testing scenarios of early vertebrate evolution," *Proc. Roy. Soc. Lond.*, ser. B, 269, no. 1486 (2002): 83–88.

31. M. A. Purnell et al., "Conodonts and the first vertebrates," *Endeavour* 19 (1995): 20–27; M. A. Purnell, "Large eyes and vision in conodonts," *Lethaia* 28 (1995): 187–88.

32. M. A. Purnell, "Armed to the teeth," *Rockwatch* 17 (1997): 10–11.

33. Aldridge and Purnell, "Conodont controversies"; Gee, *Before*, xvii; H. Gee, "What remains, however improbable . . . ," *Nature* 377 (1995): 675.

34. Benton, *Basic Palaeontology*, 197–98.

35. P. C. J. Donoghue, P. L. Forey, and R. J. Aldridge, "Conodont affinity and chordate phylogeny," *Biol. Rev.* 75 (2000): 191–251.

36. P. A. Pridmore, R. E. Barwick, and R. S. Nicoll, "Soft anatomy and affinities of conodonts," *Lethaia* 29 (1997): 317–28; P. C. J. Donoghue, M. A. Purnell, and R. J. Aldridge, "Conodont anatomy, chordate phylogeny and vertebrate classification," *Lethaia* 31 (1998): 211–19.

37. G. I. Buryi and A. P. Kasatkina, "Functional importance of new skeletal elements ('eye capsules') of euconodonts," *Albertiana* 26 (2001): 7–10.

38. A. Blieck, "Comments," *Ordovician News* 24 (2007): 8

39. A. Blieck et al., "Organismal biology, phylogeny and strategy of publication: Why conodonts are not vertebrates (abstract)," Third International Conference Geologica Belgica 2009, Ghent University. The paper was published in 2010 in *Episodes* 33:234–41.

AFTERWORD

1. B. Malinowski, *Argonauts of the Western Pacific* (1922; reprint, London: Routledge, 2002), 18. On the ethnographic study of science, E. Gellner, *Postmodernism, Reason and Religion* (London: Routledge, 1993); S. Franklin, "Science as culture, cultures of science," *Annual Reviews in Anthropology* 24 (1995): 163–84; H. Gusterson, *Nuclear Rites* (Berkeley and Los Angeles: University of California Press, 1996); S. Traweek, *Beamtimes and Lifetimes* (Cambridge: Harvard University Press, 1988). On constructive and dynamic disciplinary cultures, P. L. Berger and T. Luckmann, *The Social Construction of Reality* (London: Allen Lane, Penguin Press, 1967); R. Wagner, *Symbols that Stand for Themselves* (Chicago: University of Chicago Press, 1986); S. J. Knell, "Road to Smith."

2. On situated truths, M. Weber, *Economy and Society* (New York: Bedminster Press, 1968). This is oppositional, of course, to views such as those of Evans-Pritchard, who wrote in 1937, "Witches, as the Azande conceive them, clearly cannot exist"; if my conodont workers believed it, then for science it exists.

3. J. Deetz, *In Small Things Forgotten* (Garden City, N.Y.: Doubleday Natural History Press, 1977), 7. The literature in anthropology and museum studies is extensive on this subject. Material culture is rarely discussed in studies of science culture, and when it is, it rarely engages with the rich literature in these other disciplines.

4. On intentions, Berger and Luckmann, *Social Construction of Reality*, 34. Also E. H. Gombrich, *Art and Illusion*, 5th ed. (Oxford: Phaidon, 1977), 53, on mental sets. On perception, and naïve, scientific, and commonsensical readings of the real world, see P. F. Strawson, "Perception and its objects," in G. McDonald (ed.), *Perception and Identity* (London: Macmillan, 1979), 41–60; J.-F. Lyotard, *The Post-Modern Condition* (Manchester: Manchester University Press, 1986), 76,

argued that eclecticism is foundational to postmodernism but is only so as an overt performance. Implicitly it exists in all knowledge making. E. Wenger et al., *Cultivating Communities of Practice* (Cambridge, Mass.: Harvard Business School Press, 2002) would refer to this as tacit knowledge.

5. B. Latour, *Science in Action* (Cambridge: Harvard University Press, 1987) on black boxes.

6. Berger and Luckmann, *Social Construction of Reality*, 35ff.

7. It is noteworthy that connoisseurship in art history has roots in the natural sciences as seen in the art historians Giovanni Morrelli and Bernard Berenson.

8. See, for example, N. J. Rapport, *Diverse World Views in an English Village* (Edinburgh: Edinburgh University Press, 1993) on individual worldviews and constellations of knowledge; and C. Geertz, *Local Knowledge* (New York: Basic Books, 1983) on overlapping spheres of operational knowledge.

9. R. K. Merton and E. Barber, *The Travels and Adventures of Serendipity* (Princeton: Princeton University Press, 2004).

10. Sweet, *Conodonta,* 170 (ref. ch. 6, n. 8).

11. Raup, *Nemesis,* 119 (ref. ch. 10, n. 4). On boundaries, B. C. Smith, *On the Origins of Objects* (Cambridge, Mass.: MIT Press, 1996); T. Ingold, *What Is an Animal?* (London: Unwin Hyman, 1985).

12. I. Parker (ed.), *Social Constructionism, Discourse and Realism* (London: Sage,1998), xii. On Goethe and Morrissey on Husserl, see A. I. Tauber (ed.), *Science and the Quest for Reality* (Basingstoke, UK: Macmillan Press, 1997), 399. On Russell, see R. E. Aquila, *Intentionality: A Study of Mental Acts* (University Park: Pennsylvania State University Press, 1977), 96; A. Schultz, *Collected Papers I:*

The Problems of Social Reality (The Hague: Martinus Nijhoff, 1963), 3.

13. By denying the object independent agency of any kind, this approach prevents such things as Callon's oysters acting. M. Callon, "Some elements of a sociology of translation: Domestication of the scallops and the fishermen of St. Brieuc Bay," in J. Law (ed.), *Power, Action and Belief* (London: Routledge and Kegan Paul, 1986). Critiqued by H. M. Collins, and S. Yearley, "Epistemological chicken," in A. Pickering (ed.), *Science as Practice and Culture* (Chicago: University of Chicago Press, 1992), 301–26.

14. This lies at the heart of the correspondence theory of truth in which the facts of the object *seem* to support the theories or interpretations built from it. See D. Gooding, *Experiment and the Making of Meaning* (Dordrecht: Kluwer, 1990). For conodont workers, lookalike things acted as interpretive lenses aiding construction.

15. For a more wide-ranging variant of this essay, see S. J. Knell, "The intangibility of things," in S. Dudley (ed.), *Museum Objects* (London: Routledge, 2012).

16. S. Jones, "Negotiating authentic objects and authentic selves," *Journal of Material Culture* 15 (2010): 181–203, 181.

17. This is seen, for example, in correspondence between Jimmy Steele Williams, USGS, Washington to Branson, 14 April 1931, Edwin Branson Folder, Williams Papers.

18. S. E. Fish, *Is There a Text in This Class? The Authority of Interpretive Communities* (Cambridge: Harvard University Press, 1990).

19. R. K. Merton, *Sociological Ambivalence and Other Essays* (New York: Free Press, 1976) for a discussion of norms, expectations, and actualities in scientific behavior.

20. See the opening quote in chapter 1. Gombrich, *Art and Illusion,* 176, 190–91.

Index

SIMON J. KNELL, Professor of Museum Studies at the University of Leicester, is renowned for his innovative studies of fossils as scientific and cultural objects. Previously a museum geologist, Knell's publications include *The Making of the Geological Society of London*; *The Culture of English Geology, 1815–1851*; and *The Age of the Earth: From 4004 BC to 2002 AD*.